THE Story OF Time

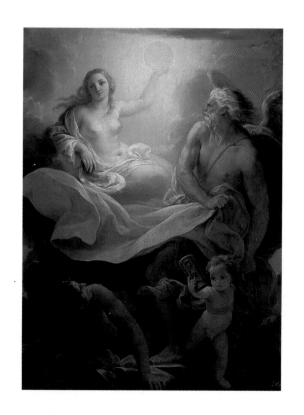

THE Story OF Time

Kristen Lippincott

with

Umberto Eco

E.H. Gombrich

and others

MERRELL HOLBERTON
PUBLISHERS LONDON
in association with
National Maritime Museum

THE Story OF Time

Contents

Any international exhibition is achieved only through the cooperation and support of a number of institutions and individuals. The most obvious debt is to those individuals and institutions who lend objects from their own collections. Without the benevolence of our lenders, exhibitions such as this one would not happen. An exhibition of any scale is also dependant on the work and ideas of a great number of scholars and curators. When that exhibition is as interdisciplinary and as cross-cultural as *The Story of Time*, that debt increases dramatically. The range and depth of learning displayed in this catalogue falls well beyond the scope of any individual. The quality of the exhibition is a direct reflection of a spirit of intellectual generosity on the part of a number of people who shared their time and expertise with us. In the first instance, I would like to thank all of the essayists who contributed to this catalogue. Thank you for your promptness, thank you for your doggedness. In particular, I would like to thank those authors who also provided help with the catalogue entries, especially Prof. Michael Loewe, Silke Ackermann, Ken Arnold, Elly Dekker and Joy Hendry. As for the numerous other individuals who have helped with the preparation of the catalogue, I apologize for merely listing them – but my thanks is no less hearty: from the Royal Collection: Hugh Roberts and Caroline de Guitaut; from the British Museum: Robert Anderson, Richard Blurton, John Cherry, Tim Clark, Anne Farrer, Anthony Griffiths, Victor Harris, Tim Healing, Peter Higgs, Jonathan King, Robert Knox, John Leopold, John Mack, Carol Michaelson, Richard Parkinson, Neil Stratford, David Thompson, Christopher Walker and Dyfri Williams; from the British Library: Beth McKillop, Pamela Porter, Anne Rose, Frances Wood and Susan Whitfield; from the National Gallery: Nicholas Penny and Gabriele Finaldi; from the Science Museum: Neil Brown and Derek Robinson; from the Tate Gallery: Sir Nicholas Serota, Stephen Deuchar and Jeremy Lewison; from the Victoria and Albert Museum: Philippa Glanville, Rose Kerr and Deborah Swallow; from the Warburg Institute: Nicholas Mann and John Perkins; from the Kunsthistorisches Museum in Vienna: Sylvia Ferino Pagden and Karl Schütz; from the Gemäldegalerie, Staatliche Museen zu Berlin: Jan Kelch and Erich Schleier; from the New York Public Library: Mimi Bowling and Alice Hudson; and from the Spanish Embassy in London, His Excellency Señor Alberto Aza and Señor Ramón Abaroa. I also wish to thank: Prof. Silvio Bedini, Koenraad van Cleempoel, His Excellency Mr David Colvin (British Embassy, Brussels), David Cordingly, C.P.E. Dercon (Museum Boymans-Van Beuningen), John Gleave, Peter Gosnell, Stefan Graeser (Naturhistorisches Museum, Basle), Peter Haward, Anita Herle (University Museum of Archaeology and Anthropology, Cambridge), Hester Higton, Rebecca Hossack, John Laverty (The National Physical Laboratory), Jay Levenson (The Museum of Modern Art, New York), Jennifer Marin (The Jewish Museum, London),Viscount Midleton (The Manor House Museum, Bury St Edmunds), Robert H. Miles, Alison Morrison-Low (National Museums of Scotland), Michael O'Hanlon (The Pitt Rivers Museum), Philippe Peltier (Musée National des Arts d'Afrique et d'Océanie), Hartmut Petzold (Deutsches Museum), Jeffrey Quilter (Dumbarton Oaks Research Library and Collection), Prof. Rudolf Schmidt, Martin Schøyen, Lorenz Seelig (Bayerisches National Museum), Desmond Shawe-Taylor (Dulwich Picture Gallery), Anthony Shelton (Horniman Museum and Gardens), Carlene Stephens (National Museum of American History), Kay Sutton, Marie-Hélène Tesnière (Bibliothèque Nationale de France), Dodge Thompson (The National Gallery of Art, Washington, D.C.), James Watt (The Metropolitan Museum of Art), Dietrich Wildung (Ägyptisches Museum und Papyrussammlung), and Carla Zarrilli (Archivio di Stato di Siena). From the staff of the National Maritime Museum, I wish to thank the Project Manager for the Exhibition, Hélène Mitchell – who certainly got 'the pants on the baby' – and Jonathan Betts and Gloria Clifton for writing many of the trickier horological and scientific instrument entries. Also, thanks go to Jane Bendall, Suki Biring, Maria Blyzinsky, Nick Booth, Lucy Cooke, Chris Gray, Sophie Lee, Alasdair MacLeod and Vicky Norton for all their help and for the long hours they put in to ensure that the exhibition was a success. Many thanks, too, to the exhibition design team, John Ronayne, Hermann Lelie and Isabel Ryan; and to Karen Wilks and the team at Merrell Holberton for helping to produce this catalogue. Finally, my most heartfelt thanks must be reserved for my husband, Gordon Barrass, the most supportive partner imaginable, who was not only kind and understanding throughout, but spent considerable time and effort trying to track down elusive bits of Chinese poetry, ancestor scrolls, talismans and images of Shou Lao: *chang ming bai sui*!

Kristen Lippincott
Director, Royal Observatory, Greenwich

On the eve of the third millennium, it is entirely appropriate that the National Maritime Museum and the Royal Observatory at Greenwich should host a major international exhibition on the Story of Time. In many people's imagination, 'time' and 'Greenwich' are nearly synonymous. Much of Britain's maritime supremacy over the past three centuries depended on an ability to send her ships to the furthest reaches of the globe and to bring them home safely again. The key to successful navigation was accurate timekeeping. Indeed, the Royal Observatory was originally founded to aid in navigation, and it is largely as a result of its success in this endeavour that it was awarded stewardship of the Prime Meridian of the World – Longitude 0° – in 1884. Today, all time and space on Earth is measured from this co-ordinate. When the new millennium begins, it will do so at mean midnight, Greenwich Mean Time, at the Royal Observatory.

We are most appreciative of all the support we have received in the preparation of this exhibition. Our primary sponsor, J.P. Morgan, has been extremely generous, imaginative and constructive throughout. This exhibition could not have been realised without their foresight and commitment to quality. The additional support from *The Times*, Parmigiani Fleurier and the National Physical Laboratory has also been invaluable, with each bringing something unique to the exhibition. We would also like to acknowledge a number of gifts from donors who wish to remain anonymous. The success of the exhibition rests largely upon their willingness to invest in Greenwich and a desire to contribute to the Story of Time.

Richard Ormond
Director, National Maritime Museum

FOREWORDS

THE TIMES

PARMIGIANI FLEURIER

For more than 200 years, since it was first published in 1785, *The Times* has been acutely aware of the time. A daily newspaper, put together against the clock, is haunted by the fear of a deadline missed, an edition lost, and readers dissatisfied. With one eye on the clock and the other on the story, *Times* journalists today have at least this in common with their illustrious predecessors. The technology may have changed, but the demand for timeliness and accuracy has not.

This makes us the most natural of partners for *The Story of Time*. It is true that clocks have an even longer history than Britain's oldest daily newspaper, and that techniques for the precise measurement of time had reached an extraordinary perfection by the time of our first publication. But since then we have delighted in reporting the key advances in time-keeping, some of which are included in this exhibition. We also expressed support in a leading article for the establishment in 1884 of the Greenwich Meridian as the Prime Meridian of the world.

The leader was sober in tone and eschewed vulgar chauvinism. "In the interest of geography, navigation and science generally" *The Times* opined, "it is well that there should be a common meridian". Yet the fact that this meridian runs through Greenwich is a permanent reminder of Britain's enormous contribution to timekeeping, navigation, exploration and the establishment of a reliable method for estimating longitude while at sea. Perhaps my Victorian predecessor was so confident of Britain's place in history that he had no need to labour the point. I am delighted that in addition to our sponsorship of The Story of Time, we are also sponsoring the Prime Meridian Line at the Royal Observatory Greenwich – the home of time.

We are committed to initiatives, such as this exhibition, which not only celebrate what has been achieved, but look forward to new achievements in the next millennium. A newspaper reports what has happened, but must also look forward to events still folded in the womb of time, waiting for the clock to give them birth – time future, as well as time past. At a moment when everybody is especially conscious of time, this exhibition celebrates it in all its dimensions, and we are happy indeed to be associated with it.

Peter Stothard
Editor

Parmigiani Mesure et Art du Temps is delighted to be associated with this landmark exhibition taking place in Greenwich, the acknowledged home of world time.

Fleurier, where our company is based, is situated in the Jura Mountains around Neuchâtel – one of the acknowledged homes of world horology. It is where Michel Parmigiani and a 70-strong team of master watchmakers, designers, engravers, casemakers, enamellers and goldsmiths, have become the inheritors of a 450-year-old tradition, and where, for the last 10 years, they have been producing some of the most exquisite and unique timepieces in the world.

Our company is built on three pillars of excellence. The first is the restoration of antique watches and ornamental clocks, the second is the manufacture of commissioned, one-off timepieces and movements, and the third is the creation of our own exclusive collection which is sold under the marque *Parmigiani Fleurier*.

In 1995, The Sandoz Family Foundation made a long-term investment in Parmigiani Mesure et Art du Temps in order to support its aims of promoting outstanding creativity, innovation and artistic excellence, within an entrepreneurial business environment. Our sponsorship of *The Story of Time* is a further expression of the philosophy of this partnership and we are proud that a piece from our new collection has been included in the exhibition, as have two pieces from the Sandoz family's own collection (now donated to the Château des Monts, Le Musée de l'Horlogerie in Le Locle, Switzerland).

The Story of Time will undoubtedly awaken new interest in the art and science of horology amongst the many thousands of people who will visit Greenwich over the coming months, but we hope it will have particular appeal for younger visitors brought up in a digital age. We feel privileged to be part of time's tradition and of its evolution.

Michel Parmigiani, President
Emmanuel Vuille, General Director

Parmigiani Mesure et Art du Temps

JPMorgan

Time preoccupies us all, never more than at the dawn of a new millennium. What better way to usher it in than with the story of time? And what better place to tell that story than at Greenwich, where each day officially begins?

The Story of Time shows how human beings have measured and perceived time through the ages. Drawing on both science and art, the exhibition is at once fantastical and empirical, questing and calming, a culmination and a curtain raiser.

J.P.Morgan is the proud sponsor of this exhibition – and an apt one. We have time on our side: more than a century and a half of financial experience. This accrued knowledge and our pivotal role in international banking give our firm a compelling strategic advantage in helping clients realize their goals and aspirations in the new century.

P. I. Guymard, Universal equinoctial ring dial, 1740 (see **125**)

Times

Umberto Eco

"What was God doing before he made heaven and earth? ... He was preparing hell for those that would pry into such profound mysteries."[1] This joke already had a long history behind it when it was quoted (alongside a warning that it was indeed only a joke) by a writer of the utmost seriousness, as he tackled one of the most profound of all philosophical mysteries: time. The writer was St Augustine, who dedicated much of the Book XI of the *Confessions* to an exploration of time. Simply by using the joke, Augustine was already close to a notion of time that would be happily shared by today's 'big bang' theorists: that time was born at a precise moment; that only from the 'big bang' onwards does it make sense to talk of a 'before' and an 'after'; and that it is nonsense to ask what was happening 'before' the birth of time.

Long before Augustine posed his question, the Greek philosophers had also tackled the problem of time. The most popular definition had been Aristotle's (*Physics*, IV, 11, 219, b1): "time is the calculable measure of motion with respect to before and afterness". The Stoics, for whom time was the 'interval' in the motion of the world,[2] did not stray far from Aristotle. Locke corrected him in part by saying that time was not necessarily the measure of movement, but rather "any constant periodical appearance of ideas", so that, even if the Sun did not move in the sky but instead simply intensified and faded in brightness, the rhythm of the alternation could nevertheless serve perfectly well as a parameter for the measurement of time.[3] Locke's correction was a good one, since it legitimized today's non-mechanical instruments of time measurement such as quartz or atomic clocks. But he still conceived of time as order and succession, and neither Leibnitz nor Newton would alter that fundamental conception. Indeed, nothing was to alter it from Kant through to Einstein. Time retained the order of a causal chain, although it is worth noting part of Hans Reichenbach's definition of relativity (on which more below): "the theory of relativity presupposes merely an order of time, not a direction".[4]

If time was conceived as the precise measure of an ordered succession of states, it should come as no surprise to find that its first criterion of measurement in every known civilization should have been the movement of the stars (which is both movement and

Time is the measure of "any constant periodical

return, in other words a 'constant periodical appearance'). Even if there were nothing more to time than this, it would still be interesting to ask why, for so many centuries, men measured out the years, the months and the days, but came so late to measuring

We can measure time, but this gives us no guarantee that we understand what time is

hours and minutes. The main stumbling block was the need for mechanically precise instruments to measure them. We only have to look at the vicissitudes of the various calendar systems to realise how hard it was even to divide the year up into an exact number of days. For millennia, the only reliable clock was the cock's crow and, in a predominantly agrarian economy, the only measures needed to regulate individual and social lives were sunrise and sunset, and the passing of the seasons. For millennia, the notion of punctuality was a vague one; at best, the days were split into measurable parts by liturgical rhythms or by the sounding of bells.

Nowadays, we are all children of a clock civilization, but we are still sometimes remarkably imprecise in how we measure time. The briefest survey of newspapers and bookshops in 1999 shows just how many books – some serious and some frivolous – are appearing as the millennium draws to a close to tackle the fraught issue of whether the last thousand years should end on 31 December 1999 or 31 December 2000.

It is astonishing that we do not all agree at once: the millennium must, of course, finish on 31 December 2000, just like the first decade in the decimal system ends with the number 10 and the next begins with the number 11. Bibliophiles know it better than most: once it had been decided that 'incunabula' were to be defined as books printed by the end of the fifteenth century, it was clear that any book printed up to 31 December 1500 (and not 31 December 1499) would be included. But those round numbers loom very large (those two zeros are the root cause of all the trouble of the 'millennium bug'). The same arguments were had at the turn of the seventeenth and eighteenth centuries, the eighteenth and the nineteenth, and the nineteenth and the twentieth. And I am sure that we will cover the same ground once more in December 2999. You cannot fight it: popular sentiment always wins out over common sense and science, so, just as our ancestors celebrated the beginning of our century on 1 January 1900, so we shall do the same for the twenty-first century on 1 January 2000.

Even the most sophisticated people lose their heads over the mystery of counting time. In the last twelve months, I have read several

appearance of ideas" (Locke)

articles about the problem of the end of the millennium which lay all the blame on poor old Dionysius Exiguus (Dionysius the Small), who had the idea in the sixth century AD of beginning to count the years from the date of Christ's birth. Before Dionysius, dates were

calculated from the reign of Diocletian onwards or from the beginning of the world, measured with great precision, of course. Now we know that Dionysius got the year of Christ's birth significantly wrong (it should have been somewhere between four to six years earlier, so that in reality the second millennium ran out of steam in 1997 or thereabouts). But there is a curious and much less plausible tendency to attribute a second error to him: to suggest that since he had no notion of zero as a number – a concept which passed from the Indians to the Arabs and only reached the West several centuries later – he therefore began his history of Christendom from Year 1. If he had made use of a year 0, the argument goes, we would not be in any doubt today: the second millennium would finish with 1999 and the third would begin with the year 2000.

To appreciate quite how absurd this notion is, we only have to imagine what would have happened if Dionysius had been fully versed in Indian mathematics and had fixed Jesus's birthday in year 0. Twelve months on, would Mary and Joseph have said that Jesus was 'zero' years old and that he would only have become a one-year-old at the end of the following year? Obviously not: we do not and could not work out the passage of the years in that way. At birth we begin the *first* year of our lives and there is no reason to think any differently when it comes to the birth of centuries. The only point of stirring up this tortured debate yet again is to show that, for all our sophisticated mechanical and atomic clocks, when it comes to counting time we are still capable of losing our heads.

The fact is that we can measure time, but this gives us no guarantee that we understand what time is or whether it is proper to measure it metrically. Let us go back to St Augustine. At the start of his reflections on the matter, he seems at one with the Aristotelian idea of time, and indeed he notes that, unlike eternity, which is fixed,[5] a period of time is long through a succession of many movements which cannot be extended within the same period of time. In fact he says "*ex multis praetereuntibus motibus*": 'for many movements that pass beyond/ that pass us by'. What strikes him in the flow of these movements, then, is that they turn into time past. From this consideration, he begins to ponder that, whilst in eternity all is present, time is a strange phenomenon in which everything past has been as if banished from the future and everything future follows on from the past, and both past and future flow out from the present. And yet, Augustine asks himself, how can past and future exist if the past is no longer and the future is not yet? Does this leave us with an eternal present? But an eternal present would be eternity and not time. And finally, even if we take the present alone, can we say that the current month is present, when a day, an hour, a minute, a second are the only real present parts of it? As soon as he tries to pin down the duration of the present second, however, Augustine realises that even that second can be infinitesimally subdivided into ever shorter entities, and that even were the briefest of all possible units of time conceivable, it would pass so quickly from the future to the past that it would have no duration at all: *"nullum habet spatium"*.[6]

And so, every attempt to define time in terms of measurable units collapses, as Augustine openly states (XI, XIII, 29). He does not agree that time depends on the movement of the Sun, the Moon and the stars. Why not use the movement of any and every body, even (and here he anticipates Locke and thus the idea of both the mechanical clock and the atomic clock) the circular, periodic motion of a potter's wheel? But Augustine goes further still. In XI, XIII, 30, he recalls the cry of Joshua, "Stand still, O Sun" (Joshua, 10:12). At that moment the sun and all the stars were stopped in their tracks: but time *went on*.[7] We readers might wonder what this time was that 'went on' after the Sun had stopped? Perhaps the time of Joshua's consciousness (and perhaps also of his body). And, indeed, having rejected the hypothesis connecting time to celestial motion, Augustine immediately posits the alternative that time is an extension or an extending movement of the soul.

Thus, Augustine tells us, we can measure neither the past, nor the present, nor the future (since these never exist) and yet we do measure time, whenever we say that a certain time is long, that it never seems to pass or that it has passed by very quickly. In other words there is a non-metric measure, the sort we use when we think

Petrus Apianus, *Planisphere*, from *Astronomicon Caesareum* (see **030**)

of the day as boring and long or when a pleasurable hour has gone by too swiftly. And here Augustine pulls off an audacious *coup de théâtre*: he locates his non-metric measure in our memory. The true measure of time is an inner measure. Centuries later, Henri Bergson would also contrast metric time with the time of our consciousness or 'inner *durée*'. But for all Bergson's wonderful writings (such as his *Essai sur les données immédiates de la conscience*) and for all the fascination of comparing his thoughts on the matter to Proust's, it is worth recalling that Augustine got there first, that his pages on time remain amongst the most modern, precise and revealing on the subject in the entire philosophical tradition.

There is no use denying how useful clock time can be, but it is clear that it is entangled in our everyday lives (although perhaps not in science) with the time of consciousness and of memory. Here we

The future should be a place where we will go, not

should perhaps embark upon a discussion of phenomenological time (Husserl) or of Heidegger's notion of time, which was not so far removed from objective time or biological time or the physical time of entropy; that time which dictates that all living beings tend towards the void. In other words (and it did not take a Heidegger to work this out), all humans are mortal. But Heidegger took things further by attempting to juxtapose this relentless physical, biological time with a "time of projection" – the only possibility we are allowed – asking how can we live accepting what we have been, existing only for death.

This introduction has no intention of offering a comprehensive reconstruction of the problem of time. Its author *hasn't got the time* to provide anything of the sort. All it wants to do is point up some areas of confusion. And much of our confusion when it comes to talking about time is apparent in the language we use to talk about it. We may not be at all confused when we say that it is ten to nine on 21 December: clock time, like astronomical time, is comforting.[8] But we are invariably at a loss when it comes to talking about inner duration.

Does time travel in front of us or behind us? It is not as pointless a question as it might seem, since when we say, for instance, that it is 6 AM, the Sun is at a certain point in the sky up on our right and, when we say it is 6 PM, the Sun is at another point in the sky to our left. Of course, the criterion will change depending on whether we are looking north or south or towards the rising Sun. So, let us say we are looking towards the Sun as it rises: as it gradually moves across the sky the past is, so to speak, in front of us, and the hours to come are behind us. Can we therefore conclude that our culture

imagines the past to be before us and the future behind us?
It seems that in some cultures the answer is yes, because the past is
what we already know (we can see it with our own eyes) whilst the
future is what we cannot yet know. But as soon as we examine our
own Western language and the way we talk about time, we realise
that our habitual reasoning is quite the opposite: we speak of
finishing the job in the weeks ahead of us, of months left behind, of
'turning back to the days of my childhood'.[9]

Wait a minute, though. We also talk about 'the following few
weeks', implying something that is following us, coming from
behind, not from in front. So the future is behind us? And there is
more. We talk as though we think of the future as something that
sooner or later will arrive here, where we are now, whereas the past
is drifting further away from us. We say that 'the time will come

something that will come to us where we are now

(*i.e.* here) when' and that the years of our infancy have gone (*i.e.* far
away from here). The irrationality of these turns of phrase cannot
be overestimated: even if it were reasonable to see the past and the
future in spatial terms, surely the future should be a place where we
will go, sooner or later, not something that will come to us where
we are now. And similarly, we ought to speak of going away from
the past where we once were, not of a past which has gone away
from where we are now.

Derek Bickerton, in his *Roots of Language*,[10] has a clever thought
experiment which we could adapt: imagine that I have been living
for a year in a really very primitive tribe, whose language I know
very roughly (some names of objects and simple actions, some verb
infinitives, proper names but no pronouns *etc*). I am out hunting
with Og and Ug, and they have just wounded a bear which is
bleeding and has taken refuge in its cave. Ug wants to pursue the
bear into its den to finish it off. But I remember that a few months
earlier, Ig had wounded a bear and brashly followed it into its den,
where it had somehow found the strength to kill Ig and eat him. I
want to recall the incident to Ug, but, in order to do so, I need to be
able to say that I remember a past event, without knowing any verb
tenses or any notion such as 'I remember that'. So all I say is, "Eco
see bear". Ug and Og obviously think I've spotted another bear and
they take fright. I try to reassure them: "Bear not here". Now they
think I am just pulling their leg at the worst possible moment. I
battle on: "Bear kill Ig". But they reply: "No, Ig dead!" Perhaps I
should now give up and let Ug die. But I try instead a visual rather
than a verbal interpretation. Saying "Ig" and "bear" I poke myself
on the head or heart or stomach (wherever the memory is supposed
to be located). Then I draw on the ground two figures and point to

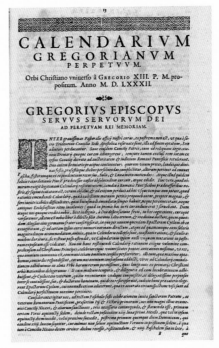

Christophorus Clavius, *Inter gravissimas ...* (see **070**)

them as 'Ig' and 'bear': behind Ig, I draw some Moons in different
phases, hoping that they take this to mean 'several Moons ago' and
finally I draw the bear eating Ig. All my efforts rely on an
assumption that my native interlocutors possess notions of memory,
and of past, present and future. But, since I have to interpret these
notions visually, I also need to know whether for them the future is
ahead of us or behind us. Otherwise, my interpretation risks
proving meaningless to them. If I place the bear killing Ig to the left
and for them the past is to the right, then all is lost, including Ug.
This is an interesting case of past and future, life and death
depending on semiotic conventions. And notice that there is nothing
at all in my conception of past and present that can tell me what my
interlocutors' spatial conception of time will be.

Of course, a scientist might object that these incidents are no more
than problems thrown up by languages and their diversity, and that
my difficulty (and that of Og and Ug) has nothing to do with a
scientific notion of time. And, indeed, the aim of the experiment is
not to suggest that our ingenuous thoughts and our flawed
languages might in any way influence science and its notion of time.
I have long felt a quasi-religious devotion to scientific (non-
ingenuous) notions of time, ever since reading Reichenbach's
Direction of Time, which tells us that the universe of our experience
contains not only 'open' causal chains (A causes B, B causes C, C
causes D, and so on *ad infinitum*), but also 'closed' causal chains (A
causes B, B causes C and C causes A). In a closed causal chain, I

could travel back to the past, meet my grandmother as a young girl, marry her and become my own grandfather.

Such journeys may never actually occur in our world, but they certainly do occur in the pages of science-fiction novels where readers are forced into thinking of time in this way, imagining worlds where 'time's arrow' can be reversed. How can we conceive of such universes, how can we imagine them, since imagine them we must in order to understand stories about them?

As an illustration, let's look at the situation in films such as *Back To The Future*. Boiling the story down to its essentials, imagine a character called Tom[1], who travels into the future and arrives as Tom[2] (*i.e.* a Tom a few hours older than Tom[1], exactly as if Tom[1] had set off from Paris and arrived seven hours later in New York as Tom[2]). But now, Tom[2] travels back in time and as Tom[3] he arrives a few hours before the original departure of Tom[1]. Tom[3] meets up with Tom[1] in the past just as the latter is about to head off to the future. Tom[3] now decides to follow Tom[1], going back to the future and arriving there as Tom[4] just a few minutes before the arrival of Tom[1] (as Tom[2]).

There are good reasons to expect that the reader might not be able to get his mind around this situation. And yet the reader does so, by identifying at all times with one particular Tom – the Tom with the highest number: all the others are seen from his point of view. In the film version of the story, the spectator identifies with the same Tom, as if he were carrying the camera on his shoulders. In short, in any encounter between Tom[x] and Tom[x+1], the 'I' and the 'eye' are always Tom[x+1].

Thus, just as with the linguistic expressions discussed earlier, where time was shown to be linked to our corporeality in our everyday language, to be imagined as near to or as far from our bodies and determined by our consciousness, so here, too, time and its paradoxes are perceived from the viewpoint of our bodily location.

The more we think about it, though, the more we come to realise that in fact all the 'clocks' used by man, at least until the invention of mechanical time-pieces, were in their way linked to our bodily location. Time was measured against the visible motion of the stars and the 'rising' and 'setting' of the Sun, that is movements that only exist in relation to our point of view (indeed, objectively speaking, it was the Earth that was moving, of course, but we did not know it and we did not really care). Even with the dawning of the age of clocks, attempts were still made to assimilate those non-anthropomorphic beasts to our bodies: the eighteenth century offers numerous examples of splendid poems describing clocks as

monsters with grinding teeth, chewing up the seconds or spitting them out like syllables. But we will surely never find a way of humanizing today's atomic clocks or even the clocks on our computer screens.

Whence our fear of the 'millennium bug' – an inhuman animal, entranced by those two zeroes and unable to *feel* the new millennium coming towards it, deceived into thinking that it has suddenly been catapulted back to the year 1900 (unable even to feel, however mistakenly, that 1900 is coming up behind it). And yet we should not be afraid. We will never leave off thinking of time from the viewpoint of our bodies. After all, watching and feeling ourselves get older every day, we are also our own time-pieces. Just try doing a few press-ups or running downstairs or jumping over a hedge, and you'll soon realise that time has passed since you were twenty years old. How lucky we are to be mortal! We can keep time under control at all times.

English translation by Robert Gordon

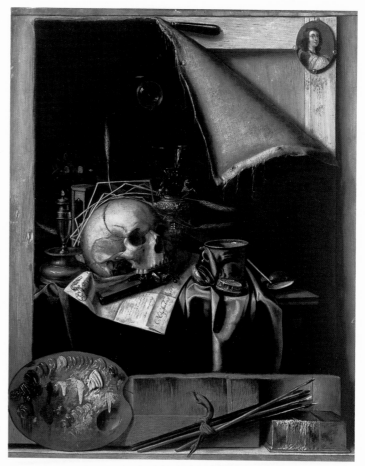

Cornelis Norbertus Gysbrechts, *Vanitas*, oil on canvas, 1664
Kingston-upon-Hull, Hull City Museums and Art Galleries, Ferens Art Gallery

Jean Dampt, *The grandmother's kiss* (see **247**)

1. "'*Quod faciebat Deus, antequam faceret caelum et terram?' Respondeo non illud quod quidam respondisse perhibetur, ioculariter eludens quaestionis violentiam: 'Alta' inquit 'scrutantibus gehennas parabat'*" (Augustine, *Confessions*, XI, XII, 14).

2. Note that in this definition by Chrysippus, 'interval' does not imply 'an empty space between two things': the Greek term is *diástêma*, which is the term used for a musical interval, that is the 'relation' between two sounds; so not a 'gap' or a silence, but rather a 'filling' heard by the ear.

3. *An Essay Concerning Human Understanding*, II, XIV, 19.

4. Hans Reichenbach, *The Direction of Time*, Berkeley and Los Angeles 1954, p. 42.

5. On the notion of eternity, Augustine drew on both Plato and the Neoplatonists, but he was also steeped in Stoic philosophy. Such links belong to an account of (approximately) twenty-five centuries of thinking about time which no preface could hope to cover thoroughly.

6. Note that, here too, a spatial term is used to indicate a period of time.

7. "*Ibat*": yet another spatial metaphor!

8. It was not so comforting for Phileas Fogg in Jules Verne's *Le Tour du monde en 80 jours*. He discovered, on his return to London, that what he thought was 21 December was, for the rest of London, one day earlier. He had circumnavigated the world eastwards, gaining a day and so arriving back on 20 December, the day before the deadline of his bet. Such is the power of Greenwich and of the 180° meridian! And yet it has only happened twice in history, once with Magellan's ships which had travelled westwards and so lost a day, and once with Fogg. Since then we have become less innocent: now it would be the epitome of kitsch to celebrate the night of 31 December 1999 on that fatal meridian, just so that you can celebrate it all a second time on the other side of the dateline.

9. For these and other interesting observations on our spatio-temporal metaphors, see G. Lakoff and M. Johnson, *Metaphors We Live By*, Chicago 1980, p. 9.

10. Derek Bickerton, *The Roots of Language*, Ann Arbor 1981, p. 270.

Indian (Jaipur, Rajastan),
Avatar of Vishnu as the Tortoise Kurma
(see 014)

THE Creation OF Time

Not all cultures define or experience time in the same way. Obviously, there are a number of phenomena in nature – such as the waxing and waning of the moon, the succession of day and night, the regular rising and setting of the stars and the pattern of the seasons – that support a view of time as essentially cyclical in nature. But there are certain facts of life, such as physical decay and the ageing process, that tend to argue in favour of time as being not only linear, but as moving in a single direction, from the past, through the present and onwards to the future. Most societies have constructed time-measuring systems that embrace both its cyclical and linear aspects. The exact proportions allotted to each, however, vary greatly. Moreover, nearly every major culture on the face of the Earth seems to have a unique understanding of what time is.

One of the best ways to explore the varying assumptions that different cultures have about time is to examine their creation mythology – in order to identify the role that time plays within those tales and to note in what ways these formulae underpin a wider understanding of how the universe operates. For example, in Judaism, Christianity and Islam – each of which turns to the opening lines of the first book of Genesis for its model of how God created the universe – the linear nature of time seems to predominate. In other cultures, where the gods tend not to be abstracted from the world of man, it is the cyclical nature of time that predominates. In these cases, quite often the relationship between gods and men is conditioned by a mutual understanding that the continued existence of the world itself – including the continuity of time – depends on man's willingness to play his part. The Navajo of southern North America, for example, believe that only man can disturb the balance of nature. In such societies, certain rituals enacted at specific times during the year help to ensure the continuity of time. Most of these rituals contain an element of re-enacting the original sequences of the creation of the world.

The Creation of Time

The Early Christians were not only the inheritors of the narrative structure of the Old Testament, they were also the heirs to the Greek tradition of Socratic enquiry and the belief that all questions can be answered if the intellect is sufficiently rigorous. One of the issues that occupied their minds for over a thousand years concerned the creation of time. For the early Church Fathers, the nub of the issue was: when God created the universe, did He create it in time, or did He have to create the matrix of space and time before he created matter? One of the main problems, it seems, was the sequence of events related in the opening lines of Genesis. For, as the Venerable Bede explains, if the world began on 19 March, time *per se* could not have begun until 23 March, when God created the Sun. Or, as one later commentator argues, "... before the creation of the Sun, there was neither time nor hour, because there was nothing which might make a shadow, that means by which time or hour would have been measured".

Following centuries of heated debate, the Fourth Lateran Council in 1215 established the official Church doctrine that God created "all things spiritual and corporeal, angelic and mundane" simultaneously out of nothing. The next problem, however, was trying to establish exactly when the world had been created. For, if the whole of everything in the world was created *in* time, then there must have been a specific time when the universe started.

The Greek and Roman philosophers had also believed that there must have been a particular disposition of the stars and planets at the beginning of time. This configuration was known as the *thema mundi* and was represented by a kind of horoscopic chart, in which all of the planets were placed 'in' specific signs of the zodiac at the moment the universe was created. The structure of the *thema mundi* was not entirely consistent across the different cultures of the Mediterranean. Those cultures that began their religious year in the spring tended to believe that, when the universe was created, the Sun was in Aries. Those that began their year in the summer, such as the Egyptians, tended to feature the sign of Leo in their *thema mundi*.

Despite these controversies, the history of the illumination of the first chapters of Genesis is relatively uneventful. Quite early on, artists discover that the best way to illustrate the first six days of the Genesis cycle is to use what one might call the 'comic strip' format. The primary action of each day – creating the heavens and the Earth and light; the creation of the waters; the creation of the plants; the

001 Italian

Biblia Vulgare Istoriata
Printed book with coloured illustrations (Venice: Adam d'Amergau, 1471)
Manchester, The John Rylands Library, University of Manchester

A number of early printed Bibles present the first chapters of the Book of Genesis as a narrative sequence of events. This so called 'comic strip' device for structuring biblical illustrations first appears in large-scale decorative cycles and in illuminated manuscripts from the Early Christian period and it continues to be the most popular format in all media well into the Renaissance and beyond. In this hand-coloured version, the actions of the first six days of the Creation are spread across the double-page spread of the book, beginning with *"Fiat lux"* and ending with the Creation of Adam and Eve.

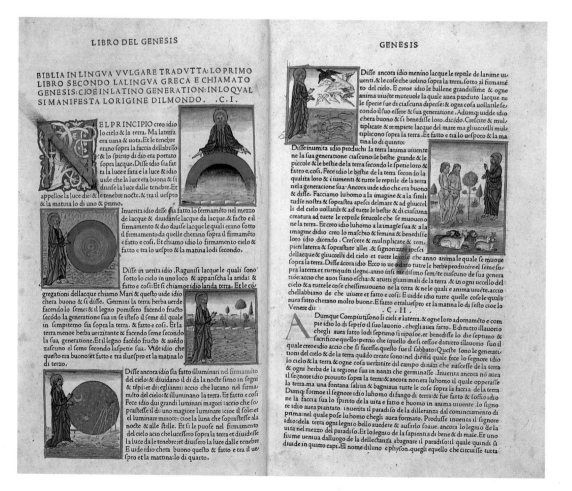

creation of the Sun, the Moon and the stars; the creation of the fishes; the creation of the animals; and, finally, the creation of man and woman – is represented with a similarity of form that echoes the repetitive cadences of the text. The central figure in each scene is God the Father and, in general, the unfolding of events can be determined by slight changes in His attributes or companions. The action is sequential. The understanding is that one 'reads' the series as progressing from one panel to another.

It is important to notice what this fact implies about one's registration of time. The rhythm of the pictures in all these early printed Bibles is based on the premise that time progresses as we move from one frame to the next. Were the order of the pictures reversed, the narrative would no longer make any sense. There is also a vague appreciation that, by moving from one day to the next, some sort of progress is being made. The implication is that the Creation of Man is the last and best of God's acts. Simple as these pictures may be, they belie a whole web of understandings about Western notions of time – how events move and develop within time; how time itself seems to move forward with differing actions; how one event seems to grow from a previous event, and so on. As everything that exists is a manifestation of the will of God, nothing exists without reason.

002 Italian

Pontificale secundum ritum sacrosanctae Romanae ecclesiae ...
Printed book (Venice: Luc Antonio Giunta, 15 September 1520)
London, The British Library
Dept. of Printed Books

The Christian Church inherited a notion that the universe began at the spring equinox, when the Sun appeared to be set against the zodiacal sign of Aries, the Ram. Numerous versions of the *thema mundi* appear in paintings, sculptures, manuscripts and printed books. In the woodcuts accompanying this volume describing the sacred rites of the Catholic Church, God the Father is depicted placing a small, smiling image of the Sun directly on the constellation of Aries. The assumption is that anyone reading this book would have been conversant with the imagery of the *thema mundi* and would have understood this as an image of God the Father creating the universe *in* time.

003 Spanish (Sant Pere de Roda)

The Roda Bible
Pen and ink on vellum,
first half of the 11th century
Paris, Bibliothèque Nationale de France [Ms Lat. 6, fol. 6r]

The Roda Bible shows a compressed version of the first chapters of Genesis – from the Separation of Light and Dark to the Temptation of Adam and the Fall. The Separation of Light and Dark is conveyed by two personifications: a dark-skinned figure of Night (*Nox*) and lighter-skinned Day (*Dies*). Below Night there is an unusual personification of the Abyss – depicted as a dark-faced, watery being. The large circle in the centre of the page may represent the description of God having "set a compass on the face of the Abyss" (Proverbs 8:27) – the divided circle being a shape that the compass marks out. It may also indicate the division of the universe into the four elements – the red and blue tones representing air, fire, earth, water.

004 **French (Paris)**

Cosmos, from Honorius of Autun,
Clavis Physicae
Tempera and ink on parchment
Mid-12th century
Paris, Bibliothèque Nationale de
France [Ms Lat. 6734, fol. 3v]

For most Christian philosophers, the passage from Revelations (21:5–6), in which Christ says, "I am the Alpha and the Omega, the beginning and the end", signalled that the end of time was implicit in the creation of time and that the universe was ordered according to a plan, some aspects of which could be understood by the mind of man. In the early 12th

century, Honorius of Autun, in an image entitled *Cosmos* in his book *Clavis Physicae* ('The Key to Nature') tried to explain how the whole history of the world is set in motion by the guiding principles of Goodness, Justice, Virtue, Reason, Essence, Life and Wisdom. These principles determine the way in which Unformed Matter can be acted upon by both Time and Place (*Tempus* and *Locus*) to create cause and effect (*Effectus causarum*). The final image in the miniature is of God closing the curtains on his creation. As God is the creator of all things, so the end of time falls within his plan of the cosmos.

One of the fundamental components of Western man's approach to the uncertainties of his world is his desire to structure, to number, to measure and to name. Images of a creator-god who fashions the universe through the act of measuring seem to be a cultural constant throughout the ancient Mediterranean. In his *Timaeus*, for example, Plato describes the 'Demiurge', a creative force that forms the rational from chaos through the act of mensuration. In the Book of Wisdom, in the Old Testament, God the Father's hand is described as "the hand that from formless matter created the world" (11:18), and the Prophet Isaiah (40:12) asks: "Who was it that measured the water of the sea in the hollow of his hand and calculated the dimensions of the heavens, gauged the whole earth to the bushel, weighed the mountains in scales, and the hills in a balance?"

Accordingly, in the great illuminated Bibles of the Middle Ages, God the Father is frequently shown creating the world by measuring – using either a compass or a pair of scales. A number of scholars have linked these images of 'God the constructor' directly to the twelfth-century rediscovery of certain influential Platonic texts. For example, the creator-architect described in Philo's commentary on Plato's writings strikes a particular chord. Like God, Philo argued, the architect creates something tangible from nothing but his own ideas.

For the early nineteenth-century Romantic artist and poet William Blake, however, this image of a god who creates by measuring was an anathema. His well known image from *Europe, a Prophecy* seems to have been based on a passage from Milton's *Paradise Lost*, in which it is Christ, the Son of God, who is responsible for creating the world through mensuration:

> "... in this hand
> He took the golden compasses, prepared
> In God's eternal store, to circumscribe
> This universe, and all created things.
> One foot he centered, and the other turned
> Round through the vast profundity obscure,
> And said, 'Thus far extend, thus far thy bounds,
> This be thy just circumference, O World'" (VII, 224–31).

Blake saw this limiting god not as a force for good, but as an evil daemon. He was the god of the pagan philosophers, like Plato, who strove to impose limiting mathematical blinkers on the "eye of the Imagination". For Blake, such representations of the 'rational' tradition were the enemy, who were

> "Fixing their Systems permanent: by the mathematic power
> Giving a body to Falsehood ..."
> (*Jerusalem*, chap. 1, plate 12, ll. 12–13).

Cy comence la bible en francois translatee selon les hystoires escolastres. Et premierement le liure de genesis du quel le premier chapi   mencement par le quel et ou quel le pere

005 French

Historia Scholastica, by Guyart des Moulins
Tempera, gold and ink on vellum, 1411–12
London, The British Library
[Ms Royal 19. D. iii , part 1, fol. 3v]

There are more than forty manuscript illuminations in which God the Father is depicted holding a dividing compass in his hand. In Proverbs (8:27) God is said to have "set a compass on the face of the Abyss". This origin of this type can be traced to the so called *Bible moralisée* – a version of the Bible in which some of the more difficult sections of the text were 'moralized' or provided with allegorical glosses to explain the 'deeper significance' of these passages to a lay audience. These glosses, in turn, formed the basis of a new type of illustration. In a fewer number of examples, God is shown holding a pair of scales in reference to the Book of Wisdom (11:21), which says He created the universe according to "measure, number and weight". In some illustrations, we see God's hand with one or two fingers extended – a pictorial device to indicate the act of numbering.

Also

• *The Montpellier Bible*, 12th century, London, The British Library [Ms Harley 4772, fol. 5r]

• Henricus Engelgrave, *Lux Evangelica sub Velum Sacrorum Emblematum ...* (Cologne 1655–59), The British Library, Dept. of Printed Books

• Albertus Magnus, *Philosophia d. Alberti. M. Illustrissimi Philosophi & Theologi: ...* (Brescia: Battista de Farfengo, 23 June 1493), The British Library, Dept. of Printed Books

006 William Blake

Frontispiece to *Europe, a Prophecy*
Relief etchings, colour printed, with wash and watercolour, 1793
Manchester, The Whitworth Art Gallery, The University of Manchester
[D.1892.32]

Blake's image of a bearded, godlike figure, holding a compass against the chaos of the forming universe, is perhaps the best known of all his engraved works. Unfortunately, it is repeatedly misinterpreted. As is clear from Blake's own words, this long haired, bearded figure is not the God of the Bible, but Urizen, one of his own mythical creations. For Blake, Urizen (an obvious play on 'your reason') symbolized the imposition of the rational methods of mathematics on chaos. In this case, however, the act was not one of benevolence; it was intended as a cruel and unwanted restraint to the otherwise free forces of the imagination. Or, to quote Anthony Blunt's formulation, for Blake, Urizen's role was to "crush man's sense of the infinite, and to shut him up within the narrow wall of his five senses". Blake's Urizen is based on his deep revulsion against Greek philosophy and, above all, against Plato – the man who expelled poetry and the arts from his *Republic* because their imagination not only conflicted with, but undermined the rational truth of facts that can be established by weight, measurement and number.

Cultures that have grown within an unpredictable or hostile environment often see the universe as being similarly fraught with dangers. The cosmos itself is characterized by ongoing conflict between opposing forces. Man's role in this process is to appease and support the gods in their battles, in the hope that – if he performs well – he will be allowed to survive. This idea is typical of cultures for whom cyclical rituals are paramount.

The principal deity of ancient Mesopotamia was Marduk, who won his place through defeating the forces of chaos and anarchy. These battles are recounted in the Mesopotamian creation stories known as the *Enuma Elish*. A fundamental part of the cycle of cult festivals – especially those that surrounded the arrival of the Mesopotamian New Year at the vernal equinox – was the re-enactment of Marduk's struggle against the watery goddess Ti'amat. As the hierarchy of the Babylonian state was seen to reflect the celestial pantheon, the king took on the role of Marduk and performed sacrifices to ensure the continued conquest of the omnipresent forces of Ti'amat. When the king was unable or unwilling to take part in these ceremonies, as happened during the reign of the last ruler of the Neo-Babylonian Empire, King Nabondius (555–539 BC), the commencement of the New Year was postponed. Time – literally – stood still for the kingdom. Or, if one

were a Neo-Babylonian, one might have seen these events as proof: the king's failure to commemorate the New Year did, indeed, precipitate the end of civilization as the Neo-Babylonians knew it.

The Aztecs also believed that their participation was needed by the gods in order to perpetuate time. Each of the first four creations had failed because its inhabitants did not fulfil the expectations of the gods. If those of the current world, the 'fifth sun', did not keep faith with the gods, they, too, faced annihilation. In this case, keeping faith with the gods involved human sacrifice to provide them with a constant supply of invigorating blood.

Like many other cultures, the Navajo draw strength from repeating the stories of their creation cycles, particularly during times of uncertainty or change. By reciting the creation stories, the Navajo 'touch base' with the essence of who they are and where they have come from. They believe that the universe is a very delicately balanced whole, in which violently opposing forces are brought into a tenuously held harmony. Only man has the ability to destroy this balance and, by extension, it is man's duty to ensure that this balance is respected and 'cured' should things begin to swing out of kilter. This 'curing' takes place by revisiting the stories of their own creation and the primordial harmony of the world.

007 Fred Stevens ('Grey Squirrel')
The Whirling Log
Sandpainting (powdered sandstone and charcoal in resin), 1966
London, The Horniman Museum and Gardens [16.9.66/1]

The Whirling Log, or *Tsil-ol-ni,* story is generally told during the seventh day of the Navajo curing ceremonial. A young hero travels downriver to a large lake, where he meets the gods, who tell him the secrets of farming and give him seeds. In the painting by Fred Stevens, we see four logs forming the shape of a cross in which male (black with a round face) and female (white with a rectangular face) gods or *yei* stand alongside each of the four sacred plants of the Navajo: corn, whose ears and leaves have been realistically rendered, squash, grey bean and black tobacco. At one side of the painting, there is a white-coloured *yei*, 'bringer of the dawn', who represents the east. At the other, there is the yellow 'twilight bearer', who represents the west. At either side is a traveller *yei* called

B'ganaskiddy. His apparent hump-back is actually the rendering of a deerskin bag, decorated with five eagle plumes. The whole of the composition is encircled by the protective hermaphrodite *yei,* known as the rainbow god. It protects the painting on the three sides from which evil might enter – the north, the south and the west. In using the events of their own creation story to re-establish harmony in the tribe or its members, the Navajo are amongst those cultures who see time as a flexible medium. Events of the past – such as their own creation – can be called into the present as a vital and curative force.

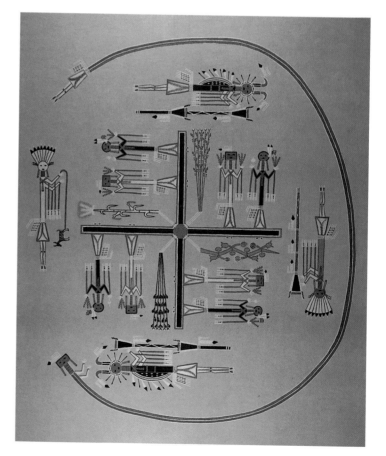

008 Babylonian
'Creation Tablet I'
Baked clay, 10th century BC
London, The British Museum
Dept. of Western Asiatic Antiquities
[WA 93014]

The creation story of the Babylonians, the *Enuma Elish,* appears to date from the first Babylonian dynasty (2225–1926 BC).'Creation Tablet I' recounts the first moment of creation and the birth of Marduk, who brings order out of chaos by defeating Ti'amat.

Also

• Neo-Babylonian, *Tablets outlining the ritual of the New Year, ca.* 1600 BC, London, The British Museum, Dept. of Western Asiatic Antiquities [WA: DT 15, DT 109 and DT 114]

009 Aztec
Hieroglyphic symbol for the 'Sun of Movement'
Stone, mid-14th century to 1521 AD
Berlin, PKB Museum für Völkerkunde
[IV Ca 6738; Uhde 1861]

The Aztecs were one of those cultures that believed that a series of cosmogonic episodes had led up to the present era, which was the fifth epoch or 'Sun' to have occurred. If mankind fails to worship the gods properly, the present epoch, or the 'Sun of Movement', will end with 'movement' – or earthquakes.

010 Northwest Coast of North American (Nisha Indians)
Raven rattle
Carved and painted wood, 19th century
London, The British Museum, Department of Ethnography
[1976.AM.3.19]

Among many tribes of the Pacific Northwest of North America, the raven is known as the world creator because he stole light from the Chief of Heaven and released it for the benefit of mankind. In most representations he is shown carrying a small box in his mouth – the box in which he carried sunlight to mankind. Rattles, such as this one, are generally carried by chiefs during the potlatch ceremonies. Through the rattle, the power of the raven can be called upon to help the tribe. The invocation of a creator-god during ritual ceremonies in times of transition underlies the widespread belief that by returning to one's temporal origins one can rediscover 'natural' strengths.

011 Aztec
Xipe Totec
Volcanic stone, *ca.* 1500
Washington, D.C., The National Museum of the American Indian
[16/3621]

Xipe Totec was the god of springtime, who presided over the renewal of vegetation and the coming of the spring rains. His principal festival occurred in March at the end of the dry season. At the beginning of the festival, a number of gladiatorial games provided the requisite victims for sacrifice. These victims were then flayed and their skins were worn by Xipe imitators for one Aztec 'week' (20 days), during which they begged for alms.The symbolism of donning a flayed skin has been explained as representing the act of germination of the seeds. The flayed skin is like the dead husk of a seed enclosing its live germ.

012 South Indian

Shiva Natarāja ('Dancing Shiva')
Bronze, 18th century
London, The Victoria and Albert
Museum [1062-1873]

The Lord Shiva is shown in his manifestation as *Nrtya-mūrti*, the cosmic dancer. He is the embodiment and manifestation of the eternal energy, set within five categories: 1) creation through unfolding and pouring forth, 2) maintenance, 3) destruction or taking back, 4) veiling his transcendental essence through the garb of appearances and 5) bestowing grace through the manifestation that accepts a devotee. His form, balanced for all eternity between movement and stasis, embodies the very essence of the god. He is duality reconciled. In one of Shiva's four hands, he carries a double-headed drum or *dumaru* that signals the creation – as sound was the first element to evolve in the unfolding cosmos. The *dumaru* itself has become a ritual object for one

form of Tibetan religion in which aspects of Tantric Hinduism and Buddhism have become fused. According to Tantric belief, the two triangular halves of the drum represents the two principles of the *linga* (or the erect phallus) and the *yoni* (or the 'triangle of nature' and the 'receptacle of the seed'). As a ritual object, Shiva's drum is constructed from two human skulls, halved and placed together, crown-to-crown. Graveyard symbolism is an extremely important aspect of 'Tibetan Tantra'. By flicking one's wrist and rotating the drum, small pellets attached to flails beat upon the stretched skin of the drumhead to make a sound which imitates Shiva's drumming that initiated the creation of the world.

Also

• Tibetan, *Hand drum* or *dumaru*, 19th century, London, The Science Museum [Wellcome Coll. 15752]

013 India (Bengal)

Kālī striding over the corpse of Shiva
Painted clay, late 19th century
London, The British Museum,
Dept. of Oriental Antiquities
[OA 1894.2-16.10]

The goddess Kālī is recognized as an aspect of Shiva. As such, she is his consort – the destructive, all-pervading 'power of time'. She is shown with black skin and a terrible, fang-filled mouth. She wears a garland of human skulls and often a crown of skulls as well. She usually carries a severed head and a flail in two of her four hands. Despite her appearance, she is the destroyer who activates the creative powers of Shiva. As here, Kālī is often depicted

striding over the murdered form of her dead consort. The goddess, in this form, represents what one might call the active principle of reality. Shiva is inert, passive and without qualities. As she treads on him (or, in an alternative version, as she sits upon him to begin sexual union), he is brought back to life and begins to swing his drum, thus calling all life into being again. The cyclical life-death-life-death nature of this pairing is fundamental to Hindu belief.

Also

• Indian (Orissa), *A Cosmological Globe of the Original World with Mount Meru*, 19th century, London, The Victoria and Albert Museum [IM.499-1924]

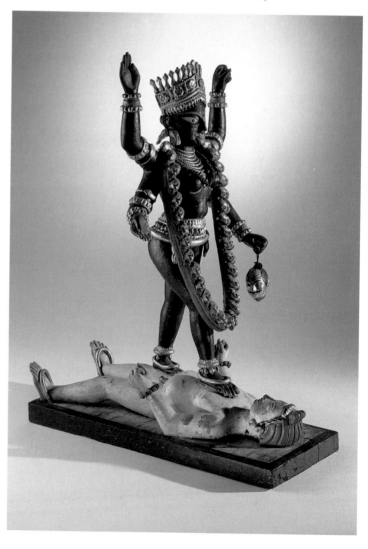

There are certainly cyclical aspects of time in Hinduism, but it is not time that binds man to god. The fundamental cycle is the process, the *dharma*, of birth, growth, decay and death. One aspect of *dharma* does not and cannot exist without the other; and many of the most powerful images of the Hindu gods convey this concept. The idea is embodied in the figure of the Lord Shiva himself, who is not only the principle of disintegration (*tamas*), but also the principle of enjoyment (*ananda*), which is fundamental to the perception of time. This duality can be expressed in beauty – as in the manifestation of dancing Shiva Natarāja – or it can be so magnificent as to invoke terror in the beholder. In the *Bhagavad-gītā*, for example, when Vishnu appears to the prince Arjuna as the Creator and Destroyer, he explains his terrible form by saying: "I am the immortal and also death .... I am the Death that seizes all, the origin of all that shall be .... Know I am Time, that makes the worlds to perish when ripe, and comes to bring on them destruction".

In Hinduism, man's role in life is to perform as well as he may in the state of life into which he has currently been born. And, as man is subject to continual rebirth, so the world in which he lives will continue to be reborn again and again. For the Western mind, used to delineation, the all-inclusive and ever-evolving system of Hindu polytheism is often difficult to grasp. Whereas most monotheistic religions look towards finding that one, unassailable answer to all questions, in Hinduism there is an inherent complexity. There are numerous different stories relating to the manner and sequence in which the world was created. The fundamental principle seems to be that the origin of the universe lies in the cosmic mind – its desire to create and to act. The manifestations of this action, however, are theoretically endless. Moreover, as the world is subject to eternal re-creation, gods such as Vishnu might take on another manifestation to create a new world. Consequently, time itself is largely relative. It does not possess an independent reality of its own.

014 **Indian (Madurai)**

Vishnu as the Supreme Creator and Destroyer of the Universe
Gouache on stiffened canvas, late 18th century
London, The British Library, Oriental and India Office Collections
[Additional Ms 15504 (A)]

This image was copied from the 18th-century wall paintings in a temple of Vishnu-Perumal in southern India. It depicts Vishnu as he appeared to the hero Arjuna in the *Bhagavad-gītā* (chap. 12). All of the worlds that ever were and ever will be are contained within his body. From every snaky hair on his body, a world is suspended. Each world will undergo a certain number of dissolutions and renovations until the whole universe is destroyed in 311,040,000,000,000 human years – or in 100 years of Brahmā's lifetime (one year in a human life equals 24 hours in the life of a *deva* (god); and 1000 years of the *deva*'s time is equal to one day in Brahmā's lifetime).

Also

- Indian (Jaipur, Rajastan), *Avatar of Vishnu as the Tortoise Kurma and the team of gods churning the Ocean of Milk, ca.* 1800, London, The British Museum, Dept. of Oriental Art [OA 1940.7–13.026]

**015 Australia, Bininyiwui
(Djambarrpuingu clan)**

Rivers, palm trees and stones
Pigment on bark, 20th century
Paris, Musée National des Arts
d'Afrique et d'Océanie
[MNAO 64-9-74]

There is no single, shared creation
myth amongst the Aboriginal people.
Though there may be common
elements in the body of stories each
clan or tribe inherits, the more
important ties are to the specific
ancestors from which each of these
different groups is descended. Before
man was born, the ancestral beings
emerged from the Earth and gave
shape to the world. Most of the
ancestors adopted the forms of well
known bush animals, such as the
kangaroo, possum or witchetty grubs.
And, as they lived, the ancestors
travelled along certain paths,
burrowed in the ground, created pools
and so on. This painting depicts the
rivers and palm trees that were
created by the primordial serpent
Yurlunggur. As he chased the mythical
sisters Wawilik, the force of his body
slithering through the mud carved
river beds and threw up rocks and
boulders. The zigzag lines running
from top to bottom are the rivers, the
circles are the stones and the palm
trees along the sides of the rivers are
depicted as triangular broom-like
growths. As a descendant from a
particular ancestor, each tribe
member is instructed not to harm, kill
or eat the flesh of his or her particular
ancestral species – because it is from
the collective energy of that species
that the individual and the tribe draw
their strength. Although each member
of a clan will call upon the strength of
his or her ancestor, only certain
members of the group are charged
with 'remembering' the stories of the
ancestor. These stories are recounted
at all group gatherings and form the
subject matter of the so called
'Dreamtime' paintings.

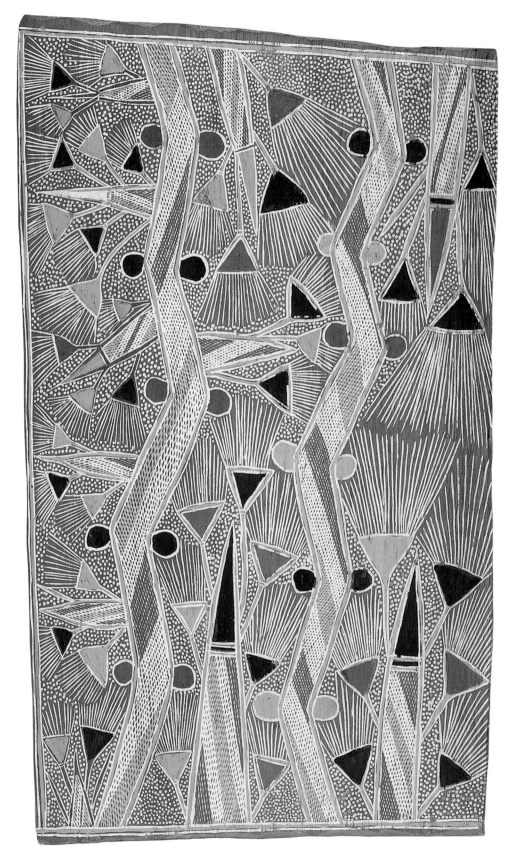

● There are a number of cultures in which the events of the world's creation are not relegated to a historical past, but exist as part of what has been termed 'a living past'. In both Maori and Aboriginal Australian views, the past is something that exists side by side with the present.

The past is the active principle that explains why things are how they are. As such, the stories attached to the creation of the earth have a tangible reality in the fact that they inform man how to live now. The purpose of the tribe is to keep those stories current and vital. To this end, the idea that one's ancestors existed only in the past is nonsensical. For the Aboriginal Australians, the vitality of their ancestors is evident in the features of the landscape which surround them. A mountain, a sink-hole or a river are not places where the ancestors existed, but places where they exist.

This animation of the past goes one step further with the Maori. If one were to make comparisons between Christian and Maori time, it would seem that the Christian concept of the present is contingent on the future, while Maori time is based on a firm integration of the past with the present. To quote the New Zealand historian, J.C.H. King, "History is the umbilical cord that connects us to the past and nourishes the present by making the present understandable. Unless we know where we come from – as individuals and as a nation – we cannot know where we are." In this way, for the Maori, the past is the vital part of both the present and the future. The term used to describe the past is *Ngā wā o mua*. It means 'the time in front of us'. Since the past is knowable, it is always 'in front of us'.

016 Maori

Pare or door lintel
Carved wood, 19th century
Dublin, National Museum of Ireland
[1896.747]

Since the early years of the 19th century, the meeting house has become the central and most important architectural feature of every Maori settlement. The structure of the building is generally believed to have symbolic associations with the body of the chief and with the tribal ancestors. The ridge pole of the building, for example, is regarded as the backbone of the tribal ancestors. The front of the house and the porch are associated with the Maori past; the interior of the house symbolizes the present and the future. As the Maori meeting house is a sacred space, the threshold has particular significance as a boundary. The most important aspect of the threshold is the *pare* or door lintel, which is often carved with aspects of Maori creation mythology. There are two main types of *pare* decoration. The first consists of a central female figure (usually

Papa or 'Mother Earth') flanked by two ancillary figures shown in profile (so called *manaia* figures). Papa is often shown giving birth to Hinenui-te-Pō, the goddess of death. One interpretation is that Papa gave birth to death in order to punish humankind and withhold from it the gift of immortality. Others, however, see Papa as a protective figure, whose genital display removes the potentially destructive powers of visitors to the settlement. The second type of door lintel (shown here) has two or three male figures, standing with their arms upraised and separated by large spirals. These figures have been identified as a representation of Tāne Mahuta, god of the forests, and his brothers, who are shown in the act of separating the primal parents of Sky and Earth (Mother Earth being shown as a decorated horizontal band, with a central omphalos-type knot and two spirit-heads at either side). By doing this, they allow light – depicted as large spirals – to penetrate the world of the Maori.

Nicolas Poussin, *Phaëton asking for the chariot of Apollo* (see 079)

THE Measurement of Time

I n the Old Testament, one of the first tasks that God the Father assigns the newly created Adam is the naming of the beasts: "each one was to bear the name the man would give it". As social historians, psychologists and anthropologists will confirm, naming is one of the tools that man uses to circumscribe objects and to make them his. Numbering is another tool. It helps man to define and structure the world around him. Both naming and numbering – setting identity and scope – are an essential part of what makes us human.

The history of man's attempts to structure time ranges from the prehistoric to the present day. It is a history that follows the erratic trail of a number of different needs, means and motives. The earliest motive was survival. By counting, one was able to understand the rhythms of nature. The most basic need to count accurately, though, seems to have been stipulated by religion. The patterns of nature were believed to reflect the will of the gods. By counting, one could begin to approach the divine. The means – the structure of the numbers themselves and how those numbers were applied to observed phenomena – forms the basis for the story of a bewildering array of myths and maps, tallies and timekeepers, calendars and clocks.

The Sun

The Sun, our closest star, is used as the primary timekeeper in almost every culture on Earth. Its apparent presence or absence defines the most basic division of the day into hours of light and darkness. Our inclined orbit around the Sun creates the time segment we call a year and causes the alternating rhythms of the seasons. Personifications of the warming Sun as a benevolent deity – a healer, a provider and a friend to man – are nearly universal.

The iconography for the Sun god in both the East and the West has been remarkably consistent from antiquity to the modern day. Sun deities are usually male and are often eternally youthful. The Sun is regularly portrayed as either the consort or the brother of the Moon. He is usually depicted standing in his solar chariot, drawn by two or four horses, and is regularly depicted with streams of light issuing from his head like a halo. Often he is accompanied by the Dawn, and sometimes by personifications of the hours. The extent to which the iconography of the god remains unchanged can be gauged by comparing a fifth-century BC red-figure vase of *Helios in his quadriga* (**018**) with Giuseppe Bazzani's painting of *Phoebus Apollo* completed some time around 1750 AD (**022**).

In Babylonian mythology, the Sun god Shamash holds a position within the divine pantheon not unlike that of the Greco-Roman Sol-Apollo. He is not the most powerful deity, but he forms part of a celestial triad with Ishtar (the planet Venus) and Sin, the god of the Moon. Like Apollo, Shamash drives the chariot of the Sun across the heavens each day.

In India, the image of Sūrya, the Sun god, riding through the skies on his chariot is established in the earliest Vedic texts. During this period, the Sun was established as the supreme soul, the creator of the universe and the source of all life on earth. Or, to quote the *Brhavishya-Purāna*, "The Sun is visible divinity, the eye of the world, the maker of the day. Eternal, no other deity can be compared to him. He is the source of time. The planets, the stars, the spheres of the elements, the divinities of life, the lord of the winds and of fire and all the other gods are but a part of him."

Sūrya is at the centre of all creation and occupies a place between the 'manifest' and 'unmanifest' spheres. Interestingly, this apparently Indian formula is actually a rewording of Aristotelian theories on the structure of the cosmos and reflects the profound influence of Greek thinking on the Hindu sciences. In essence, this placement of the Sun mirrors the so called Ptolemaic world order, in which the Sun nestles between the outer planets and the sublunar (manifest) worlds of change and decay. As Hinduism developed, the primacy of Sūrya was supplanted by the greater gods. Nevertheless, a number of the most important early temple complexes are dedicated to the god.

017 **Eastern Indian (Bihar)**
Sūrya, the Sun god, with his charioteer Aruna
Black basalt, 9th–10th century (Pala period)
London, The Victoria and Albert Museum [IM. 109-1916]

Sūrya is often depicted standing on a lotus and holding two lotus blossoms in his hands. Usually, he is shown in his solar chariot, drawn by a team of seven bright horses driven by his legless charioteer, Aruna. He is also regularly accompanied by his four wives: Samjñā (Knowledge), Rājñī (Queen), Prabhā (Light) and Chāyā (Shade); and by his two consorts Usahas (Dawn) and Pratyūsha (Twilight), who, in this case, are shown shooting arrows 'before' and 'after' the god. His dress is not Indian. This reflects the fact that the iconography for Sūrya was probably originally Greek, having reached India through Iranian intermediaries. Unlike the great gods of Hinduism, such as Shiva or Vishnu, whose perpetually divine essence appears on Earth through different manifestations or *avatāras*, the lesser gods die and are reborn with each successive re-creation of the world.

018 Greek (Apulia?)

Red-figure krater with Helios rising
from the sea in his quadriga
Terracotta, 420 BC
London, The British Museum, Dept.
of Greek and Roman Antiquities
[GR 1867. 5–8.1133, Vase E 466]

The overall subject of this vase is the
dawning of a new day. In the middle
of one side of the vase, Helios-Apollo
is shown rising from the sea in his
quadriga, drawn by four winged
horses. Behind him is Selene, the
goddess of the Moon, shown riding
side-saddle with her mantle drawn
over her head. On the other side of
the vase, there is a scene of the
winged and crowned Eos (the Dawn)
trying to seize the handsome hunter
Cephalos. The young boy shown to
the right of this pair appearing to flee
is Eosphoros or Lucifer – the Morning
Star – who disappears from the night
sky with the arrival of the bright glare
of the Sun. Cavorting in the waves
beside and below the Sun god, there
are a number of nude young boys.
Some are shown about to dive into
the sea and others are shown

emerging from it. These have been
identified as those stars which rise
and set with the dawn. Their
appearance on this *krater* underlines
the importance of these stars as
timekeepers in the Greek reckoning
of time. References to the rising and
setting of stars at dawn and at dusk
are one of the basic components of
all the astronomical and calendrical
texts of the period.

019 Babylonian (Sippar)

Tablet representing the worship of
the Sun god
Schist, *ca.* 870 BC
London, The British Museum, Dept.
of Western Asiatic Antiquities
[WA 91000]

The tablet is sculpted with a scene
representing the worship of the Sun
god at the Temple of the Sun in
Sippar. It shows the god, Shamash,
seated within his shrine at the right
of the composition. In his right hand
he holds a disc and a bar, symbols of
the Sun and its circuit. Above his
head, there are three circular marks
that represent the three major
planetary deities of the Babylonian
cosmos – the Moon, the Sun and
Venus. Outside the shrine, there is an
altar table upon which an image of
the Sun has been placed. It appears
to be held in place by a system of
ropes held by two celestial beings
perched on the roof of the shrine. In
front of the shrine, there are three
figures. The first of these represents
the high priest, who leads the king by
the hand to the shrine. The pattern of
wavy lines at the bottom of the relief
is meant to be the celestial ocean
with the four small roundels
indicating the four cardinal points.

020 Romano-Egyptian

Harpocrates, the young Horus
Painted limestone, 4th–5th century AD
Oxford, The Ashmolean Museum
[1971–993]

The Egyptians seem to have had
multiple versions of various gods that
they recognized as incarnations of the
life-giving principle of the Sun. Amon,
one of the four great creator deities,
became recognized as the national
god of Egypt when his name and
identity were connected to the solar
deity, Ra, in the second millennium
BC. Atum, another of the creator
gods, was also called Ra-Atum and
personified the evening Sun. In later
myths, the figure of Ra became fused
with that of Osiris the god of the
underworld, since the Sun was
believed to journey through the
underworld every night. The Sun god
is also associated with the falcon-
headed Horus, who is often depicted
carrying the solar disc on his head.
Harpocrates, the child Horus, is often
cited as the god of the first two hours
after dawn, a variant of 'Ra, the
Horus of the Horizon'. As he was
born from the marshes of the Nile,
the lotus blossom is his attribute – a
symbol of the rising Sun. The date of
his birth coincided with the winter
solstice in late December and the
return of longer days.

In China, the Sun and Moon were husband and wife. Together they symbolized the duality of the universe: the Sun as *yang*, or the male principle, and the Moon as *yin*, the female principle. Both gods were elevated to immortality quite late. The god of the Sun himself started life as a mortal, known as Shen Yi, 'the divine archer'. When his wife, Heng E, accidentally ate the pill of immortality, she floated into the heavens and took up residence in the Moon. The Immortals then took pity on Shen Yi and offered him the Palace of the Sun and bestowed the gift of immortality. He was given a bird to help him in his task of maintaining the regular schedule of the Sun's risings. The only restriction that the gods imposed was that, whereas he would be able to visit his wife in her palace, she would never be able to visit him in his. Shen Yi agreed to visit Heng E on the fifteenth day after every New Moon. It is this regular meeting of husband and wife and the joining of their *yin* and *yang* that causes the Moon to shine so brilliantly when it is full.

For the North American tribes of the Great Plains, the Sun Dance formed one of the most important foci of the ritual year. By the time the Western anthropologists had arrived to study the Sun Dance, the dance had evolved to a point where each tribe had its own variant and its own alleged purpose for the ceremony.

Nevertheless, it does seem that there must have been a common origin for the dance, as well as a common purpose: the dance probably originated as a re-creation ceremony – a ritual that made everything new again by rejoining the tribe to the circumstances of its own creation. This suggestion is supported by the fact that most of the dances coincided with the early summer months, when the land was bountiful and the tribes were at their strongest.

Whereas the Sun is certainly the source of all life on Earth, it is perhaps not surprising to discover that, in those northern countries where the latitude is so high that the people can go for months without daylight, the Sun often plays a relatively marginal role in the pantheon of gods. In the case of the Inuit, for example, the primary celestial god is the male Moon spirit, Tarqeq, who is concerned with fertility and moral propriety and controls the patterns of animal migration. Similarly, in Norse mythology, the god of thunder (Donar or Thor) and the god of battle (Wodan) are much more important than the Sun.

021 Attributed to Mató-topé (North American, Mandan Tribe)
The Sun and warriors
Painted bison robe, second quarter
19th century
Berne, Bernisches Historisches
Museum [NA 8]

Mató-topé was a chief of the Mandan tribe, one of the small village tribes settled in the Missouri River basin. He is renowned for his paintings on robes and skins. This painted bison skin combines the symbolic – in the highly stylized depiction of the Sun in the centre – with events taken from the life of the tribe. It possibly relates to aspects of the Sun Dance ceremonials. As one central feature of the ceremony was a celebration of prowess of the young warriors, the image concentrates on scenes of warfare and hand-to-hand combat. The groupings of marks scattered throughout the composition appear to be what are called 'counting coup'. Each horse-shoe mark and each dot indicates a horse or man captured or killed during battles fought that year.

023 Chinese

Shen Yi (the Sun god)
Porcelain, 19th century
London, The Victoria and Albert
Museum [HMC CD 80/2]

The immortals who inhabit the Sun
and the Moon are not ranked very
highly in the Chinese pantheon. In
many ways, they are merely
functionaries who carry out the never-
ending task of ordering time. Shen Yi
is depicted as an old man, dancing,
and holding his solar disc above
his head.

Also

• *Heng E* (the Moon goddess),
19th century, London, The Victoria
and Albert Museum [HMC CD 80/1]

022 Giuseppe Bazzani

Phoebus Apollo
Oil on canvas, *ca.* 1750
Amherst MA, Amherst College, The
Mead Art Museum [Gift of the
Samuel H. Kress Foundation. K. 1270
(1961–75)]

Bazzani's exuberant painting of a
handsome and youthful *Phoebus
Apollo* shows the dramatically
foreshortened Sun god driving a two-
horse chariot or *biga*. According to
tradition, one horse should be dark
and symbolize the night and the other
should be light, symbolizing the day.

Also

• Niccolò Pellipario (after engravings
by 'GGF'), *Majolica with Sole* (The
Sun), after 1533, London, The
Victoria and Albert Museum [C.
2272–1910]

024 Islamic (Iznik)

Plate with Leo and the Sun
Blue and white earthenware,
17th century
London, The British Museum, Dept.
of Oriental Antiquities [0A 1983.G123]

It was part of astrological doctrine
that each planet was said to be
'happiest' when it was located in or
lay in harmonious aspect to the
zodiacal sign it occupied when the
universe was created. The house (or
domicile) of the Sun, for example,
was the zodiacal sign of Leo. The
image of the 'Sun in Leo' found
particular favour in the Middle East
and India. The iconographic formula is
always the same: Leo, the lion, is
shown standing in profile, carrying
the disc of the Sun or a man whose
head is formed by the disc of the Sun
on his back. This image was
subsequently adopted as the royal
emblem of the Seljukian sultans and,
until 1979, appeared as the national
emblem on the state flag of Persia.

Nineveh, Fragment of an ivory prism for calculating the lengths of hours throughout the year, 8th century BC, London, The British Museum [WA123340]

From Observation to Record

Astronomy in Prehistory and the Early Civilizations

John North

Stonehenge, silhouette of the Heel Stone at midsummer sunrise

The islands of Britain and Ireland are extremely rich in prehistoric archaeological remains. In particular, they are the home to some of the most intriguing rings of large stones (known as megaliths) to be found anywhere. The precise number of these sites depends on how we define them, but seems to be of the order of a thousand, with most of them having been constructed during a period between three and five thousand years ago. During this same period, there were also comparable rings made from timber posts. Unfortunately, most of these timber monuments have long since disappeared, but there is evidence to suggest that there must have originally been at least several hundred of these wooden structures as well.

While some archaeologists tend to be deeply suspicious of those who propose complex astronomical explanations of such prehistoric monuments, it is generally accepted that many (perhaps the great majority) were arranged along an axis that, in some way, pointed towards either the rising or the setting of the Sun at the time of the midsummer or midwinter solstice. For example, there is ample evidence that the best known of all megalithic monuments – Stonehenge – is aligned towards the midwinter solstice. The traditional view is that priests or persons unknown stood in the centre of the monument and looked to the north-east, over the Heel Stone, over which the Sun supposedly rose at the time of the summer solstice. Actually, it has been known for a long time that midsummer sunrise has never been precisely over the Heel Stone and that there is a far better case that the focus of attention was the setting midwinter Sun, and that it was observed from the Heel Stone by someone looking in precisely the opposite direction to that generally supposed. Perhaps observation was done in both directions, according to season. There are certainly monuments at which the general plan suggests a summer observance, just as there are others at which a midwinter observance was made. Moreover, with some megaliths, it seems to be the extremes of rising and setting of the Moon that were the focus of attention. Each case must be argued on its own merits, but the general principle that most megalithic structures were aligned according to risings and settings of one of the principal luminaries seems to hold good.

The rising and setting of the

Although astronomy is the best route to an understanding of the purpose of these stones, it is not useful to suppose that Stonehenge or the megalithic or timber rings generally were, in any real sense, astronomical observatories. They were places of religious observance with reference to the apparent movements of the Sun and Moon that were already well understood. For example, the rising and setting of the Sun at its most northerly (at the time of the summer solstice) would probably have been viewed as a climax in the enjoyment of life within the human cycle. The rising and setting of the Sun at its southernmost point during the time of the winter solstice would have marked in a sense both the death of the old year and a beginning of hope for a 'new' year. The setting of the midwinter Sun at Stonehenge was, therefore, a symbol of death twice over. Observed from the right-hand side of the Heel Stone, through the central gap in the monument, the brilliant light of the setting Sun against the bright background of the western sky would have produced a memorable silhouette of the monument. A somewhat similar extreme of the setting Moon could have also been observed from the Heel Stone, to the left of the silhouette.

Even if the megaliths were 'octile' calendars, this does not mean (as so many have claimed) that they were 'extremely sophisticated calendars'. Nevertheless, the fact remains that, with or without an unproven and unprovable calendar function, such monuments were extraordinarily sophisticated. In their own way, they were far more ingenious than many of the calendars of later history.

Since, in a number of prehistoric and early historical cultures, celestial observation was often intimately associated with the rituals of death, it seems likely that there was a religious motive for building these monuments. The first indicator of this is the huge amount of energy it must have taken to build these structures. The

Celestial observation was often intimately associated with the rituals of death

The broad principle that observations were made of the Sun and Moon at the extremes of their places of rising and setting on the horizon can hardly be challenged. It is a custom or tradition that has passed down to modern times, within agricultural and religious communities. But it is all too easy to colour our views on the motivation of prehistoric communities by reference to later practices. At a time in history when we are obsessed with our own calendar, it is natural to see prehistoric attention to the cycles of the motions of the Sun and Moon as in some loose sense 'calendrical'. The question that must be asked, however, is what such a claim might imply.

We can think of the orientated stone circles as calendars in the sense that they helped to determine the beginning, and perhaps the length, of successive years – assuming that people took the trouble to count the days between extremes of rising or setting. But a year-marker (if that is all that it is) is hardly a calendar in the fullest sense. A calendar surely entails a subdivision of the year into its parts, as in the case of the great calendars of antiquity. In fact, most early calendars involve a strong lunar component, and many try to take into account the almost irreconcilable motions of both Sun and Moon. While it would be rash to say that this was not done by the builders of the megalithic monuments, there is no real evidence that it was. Some scholars, for example, have claimed that certain

second is the time that was expended – at least in relation to the small human populations of the time. Historically speaking, a monument like Stonehenge did not suddenly appear out of the blue. It was the culmination of perhaps two thousand years of earlier tradition in applying simple astronomical and geometrical techniques to the building of monuments of a very different sort. One example of these earlier structures is that of the so called 'long barrows'. These were elongated mounds of earth or stone, usually covering a chamber at one end (or in the middle), in which members of the community were buried. Some British long barrows date back more than six thousand years before the present. There are many variant types, but most long barrows were flanked by ditches, and there are reasons for thinking that these ditches were used to observe the rising and setting of particular stars over the mounds. In this way, the mounds themselves created an artificial horizon. Indirect support for this idea can be found in the fact that long barrows, often several kilometres apart, were lined up in threes and fours in a network covering the terrain in such a way that the lines joining them were directed to the risings and settings of the same bright stars as were viewed over the barrows. The risings and settings of bright stars seem also to have been observed over monuments of other types from this early period, such as, for example, structures like the Stonehenge *cursus* (a strip flanked by ditches several

Sun at its most northerly would have been a climax in the enjoyment of life

monuments appear to imply a division of the range of the Sun's rising (or setting) positions into eight parts, and, from this, they have proposed that early man divided his time in accordance with an eight-part year. There is some evidence to support this suggestion, such as the fact that, in historical times, Scandinavian farmers are known to have used a similarly crude division – crude to the extent that it will not produce 'months' containing equal numbers of days.

kilometres long at the end of which was an extremely important long barrow). The same is likely to have been true of some of the British downland chalk figures, such as those at Uffington and Wilmington. It seems unlikely that a system for collating the risings and settings of particular bright stars served any practical calendrical purpose.

2 · THE MEASUREMENT OF TIME

What is to be counted as 'prehistoric', in the sense of belonging to a time before written historical records, clearly depends very much on place and culture. Prehistory for the Egyptians or the Babylonians is much earlier than it is for northern Europeans. The earliest Egyptian pyramids date from much the same period as Stonehenge, but, by the time the Pyramids were built, the Egyptians had been keeping historical records for centuries.

The astronomical riches of the Middle East are not to be summarized briefly, but some of their qualities are worth mentioning for the similarity they seem to bear to the monuments built by the unlettered northern Europeans. All these cultures paid attention to horizon phenomena – that is, to risings and settings. In all cases, astronomical practice seems to have been the servant of religion. And, almost incidentally, astronomy gave rise to techniques of

or constellations were therefore sought that marked, by their heliacal risings, the beginnings of those thirty-six weeks. The main criterion for choosing a star seems to have been its closest possible similarity to Sirius/Sothis, especially in being invisible for about seventy days in the year.

Forget, now, the reasons for which the stars were chosen – reasons having to do with the division of the year, that is, with the calendar. We simply have thirty-six stars, or groups of stars, recognized as being of great importance, and during any night their risings will occur at moderately regular intervals. In fact, these stars generally divided the period of the night into twelve equal parts, and the night was finally regarded as having been exactly so divided. Daylight was later divided into twelve hours by analogy with the night. And in this way we were given the twenty-four hours of our day.

There have actually been two fundamentally different 24-hour systems in play

timekeeping and geometry (for instance the theory of proportions) in ways for which it was not originally intended.

There is, in all timekeeping, an element more or less forced on us by the cycle of the Sun. The biology of the living world makes the day and the year of paramount importance. The cycles of the Moon, on the other hand, have similar repercussions on human life – the menstrual cycle in women and the tides of the sea, for instance. But there are human conventions in timekeeping that are not at all as compelling. The day that we divide into 24 hours, for instance, could just as easily have been divided into ten or twenty or any other number, given a suitable procedure for measuring the divisions. The week of seven days is likewise a rather arbitrary unit, although the week very probably has its origins in the quartering of the month, as judged by the Moon's quarters. Why we measure time in hours, minutes and seconds is a much more complex problem.

The ancient Egyptians were particularly interested in the risings and settings that happened just before the rising of the Sun or just after its setting. As the Sun seems to pass round the sky in the course of a year, there will be periods when any number of stars will be 'invisible' because they are above the horizon when the Sun is also there. The time of first morning visibility after a period during which a given star rose only in daylight was an important Egyptian calendar notion. We give it the technical name of 'heliacal rising', and the Egyptian calendar is largely based on a succession of heliacal risings of selected bright stars. The brightest star, Sirius (Sothis), was the most important marker of all, but other stars were also needed as suitable time-markers. A decision was made to divide the civil calendar year into thirty-six 'weeks' of ten days each. Thirty-six stars

The ancient Babylonians also divided day and night into twelve equal parts each. Through the influence of Babylonian astronomy on later Greek astronomy, the system became customary throughout the ancient Near East, North Africa, the Roman Empire and, eventually, throughout Europe.

The shared idea of a 24-hour day, however, is not as straightforward as it might seem. Well into the modern era, there have actually been two fundamentally different 24-hour systems in play. In one system, the daylight was divided into twelve parts and the night was also divided into twelve parts, despite the fact that the length of both varies with season. This practice resulted in hours of varying length during both the day and the night throughout the year and the resulting hours are known as 'unequal' or 'seasonal hours'. By contrast, if one measures time on a water clock, or by the constant daily rotation of the heavens, hours will be more or less equivalent to our own 'equal hours'. Quite often these two different systems for counting the 24 hours of the day would exist side by side.

An ivory prism from Nineveh, dating to sometime during the eighth century BC, now in the British Museum, shows that converting between the two systems of time-reckoning (equal and unequal hours) was then clearly conceived of as an astronomical problem. The Babylonian astronomers used a system of a 24-hour day that was made up of 12 double-hours (beru), each divided into 30 parts (USH), one of which was equal to four minutes of our time. Like the Egyptians, the Babylonians had made use of certain fundamental stars (called ziqpu stars) from the horizon position of which time and the calendar were reckoned.

Stonehenge, midwinter sunset over the stones

Where, then, did the division into sixty arise? This is one of those questions to which only a very general answer can be given, but we must begin with arithmetic. The Babylonians are notorious for having given us a so called 'place-value notation'. When we write a decimal number, say 2539, we mean it to signify 9 units, plus 3×10 units, plus 5×10^2, plus 2×10^3. The place where the number is in the sequence tells us by how many times the power of 10 the number should be multiplied. The Babylonians devised an earlier form of this system, where the 'base' was not ten but sixty. Thus 2539 in this system would mean 9 units, plus 3×60 units, plus 5×60^2, and 2×60^3. We now use commas and semicolons to separate the 'places' in notation. For example, we might write 17,12,59;11,7 to mean $(17 \times 60^2) + (12 \times 60^1) + 59 + (11 \times \frac{1}{60}) + (7 \times \frac{1}{60^2})$. With a sixty-based (sexagesimal) system, the Babylonians could do amazingly complicated calculations.

This is the arithmetical system we still use for minutes and seconds, not only of time but of angular measure. For example, the map coordinate of 51° 28′ 38″N (the latitude of Greenwich) actually means 38 seconds of arc, plus 28×60 of the same unit, plus 51×60^2. Our clock that reads 9:30:45 AM actually is telling us that the time is 45 seconds, plus 30×60 seconds, plus 9×60^2 seconds past midnight.

The Babylonians also used number systems based on 24, 12, 10, 4, or even 2 units in the base, but, somehow, the supremacy of the scale of 60 pulled astronomy in line, and certainly during the Seleucid dynasty the 24 hours of the day were being regularly subdivided according to the sexagesimal scheme. The Seleucid empire was that established in the region by the followers of Alexander the Great, the Greek conqueror of the world between Egypt and India, and it was this originally Greek connection that ensured that the Middle East's great achievements in astronomy became known to Mediterranean cultures. In that way, the powerful and highly consistent Babylonian way of reckoning was handed down to posterity.

The Movement of the Sun and Moon

The Sun and the Moon are the two great timekeepers in the heavens. The rising and the setting of the Sun marks the days and nights. The waxing and waning of the Moon marks the progression of the months. For the Sun to appear in the sky against the same backdrop of distant stars, it takes one year.

In the West, the year was often personified as 'Annus'. He is generally depicted standing or seated and holding the Sun and the Moon in his hands. He is represented as a young man, surrounded by a large ring or band containing depictions of the twelve signs of the zodiac. The pictorial source for this image seems to have developed from three different iconographic traditions. The first is the conflation of Annus with the figure of the Sun, who, by his yearly transit through the twelve zodiacal signs, defines the year. The second is an association with the Greek god Aion, who represents the concept of infinite time – he is the cause that keeps the year ever spinning into the future. Finally, there is also the punning device between the name of Annus, and the Latin words for 'ring' or 'band', *anus* and *anulus*.

Considering the apparent movement of the stars and the planets, it makes sense to assume that the Earth is stationary and that the Sun and the Moon travel overhead through the visible sky from the eastern to the western horizon. The representation of both the Sun and the Moon as riding through the sky in a chariot or boat is widespread. How the Sun got back to the eastern horizon was more difficult to explain. The ancient Egyptians believed that Nut, goddess of the night sky, swallowed the Sun every night. After passing through her digestive system it was 'reborn' with every new dawn. The Babylonians had myths of planetary gods in chariots, but their inherently mathematical outlook suggested that the stars and planets must move in circular orbits above and below the earth. The ancient Greeks seem to have been the first to propose a picture of the universe in which a central, stationary Earth was enveloped by a series of spheres along the surface of which the planets travelled. The outermost of the spheres was the so called 'crystalline sphere of the fixed stars'. This understanding of the structure of the universe – the 'Ptolemaic universe' as it came to be known – remained virtually unchallenged for nearly 2500 years.

The planets (or 'wandering' stars, as the Greeks called them, because they did not appear to circle round the single pivot of the northern celestial pole, but to speed up, slow down, even stop or reverse) seemed to move within a relatively narrow band of the sky, approximately 12° wide. As early as the second millennium BC, the ancient Babylonians had defined the boundaries of this band and identified and named some of the star groupings through which the band seemed to pass. The Babylonian zodiac of twelve signs of 30° each was fully established by the fifth century BC. The list of zodiacal constellations that we use today reflects a Greek adaptation of the original Babylonian system, with the slight modification that took place during the first century AD, when the 30° marking the claws of Scorpio were redesignated as the constellation of Libra.

Another system for celestial timekeeping was based on the movement of the Moon against the background of the fixed stars. Not only did the ancient Babylonians divide the yearly course of the Sun into twelve segments of 30°; they also divided the monthly orbit of the Moon into 28 different sections called 'lunar mansions'. Each 'mansion' equalled approximately one day of the lunar cycle or about 12.2° of the Moon's 360° circuit. In the same way that the animals of the zodiac mark those constellations through which the Sun travels, the lunar mansions are marked by groups of bright stars through which the Moon passes each month. For example, the first lunar mansion encompasses the two bright stars in the head of Aries, the second covers the small stars in the belly of Aries, the third covers the Pleiades, the fourth is located around the 'Eye of the Bull' (Aldebaran; α Tau), and so on. Like zodiacal signs, each of the lunar mansions was also believed to harbour specific astrological powers.

025 German

Scivias (The Salemer Codex)
Pen and ink on parchment, 1200
Heidelberg, Universitätsbibliothek
[Ms cod. SAL. X. 16, fol. 2v]

During the late classical period,
authors began to add diagrams to
their texts in order to help explain
potentially complicated ideas to their
readers. In this diagram, the
personification of the yearly cycle of
life – Annus – is depicted at the
centre. The elements, the seasons
and the months surround him in a
merry temporal dance.

Also

• Missal, *ca.* 1300–10, St Florian bei
Linz, Augustiner Chorherrenstift
[Ms CSF. XI. 221, A, pg 1r]

• Thomas von Cantimpré, *De Naturis
Rerum Visibilium, ca.* 1300, Munich,
Bayerische Staatsbibliothek [Ms
Cod. Lat 2655, fol. 104 v-105r]

• Italian, *Supplicationes Variae*, late
13th century, Florence, Biblioteca
Medicea Laurenziana [Ms Plut.
XXV.3, fol. 14v]

026 Abd al-Wahid

Plate with the twelve zodiacal signs
Glazed blue and white earthenware,
AH 971/1563–64 AD
Berlin, PKB, Museum für Islamische
Kunst [I.1292]

The Greek iconography of the twelve
signs of the zodiac travelled
eastwards into Persia, India and
throughout Islam. Here each of the
twelve signs of the zodiac is depicted
following the Sassavid (Persian-
Islamic) tradition, but is clearly
recognizable to Western eyes: Aries
(the ram) starts the cycle; the Gemini
are a single-bodied creature with two
heads; Leo is depicted as 'the house
of the Sun' (see 024). Virgo is shown
cutting wheat. Sagittarius, the archer,
is taking aim at a dragon that has
sprouted from his own tail.

027 Chinese

*Mirror with the 'five directions' and
the twelve signs of the 'zodiac'*
Bronze, Tang Dynasty (618–907 AD)
Toronto, The Royal Ontario Museum
[928.12.2]

From perhaps as early as 400 BC,
mirrors were regularly buried with the
dead in China. Of various styles, they
were intended to convey a message
of good fortune. This late example
shows the twelve animal symbols
which denoted the regular passage of
time – often mistakenly referred to as
the 'Chinese zodiac'. In fact, these
animals were associated with star
groupings on the celestial equator
that were related to the apparent
orbit of the planet Jupiter, which
takes 11.6 years to complete.
Different animals were assigned to
each of these '12' years, which were
named accordingly the year of the rat,
the ox, the tiger, and so on.

028 A. van Aken

*Planetarium with the Sun and the
Moon*
Brass, ivory and paper, 18th century
Geneva, Musée d'Histoire des
Sciences [649]

Mechanical simulacra of the heavens
have played a useful role in the
teaching and dissemination of
astronomical concepts from classical
Greece onwards. A planetarium like
this, for example, is geared to show
the changing relation between Sun,
Moon and Earth. A candle in the
'Sun' simulates its light.

The way in which the constellations are described and depicted changed remarkably little from their earliest depictions in ancient Sumeria two and a half millennia ago up until the seventeenth century. The continuity and consistency of astrological imagery is astounding. For whatever reason, this body of material was believed to be important and worth preserving.

The lines of transmission of astronomical and astrological learning are quite straightforward. Two equal cultures developed divergent mythologies: Egypt and Babylonia. The conquering Greeks subsumed both cultures and offered a philosophical structure whereby irregularities between the two systems were resolved. In general, the Greeks found the Babylonian system more sympathetic and completely absorbed the notion of planetary gods, the band of the zodiac and the system for measuring celestial movement. This new 'Greek' understanding of the heavens was disseminated to the boundaries of the Alexandrian empire – from Saharan Africa to the Ganges. The Romans inherited this tradition from the Greeks and spread its influence even further. Travellers along the Silk Route exported Hellenistic learning to the Far East. As a testament to the longevity of constellation imagery, it is quite possible to compare a fourth-century Greek globe (**031**) with a twentieth-century one and have difficulty finding major divergences between the two.

It was always a problem, however, how best to represent the sphere of the fixed stars in two dimensions. Two preferred options emerged. The first was to divide the sphere into two equal halves, or hemispheres, and project those hemispheres as if they were flattened circles. This is the solution offered in Albrecht Dürer's *Celestial hemispheres*, printed in 1515 (**030**). Such maps are portable, relatively easy to read and one can plot the stars against the co-ordinates of the ecliptic that run along the edge of each circle. The main disadvantage is that the constellations themselves become distorted. In the northern celestial hemisphere, for example, the massive constellation of Draco (the Dragon) occupies only a very small portion of the very centre of the map, while all the zodiacal signs, arranged around the circular edge of the chart, are exploded to unnaturally large proportions.

If one compares the individual constellations in these maps with their Greco-Roman counterparts from the Farnese *Atlas*, one can detect subtle changes in the iconography of some of the figures that attest to the cultural wanderings that these images have endured. During the Middle Ages, astronomy was kept alive by Arabic and Persian scholars, who, not overly familiar with some of the Greek myths upon which the constellation figures were based, added their own iconographic and stylistic embellishments to the tradition. When Dürer and his colleagues were searching for suitable models upon which to base their own celestial charts, the influence of Arabic iconography was still sufficiently strong to make its mark.

The second way in which one might 'flatten out' the celestial globe is by taking the northern pole as the centre point and bringing up the southern edges of the sphere – as if the whole globe were being opened out like an umbrella. These 'planispheres' are, literally, a sphere projected on to a planar surface. Planispheric projection forms the basis for that most revered of all astronomical instruments – the astrolabe. Moreover, some of the earliest examples of two-dimensional rendering of the heavens use the planispheric format – such as the Babylonian cuneiform planisphere (**029**). Nevertheless, there are major pictorial disadvantages in using such projections for constellation maps. The amount of 'stretching' that has to be done to the southern constellations creates images that verge on the nonsensical.

029 Babylonian (Nineveh)
Celestial planisphere
Baked clay, 650 BC
London, The British Museum,
Dept. of Western Asiatic Antiquities
[WA. K. 8538]

There are several late astronomical texts which provide information on the Babylonian constellations. Depictions or illustrations of how they were actually plotted by the Babylonians are relatively rare, however. In this 'map' the sky has been divided into eight sections. It represents the night sky of 3–4 January 650 BC over Nineveh. The rectangular shape at the top of the tablet has been identified as the constellation we call Gemini and the stars contained with an oval shape are the Pleiades. The two triangles in the lower right segment mark the bright stars of Pegasus.

030 Albrecht Dürer

The northern and southern celestial hemispheres
Woodcut, dated 1515
London, The National Maritime Museum [G200:1/6 A and B]

One convention that dates back to the earliest Greek manuscripts is to divide the celestial sphere into two equal hemispheres. As the main purpose for Dürer's pair of maps of the northern and the southern skies was to provide ecliptical co-ordinates for the stars, he used the ecliptic (or band of the zodiac) as the outer boundary for both maps. Since the line of the ecliptic runs at an angle inclined 23° from the celestial equator, any hemispheres divided along this line will cut the sphere of the heavens at a similar angle. This means that the central point in Dürer's northern map is not the northern celestial pole, but the northern ecliptical pole. The construction of these prints was a collaborative effort. Johannes Stabius, who is credited with arranging the map, was astronomer to the Emperor Maximilian I. Conrad Heinfogel, who 'placed the stars', was an astronomer and mathematician from Nuremberg. Dürer, also from Nuremberg, is named as the artist who 'drew the lines around the images'. This idea of great minds coming together is also echoed in the placement of the four great astronomers in each corner of the northern map: Manilus representing Rome, Ptolemy the Egyptians, Aratus from Greece and "Azophi" (as-Sūfī), the Arabic astronomer.

Also

• Petrus Apianus, *Planisphere* from *Astronomicon Caesareum*, Ingolstadt 1540, London, The National Maritime Museum [PBC 1352]

031 Giovanni Battista Passeri

View of the celestial globe held by the Farnese Atlas (from Antonio Francesco Gori, *Thesaurus Gemmarium Antiquarum Astriferarum*)
Printed book, Florence 1750
London, The National Maritime Museum [AST0006]

The so called Farnese *Atlas* is a 2nd-century Roman copy of an original Greek statue. It depicts the Greek demi-god Atlas supporting the heavens on his back, which are represented as a celestial globe decorated with the figures of the constellations. The globe is one of the earliest pictorial records of the Greek and Roman constellations. It records a vision of the sky that its remarkably close to the one recorded in the *Phaenomena* of Aratus, a poem written in the fourth century BC that is based on the astronomical texts of Eudoxus of Cnidos, written a generation earlier. Possibly more remarkable than the fact that the shapes of the Greek constellations had scarcely changed over a period of 600 years is that these figures are virtually identical to those found on a twentieth-century celestial globe.

Also

• Durgansankara Pathaka, *Sarvasiddhantatattvacudamani* ('The Jewel of the Essence of all Sciences'), 1840, London, The British Library [Ms Oriental 52829, ff. 56v–57r]

Durgansankara Pathaka, *The Jewel of the Essence of all Sciences* (see **031**)

Time in India

It is difficult, if not impossible, to make generalizations about 'Indian time' as people living in the area defined by the subcontinent represent so many different cultural backgrounds. There are at least half a dozen major religions practised in India – not to mention numerous variants of Hinduism itself. There are no pan-cultural constants in India. There is no single creation mythology shared amongst Hindu, Muslim, Jain, Sikh, Buddhist and Christian. The climate – an element that often informs a society's perceptions of time – ranges from permafrost through temperate climes to desert and sub-tropical jungle. The landscape and the occupations it yields also vary tremendously. On top of this, India has been subject to wave after wave of invasion and occupation. Indeed, many of the basic 'Indian' timekeeping components do seem to have been imported. There is evidence of Babylonian mathematical, astronomical and calendrical knowledge in India from about the sixth century BC and the Greek version of the signs of the zodiac appear some time in the second century BC. The seven-day week seems to have appeared in India via the Romans in the second century AD.

Nevertheless, there are several elements of timekeeping that do appear to be uniquely Indian. The first is the division of the 24-hour day. The predominant form for the Indian day was a 24-hour, equal-hour system in which the day began at sunrise. The period of day and night is divided into 30 *muhrtas* (each 48 minutes long) or 60 *ghati* or *ghatikas* (each *ghatika* is 24 minutes long; an hour is 2.5 *ghatikas*). Each *ghatika* is further divided into 30 *kāla* (each is 48 seconds) or 60 *palas* (each *pala* being 24 seconds), and each *pala* is further subdivided into 60 *vipala* (a *vipala* is 0.4 seconds long). Unquestionably, this 60-based, or sexagesimal, timekeeping system is Mesopotamian in origin, but the way in which it was developed appears to be indigenous.

The nineteenth-century Tibetan shepherd's time stick or clock spear provides an excellent example of how this system worked in practice (see **110**). The stick has eight sides, each of which bears a time scale calculated according to the relative amount of daylight during the different months of the year. The text on this stick explains that the shortest period of sunlight it can measure is 26 *ghatika* (or about 10.4 hours of daylight); the longest is 34 *ghatika* (or 13.6 hours of sunlight). The spear is used just like a pillar dial (see **110**). In order to find the time, a peg is inserted horizontally into a hole at the top of the stick in line with the correct month. The stick is then turned towards the Sun so that the peg's shadow falls straight down the scale. The time is indicated by the lower edge of the shadow.

The structure of the year varied widely over the breadth of the region. At the end of the nineteenth century, there were over twenty different types of calender in use. In 1953, Nehru noted that this number had risen to well over thirty. Solar, lunar and luni-solar calendars were all used. The names of the months, the starting date of the year, rules for intercalation, the era for counting the years and which phase of the Moon began the month, all varied.

Despite the fact that early Indian solar calendars are based on the Greco-Roman signs of the zodiac, the time that each solar month was considered to occupy reveals another indigenous development. Most solar calendars are based on a series of months of which the length alternates between 29 and 30 days. The Indian solar calendar, however, reflects the fact that the Earth's orbit around the Sun is elliptical. This means that the speed at which the Earth travels changes throughout the year. The Earth is slowest when it appears to pass through the sign of the Gemini. It then gains speed until it arrives at its quickest rate when passing through Sagittarius. The different length for the solar months is as follows:

Mesha	(Aries)	30.9 days
Vrishabha	(Taurus)	31.4 days
Mithuna	(Gemini)	31.6 days
Karka	(Cancer)	31.5 days
Simha	(Leo)	31.0 days
Kanya	(Virgo)	30.5 days
Tula	(Libra)	29.9 days
Vrischika	(Scorpio)	29.5 days
Dhanus	(Sagittarius)	29.4 days
Makarus	(Capricorn)	29.5 days
Kumbha	(Aquarius)	29.8 days
Mina	(Pisces)	30.3 days

The early Indian calendars also followed the Babylonian and Greek convention of beginning the year in the spring; but they did not use the vernal equinox or the first degree of Aries as a starting point. Instead the year was begun when the Sun reached a particular star in the zodiacal constellation of Pisces. Also, despite the fact that the early Indian astronomers understood the concept of the precession of the equinoxes, they made no attempts to adjust their calendars in order to accommodate them. As a result, the six seasons into which the year is divided – each of which is named according to the type of weather it typifies: *vasanta* ('spring'), *grishma* ('hot'), *varsha* ('rainy'), *sarad* ('autumn'), *hemānta* ('cold') and *shishira* ('dewy') – are now nearly six weeks out of 'synch'. The descriptive names of the months now only bear a vague relation to the weather patterns noted when the system was first begun sometime around 500 BC.

The astronomical lunar year is defined by the lunar mansions (see p. 38). These, too, betray a Babylonian origin. In the Indian system, each lunar mansion or *naskahatra* equalled approximately one day of the lunar cycle, or about 12° 27' of the ecliptic, and was defined and named in accordance with certain bright stars or stellar clusters that fell within that segment. The names of the months in the lunar year are, in most cases, derived from twelve of the lunar mansions. The starting point for the months themselves differs from region to region. The tradition in southern India is to begin the month with the New Moon. In northern India, however, it is more common to begin the month with the Full Moon. The lunar year can begin either in the spring (with the month of Chaitra and the sign of Aries) or in the autumn (during the month of Karttika, which coincides with the Sun passing through the sign of Libra).

The way in which the years are numbered varies as well. For example, in northern India, a lunar calendar of the Vikrama era is used. Its epoch (or starting point) is dated to 57 BC. According to this system, the year 2000 AD is 2057/58. In southern India, the Saka calendar is more common. Its epoch is 78 AD, making the year 2000

AD equal to 1921. The Bengali use a predominantly solar calendar of which the epoch is 1556 AD, making our 2000 AD equal to the year 444.

Mathematically, the Indians resemble the Aztec and the Maya in their ability and desire to count huge amounts of astronomical time. One reason for this is the fact that the time relating to the life of Brahmā needs to be calculated according to a different temporal scale. One month in human terms is equal to 24 hours in the life of a *pitr* ('deceased forefather'). One human year is equal to 24 hours in the life of one of the minor gods (*deva*). One thousand years in the life of a *deva* is equal to one day and one night in the life of the creator god, Brahmā. This means that one period of 24 hours in the life of Brahmā is equal to 8.76 million human years.

Life on Earth, however, is subject to a repeated cycle of creation, destruction and re-creation. Each creation cycle lasts for a *kalpa* or 'eon' and is equal to 4.32 million human years. Each *kalpa* is subdivided into four periods or *yuga*. The current creation, or *Kali yuga*, 'the age of strife', is reckoned to have begun at either midnight or dawn on Friday 18 February 3102 BC, when the Sun, Moon and the five planets were all aligned in the sign of Mesha, or Aries. This idea of the creation being marked by a great conjunction of planets in a certain sign seems to reflect a Babylonian concept that also reappears in Greek, Roman and Christian cosmological thought. The Roman writer, Seneca, recounts that since the world began when all of the planets were aligned within the same degree in the sign of Cancer, should that conjunction happen again, it would signal the end of the world. Once destroyed, though, the world would then be re-formed and a history exactly the same as that of the present world would be re-enacted in every detail. In the same way that the annual solar cycle creates similar patterns of death and rebirth, so the Babylonian concept of the Great Year is used to explain a never ending cycle of successive deaths and rebirths of the universe. The Babylonians calculated the period of the Great Year as the simple multiple of 600 and 3600, or 2.16 million years. The Indian variant of the Great Year, the *kalpa*, is twice that length.

One other aspect of Indian cosmology that appears to be wholly indigenous is the addition of two 'planets' to the canonical list of seven. These are Rāhu, the 'head of the demon', and Ketū, the 'tail of the demon', and they represent the two points where the Moon crosses the path of the ecliptic, the ascending and the descending lunar nodes. As these are the only places on the ecliptic where a lunar or solar eclipse can occur, both these demons are seen as personifications who strive to capture the Moon. The two figures were originally part of the same demon, but now they lie exactly 180° from each other, at opposite ends of the sky.

032 Unkoku

Netsuke group of zodiacal animals
Ivory, 19th century
London, The British Museum, Dept.
of Japanese Antiquities
[1953.12-17.3]

Netsuke are small ornamental
carvings in wood, bone or ivory
usually worn by men on their belts.
Quite often, a man would wear a
netsuke depicting the 'zodiacal
animal' associated with the year of
his birth. Like the Chinese, the
Japanese marked the years by the
succession of these animals, each of
which was believed to impart a
certain character to the year as well
as to all those children born within
the year. In this larger piece, the
complete series of all twelve animals
is dextrously carved into a single
composition.

Also

• Masakatsu, *Rat with kagami-mochi*,
19th century [HG 636]; Tomotada,
Ox, 18th–19th century [HG 453];
Unsigned, *Tiger*, 19th century [HG
700]; Okatomo, *Rabbit with
loquats*, 18th–19th century [HG
441]; Masatsugu, *Manjū with
dragon*, 19th century [HG 563];
Masanao, *Snake*, 19th century [HG
620]; Kaigyokusai, *Horse*, 19th
century, [HG 369]; Unsigned, *Goat*,
18th–19th century [HG 452];
Yoshimasa, *Cock*, 19th century [HG
465]; Rantei, *Boar*, 18th–19th
century [HG 418]; Tomotada, *Dog*,
19th century [HG 445]; Masatsugo,
Monkey, 19th century [HG 566]

Very early on, the ancient Egyptians had marked a set of thirty-six different stars or star clusters the rising of which marked the night time into twelve equal segments, each of which was about forty minutes each. These stars, which were later named 'decans' by the Greeks, provided a kind of celestial clock for the Egyptians. The origin of these 'decan clocks' is so old that by the third millennium BC – the period from which most of the documents survive – the system was no longer astronomically reliable. The constellations had shifted, making the real purpose of the decan clock obsolete; nevertheless, the Egyptians continued to preserve the form of the decan clock well into the period of the New Kingdom.

By this time, the Greeks had added these thirty-six celestial timekeepers to their own zodiacal band. As the celestial circle measured 360°, each of the thirty six decans was given rulership over 10° of the band, or one third of a zodiacal sign. The name 'decan' is derived from the Greek word for 'ten'. In the same way that each planet had a planetary ruler, so each decan was given planetary influence. The first decan of Aries, for example, was said to have the power of Mars; the second had the power of Venus and the third the power of Mercury. As garbled as this system might seem, the decans and the decan gods were extremely popular astrological figures throughout Europe and the Middle East.

Another system for marking the segments of the zodiacal sign was to note which non-zodiacal stars and parts of constellations rose within the 10° of that sign. These parts of constellations were called *paranatellonta* or *sunanatellonta* – again from the Greek, meaning 'rising with' or 'rising beside' a part of the zodiac sign. If one were to imagine the sky as a giant Ferris wheel, with thirty-six passenger cars, each car could be thought of as a decan, and the passengers rising in the cars as *paranatellonta* or *sunanatellonta*. In some instances, the *paranatellonta* had specific astrological fortunes attached to them. In most cases, however, they were used as timekeepers.

In the early ninth century, an Arabic astronomer named Abū Maꞏshar al-Balkhī Jafar ibn Muhammad (known to the Latin West as Albumasar) decided to write a great synthetic treatise pulling together the astronomical knowledge of the Greeks, the Persians and the Indians. In his *Kitāb al-madkhal al-Kabīr*, or 'Great Introduction', he provided a list of the *paranatellonta* that rose with every 10° of the ecliptic according to these three traditions. The resulting text and illustrations (see **034** and **035**) exemplify perfectly what happens to unconventional, para-scientific systems over a period of five hundred years. It is fascinating nonsense that bears only the dimmest recollection of its grander astronomical origins.

034 Franco-Flemish

Liber Albumazaris, by Georgius
Zothorus Zaparus Fendulus
Pen, wash and colouring on vellum,
ca. 1350
London, The British Library, Dept. Of
Western Manuscripts [Sloane Ms
3983, ff. 16v–17r]

There are about half a dozen
illustrated manuscripts of the so
called 'Fenduli Abridgement' of Abū
Maᶜshar's *Introductorio*. In the Sloane
manuscript, full-page illustrations of
each zodiac sign are followed by
three pages in which that sign's
decans are depicted. Both the top
layer (the decans according to the
Persians) and the lowest zone
(decans according to the Greeks)
depict parts of constellations that
rise alongside the first 10° of Leo.
In the top band, we see the tail of the
Dog (either Canis maior or minor), the

head of the lion itself, an ass and a
horse (possibly Pegasus), and a
houseboat of sorts with oars and a
nauclerius ('skipper') – no doubt
relating to part of the constellation of
Navis. Along the bottom of the page,
there is the left hand of Castor (one
of the Gemini), the neck and right
foot of the Bear (Ursa maior or
minor), the legs of the Lion and the
middle of the Ship (again, Navis).
The middle register, the decans
according to the Indians, has a series
of figures drawn primarily from
indigenous Indian astro-mythology.
One can only assume that they, too,
represent stellar bodies associated
with the first 10° of Leo.

Also
• Mehmet b. Emir Hasan el Suudi,
Astrological manuscript, 1582,
Paris, Bibliothèque Nationale de
France [Ms Suppl. turc. 242]

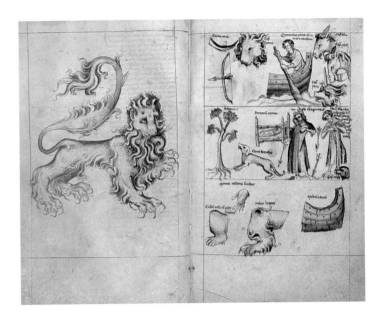

035 French

Liber Astronomiae, by Georgius
Zothorus Zaparus Fendulus
Manuscript, *ca.* 1500
Paris, Bibliothèque Nationale de
France [Fonds Smith-LeSouëf, Ms 8]

In comparing the iconography of the
constellations and zodiacal signs from
Babylonian times up to the present
day with that of the decans and
paranatellonta, one is struck by the
relative stability of the former and the
huge number of variations in the
latter. Despite the fact that the
original Egyptian decan-imagery
reappears in Greek, Roman, Persian,
Indian, Islamic and Latin sources, it
never seems to have become
sufficiently 'mainstream' to have
created a constant form. One reason
for this may be that both decans and
paranatellonta were regularly cited by
the authorities as being somewhat
mysterious. They were often
described as being part of that
shadowy world of astral magic. The
Latin astrological poet Marcus
Manilius skirts the issue of how the
decans manifest their powers by

simply stating that "nature is ever
hedged about with deep darkness,
and the truth is hidden and wrapped
in much complexity" (*Astronomica*,
IV, 303–09). Three centuries later,
Firmicus Maternus wrote that the
decans were celestial demi-gods,
whose powers were first used by the
ancient Egyptians. Their forms had
been shrouded in obscurity by the
ancients so that their secrets might
not be discovered by the profane.
This claim that the decans formed
some part of the 'hidden wisdom of
the ancients' ensured their popularity.
In particular, it may help to explain
their sudden reappearance in
illustrated manuscripts dating from
the early years of the 15th century.
For the scholars of the early
Renaissance, who were fascinated
by everything purported to be derived
from ancient Egypt (see 223), the
combination of a set of intriguing
Egyptian images, culled by an
Arabian *magus* from Indian, Persian
and Greek sources, must have been
intoxicating.

036 Italian (Florence)

Luna and her planet children
Engraving, *ca.* 1460
London, The British Museum,
Dept. of Prints and Drawings
[1845.8–25.476]

The print depicts Luna with the horns
of a crescent moon sprouting from
her head. Her chariot is drawn by two
maidens – presumably the
companions of the huntress goddess
Artemis/Diana. Her 'children' are
shown involved in various activities
relating to water: swimming, fishing
and using the water's power to drive
a mill. On the left of the composition,
there is a group of men shown

gambling, in the company of a fool
and a small monkey. Fools, perhaps
by their association with lunatics, are
often depicted as children of Luna.
The inscription makes clear that the
Moon is a feminine planet and is in
the first sphere (*i.e.* she is closest to
the Earth). She is moist, cold and
phlegmatic. She loves geometry and
her metal is silver.

037 French (Paris)

*L'influence de la Lune sur la Teste
des femmes* ('The influence of the
Moon on the head of women')
Engraved print, 17th century
Paris, Musée Carnavalet [G.2210
G.C.2 Mœurs]

In the West, the appearance of the
Full Moon has been believed to be a
signal for magic and mayhem since
antiquity. It was a time for magic
plants to be harvested, potions to be
distilled and sorcerers to meet. The
rays of the Full Moon were generally
believed to be one of the causes of
madness. The idea was that the
Moon affected the humours of the
brain adversely. Those with
'excessive moisture' in their
constitutions – and this included
women – were especially vulnerable
to the baleful influence of the Moon.
Owing also to the similarity between
the periodicity of the phases of the
Moon and the duration of the
menstrual cycle, women were
thought to be particularly susceptible
to the Moon. In this print, hapless
women, bewitched by the rays of the
Moon, have come out to dance in its
light. The changing face of the Moon
was also associated with physical
changes in the proper order, shapes
and boundaries of things and beings.
In folklore, the Full Moon brings

about physical metamorphoses of all
manner of creatures. It is also the
time when evil spirits, ghosts and
werewolves roam the countryside.
At a basic level, these superstitions
about the Moon might have
something to do with the way in
which moonlight is different from
sunlight. The Moon's weaker, silvery
light changes the perceived colour
relationships of objects, making
things appear more monochrome,
less corporeal and more ethereal.
Also, the shadows cast by the fast-
moving Moon tend to be more
erratic than those cast by the Sun.
This can lead to the impression that
the objects themselves are more
lively. Shadows seem to dance in the
pale moonlight. As odd as it might
seem, though, for human beings,
who are psychologically programmed
to make some sort of sense out of
what they perceive, it is often much
easier to interpret an unexplained,
shadowy figure as some sort of
monstrous apparition or magical
changeling than as a mere 'trick of
the light'. The question that remains
unanswered, however, is why a
Western mind might be programmed
to see evil, when an Eastern one
sees only beauty in the changing
light of the Moon.

Apart from the flux between day and night, it was the movement and changes in the Moon that first occupied the mind of primitive man. The Moon is not only the largest, but it is usually the brightest object in the night sky. The use of the Moon as a timekeeper seems to predate the use of the Sun. Almost all the very earliest calendars are lunar.

The mythology of the Moon varies greatly from one culture to the next. Many Western myths concerning the Moon revolve around the concept that the Moon is a symbol of inconstancy. This notion can be traced back to Aristotle's division of the cosmos into two parts: the indestructible world of the fixed stars and the sublunary world of the Earth, which is subject to change, growth, decay and destruction. For the early Christian Fathers, the separation of the Universe into two parts became a moral paradigm. In the supralunary world, the stars and the planets were pure and unchanging. In the sublunary world, everything was corruptible and potentially evil. The Moon itself became associated with everything that was weak, lowly, deceptive and transient.

In the Far East, however, the Moon was seen as a symbol of constancy and eternity. The Moon is the greatest of the *yin* powers in the Chinese cosmos and the manifestation of the *qi* or 'breath' of heaven. She protects women and is the patron goddess of the harvest. Her birthday is the fifteenth day of the Eighth Month – the Full Moon of the harvest festival. The Moon Festival is also a time for families to come together, and for friends to reunite and re-form the circle of friendship. There are numerous Chinese poems that use the image of the Full Moon as a symbol for friendship or to remind the reader of distant loved ones. Or, to quote the poet Li Bai, "I lift my head and watch the bright Moon, then my head drops in meditation of my hometown".

In China, the most popular image associated with the Moon – aside from the Moon Goddess herself – is the hare. The hare's duty was to manufacture the elixir of immortality for the Great Immortals of the Chinese pantheon, grinding it together with his pestle and mortar every night. In the same way that Europeans might see a 'Man in the Moon', the Chinese see an image of the hare, pounding the elixir of immortality.

The Japanese also inherited a festival of the Harvest Moon on the fifteenth day of the Eighth Month. Moon-watching, or *tsuki-mi*, became established relatively late in Japan, perhaps some time during the ninth century, but quickly became one of the most significant times in the Japanese year. In one well known verse, Saigyō, a Zen monk poet of the early Kamakura period, claims that his only regret in dying will be to lose sight of the Moon:

> Some day I may have
> To pass away from this world, alas!
> Forever with a longing heart
> For the Moon, for the Moon!

038 Kitagawa Utamaro
Kyōgetsu-bō ('The Moon-mad monk' or *'Crazy gazing at the Moon')*
Colour woodblock album, 1789
London, The British Museum, Dept. of Japanese Antiquities
[1979.3–5.0154 (JH 154)]

Utamaro's prints accompany a set of 72 *kyoko* poems on the theme of the Moon. They must have been inspired by or commissioned to celebrate the Full Harvest Moon of the fifteenth day of the Eighth Month of that year. The pictures are as quietly evocative as the poetry of the period. In one, Prince Genji is standing by the bay of Akashi, watching the Moon. In another, there is a scene of a peasant family, with the wife pointing out the beauty of the Moon to her children.

039 Matsumura Goshun
The Hare in the Moon with his pestle
Woodblock print, early 19th century
London, The British Museum, Dept. of Japanese Antiquities
[1993.4–5.01]

One of the regular features of the Harvest Festival is prints and figurines of the Hare in the Moon.

Also

• Mirror, Song Dynasty, London, The British Museum, Dept. of Oriental Antiquities [OA.1986.6–28.1]

• *Heng E and the Hare*, 20th century, London, collection of G.S. Barrass

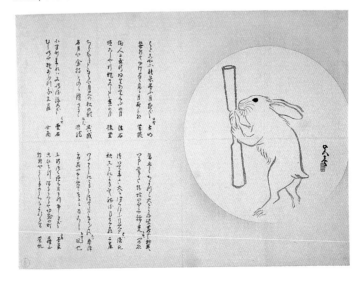

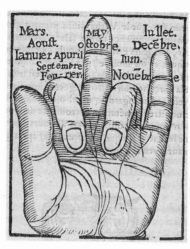

Jehan Tabourot, *Compot et Manuel Calendrier* ... (see **059**)

The Principles and Uses of Calendars

Political and Social Implications

Silke Ackermann

Anton Linden, Astronomical compendium (see **061**)

W hen champagne bottles are being opened worldwide during the first seconds of 1 January 2000 to celebrate the beginning of the third millennium in the Gregorian calendar, numerous other calendars will not indicate a new millennium at all. The date 1 January 2000 of the Gregorian calendar equals, for example:

19 December 1999 *anno Domini* in the Julian calendar;

23 Tebeth (the fourth month) 5760 *anno mundi* in the Jewish calendar;

24 Ramadan (the ninth month) 1420 *Hijra* in the Muslim calendar;

the 25th day of the 11th month in the year of the rabbit, and the 16th year of the current 60-year cycle in the Chinese calendar;

the 25th day of the month of Margasira in the year 1921 of the Saka era according to the religious lunar Hindu calendar; and

the 11th day of Pausa in the year 1921 of the Saka era according to the civil solar Hindu calendar.

Although most of these calendars are based on the same astronomical principles, each one has completely different epochs or historical or mythical dates from which the beginning of time is counted. In the case of the Chinese calendar, the counting of time is organized according to a series of cycles of 60 years.

Organizing time gives the sense of controlling time and thereby provides the feeling of being in control of an increasingly important part of everyday life and society. Since most calendars are based on astronomical cycles, *i.e.* the movements of the Sun and the Moon, they provide a link between mankind and the cosmos and can reach sacred status. Without a calendar, however simple it may be, it is very difficult to memorize the past in a structured way or to plan future events, be they farming and hunting or religious and sacred duties. Inaccuracies in a calendar, which bring it out of step with its astronomical basis, can seriously upset the life of a society, while reforming a calendar which people have become used to and which is part of their everyday lives is an extraordinary event. Its acceptance depends on political or religious forcefulness and its social and economic implications.

The cycles of the two most obvious and lightgiving celestial bodies, the Sun and the Moon, were used to reckon time over extended intervals for civil or religious purposes from as early as 10,000 BC. The periods we recognize as the day, the month and the year mirror these astronomical cycles and are the basis of most calendars. The smallest unit of any calendar, the (solar) day, can variously be defined as the interval between two successive sunrises, two successive sunsets, two successive midnights or two successive Noons, and the beginning of the day can be quite diverse in different

Organizing time provides control over an

Weeks of seven days are the only calendar unit not based on an astronomical cycle

calendars – even within one calendar – depending on religious, civil or academic contexts. At the International Meridian Conference, held in Washington, D.C., in October 1884, it was agreed that, for international purposes, the civil day would start at midnight. According to religious calendars, for example the Jewish or Islamic calendars, the day begins after sunset. Astronomers vigorously argue that the day should start at Noon, for two reasons: first, the transit of the Sun across a meridian (*i.e.* local Noon) can be easily measured and, secondly, if the day began at Noon, the nightly observations of the astronomers would not have to be split across two dates.

The 'month' is derived from the period of the Moon's orbit around the earth and the cycle of the Moon's phases. These lunar phases depend on the Moon's position relative to the Sun and the Earth, and they run in a cycle of about 29½ days' duration. The interval between two New Moons is called the 'synodic period'. Ancient astronomers realised that 235 synodic periods equalled nearly 19 solar years (the solar year equalling approximately 365¼ days). After 19 years, the so called 'metonic' cycle, the lunar phases repeat themselves on the same date. The lunar month is the essential unit of any lunar calendar. In a solar calendar, though the months are still present as units, they are often detached from the actual cycle of the lunar phases.

The greatest single unit, the year, can either be a lunar or a solar year. The lunar year consists simply of an agreed number of lunar months. The length of the solar year is determined by the period of the revolution of the Earth around the Sun, for which external references are needed. One such reference point is the vernal, or spring, equinox – that point during the spring when the length of daylight equals the length of the night. An interval between two successive passages of the vernal equinox by the Sun has a length of 365.24222 days (or approximately 365¼ days) and is called a 'tropical year'. Every 28 years the calendar dates fall on exactly the same week-days and this period is called a 'solar cycle'.

The division of the months into weeks of seven days, common to many calendars, is the only unit in the calendar not directly based on an astronomical cycle. The origins of the seven-day week are shrouded in mystery. It is a common belief that it goes back to the Babylonians, who established a seven-day cycle in an attempt to reflect the four phases of the Moon. The Babylonian calendar served as the basis of the Jewish calendar, where the seven-day week found its rationale in the seven days of Creation described in Genesis. The

Roman calendar, although it originally had an eight-day market week, gradually adopted the seven-day week in the first century AD, probably for astrological reasons. The Christian calendar, based on a combination of the Roman Julian calendar and Jewish cosmology, subsequently also adopted the seven-day week, with six working days and one rest day.

But in the Semitic languages the days of the week are numbered, not named. The modern names of the days of the week come from the Roman world of the second and first centuries BC and the Latin names of the seven planetary gods – Saturn, Jupiter, Mars, the Sun, Venus, Mercury and the Moon. In most Romance languages, these origins are still apparent, as, for example, in the French word for Monday, *lundi* (from the Latin *luna* for Moon), or in the Italian word for Tuesday, *martedì* for Mars's day. In the Germanic languages, some of the Roman gods were replaced by Germanic gods, as for example Venus by Freia, Jupiter by Thor, Mars by Tyr and Mercury by Wodan, leading to names for the days of the week such as Friday, Thursday, Wednesday and so on.

Contrary to possible expectation, the sequence of the names of the days of the week is not directly related to the relative distances of the planets from either the Earth or the Sun. Instead, astrologically, every hour of the day was given a so called 'planetary ruler'. The sequence of these 'rulers' does follow their order in the sky, starting with the outermost planet: Saturn was the ruler of the first hour of the first day, hence its name, Saturday. The second hour was ruled by Jupiter, the third by Mars, the fourth by the Sun, the fifth by Venus, the sixth by Mercury, the seventh by the Moon and so forth; the last hour of the first day was ruled by Mars. As a result, the first hour of the second day is ruled by the Sun, hence its name, Sunday, and so forth.

Many historical calendars were lunar in origin. A year in the lunar calendar usually consisted of 12 months of alternating 29 and 30 days in order to make up 354 days in total. Such lunar years, 11 days shorter than the solar year, soon 'fall behind' the longer solar years and any lunar civic or religious festivals will also move backwards through the seasons. The Islamic *Hijra* calendar is based on a 30-year cycle of 12 lunar months of 30 and 29 days alternately. Eleven years of this cycle have 355 days, the other 19 have 354 days – in this way the calendar stays adjusted to the lunar cycle and makes a complete circuit through the seasons in 33 years. Although for civil purposes the calendar is fixed, for religious purposes the new month starts with the first sighting of the lunar crescent and the days begin at sunset. The beginning point or epoch of the Islamic calendar is the date of the prophet Muhammad's emigration (*Hijra*)

increasingly important part of life and society

from Mecca to Medina on 16 July 622. This is noted as 1 Muharram 1 *anno Hijrae* and corresponds to 16 July 622 *anno Domini*. Since the Islamic year has only 354 days, 1 Muharram AH 2 corresponds to 5 July 623 AD. This shift of dates 'backwards' through the seasons means that fasting during daylight hours in the holy month of Ramadan can be very hard when the month falls in a season with long hours of daylight in great heat.

To harmonize the lunar cycle with the solar cycle, a complicated intercalation (leap) system had to be introduced. Such a calendar is called a 'luni-solar calendar' and is represented by the Jewish, the traditional Chinese and the Hindu calendars. The Jewish calendar is based on the epoch of the mythical date of the Creation of the universe and is given as 7 October 3761 BC in the Julian calendar. The Jewish calendar is heading towards its seventh millennium. It has many features of the old Babylonian calendar and a very complicated 19-year cycle with 12 common years (with 353, 354 or 355 days) and 7 leap years (of 383, 384 or 385 days).

The traditional Chinese luni-solar calendar was regarded as a sacred document which demonstrated the harmony between the imperial court and the heavens. It consisted of 29 or 30 days, the beginning of which was determined astronomically with the New Moon. Every two or three years a leap month was introduced to keep the calendar in step with the solar year. There is no epoch for the Chinese calendar; years are organized in 60-year cycles, with each year starting at a New Moon near the winter solstice. The names of each year in such a cycle are combined from one of the ten 'celestial stems' and one of the 12 'terrestrial branches' (*i.e.* the names of 12 animals, often mistakenly referred to as the 'Chinese zodiac'). After 60 years, the names repeat themselves and a new cycle starts. The initial year of the current cycle began on 2 February 1984.

The earliest purely solar calendar existed in Egypt probably already in the First Dynasty (in about 3000 BC). It was the basis of the Julian calendar and, ultimately, the Gregorian calendar. The Julian calendar was introduced by Julius Caesar in 45 BC in order to reform the old Roman calendar, which in Caesar's time was about ninety days out of step, leaving society in confusion. Every year has 365 days with every fourth year being intercalated with one extra day. The Julian year started on 1 January and consisted of twelve months with 30 and 31 days respectively, except for February, which had 28 days in common years. The fact that the old Roman calendar originally started in March (and not in January) is still reflected in the month names, such as September (the seventh month), October (the eighth month) *etc.* The new calendar was rather confusing and was only really established in the time of emperor Augustus. The gratitude of the senate was expressed by naming two months in honour of Julius

Caesar and Augustus: July and August, the seventh and eighth of the calendar and originally named *Quinctilis* and *Sextilis*. The Julian calendar started with the mythical founding of Rome (*ab urbe condita*) in what we now call 753 BC.

Unfortunately, the Julian calendar had a slight error in its calculation of the length of the year, which is actually less than $365\frac{1}{4}$ days. This accumulates to three days too many in 400 years. This affected the fall of Easter, a matter of great importance to the early Christians. In an effort to find a general rule, the Council of Nicaea in 325 had defined that Easter should take place on the first Sunday after the ecclesiastical spring Full Moon, with the spring equinox fixed at 21 March. Consequently, the calculation of the date of Easter involved a combination of the 19-year lunar cycle and the 28-year solar cycle. As the calendar drifted increasingly, the spring equinox began to fall – in the calendar – earlier and earlier in March. By the sixteenth century, the calendar was ten days in advance of the seasons. The lunar phases were equally out of step, since the epacts (a system indicating the age of the Moon at a particular date) were also calculated wrongly. Even though this was a matter of great concern in the Christian world and in spite of several efforts to revise the calendar, it was not before the end of the sixteenth century that the calendar was reformed.

Three issues had to be addressed: the removal of the additional ten days, a change in the leap rule to keep the calendar and the seasons adjusted, and the development of a new algorithm to calculate the lunar age and the date of Easter. A commission nominated by Pope Gregory XIII eventually solved all three problems. By papal edict, *Inter gravissimas*, in 1582, Thursday, 4 October was followed immediately by Friday, 15 October. New Gregorian epacts were devised and a new leap year rule was fixed, which stated that years divisible by 100 should not be leap years, except for those divisible by 400. (This is the reason why the year 2000 will be a leap year, but the year 1900, although divisible by four, was not.)

The introduction of the 'Gregorian calendar', as the reformed Julian calendar was called, was by no means a smooth process and went on over five centuries. The bishopric of Strasbourg, for example, introduced the Gregorian reform in November 1583, while the city itself only switched over in February 1682 – the confusion this must have caused can easily be imagined. Generally, Catholic areas introduced the new calendar earlier than places dominated by Protestants. In some cases, the Gregorian calendar was introduced for economical or political reasons, while the religious calendar remained Julian – as for example in Soviet Russia since 1918. In China, the Gregorian calendar was officially introduced in 1912, but the old Chinese calendar is still used for religious purposes. In

Turkey, the Gregorian calendar was adopted for civil purposes in 1927 alongside the Islamic religious calendar. Nowadays, the Gregorian calendar serves as an international standard for civil use.

England went to some length in considering the new calendar in the time of Elizabeth I, with reform ideas formulated by the famous mathematician and magician John Dee, but did not actually change until 1752. At a time of expanding trade and commerce one can only imagine how confusing the correspondence must have been between English and Continental merchants who were forced to use double dating. England did not change until 1752, allegedly triggering the so called 'time-riots'. These 'riots' with their famous slogan "Give us back our eleven days" were stylized as a symbol of popular ignorance in the age of Enlightenment, but never took place as such, although the introduction went by no means smoothly and faced considerable opposition from all classes – not just the common people. This was perhaps inevitable given the financial implications of losing eleven days and the promise at the same time that nobody would face any financial losses.

In Christian countries calendrical issues were further complicated by the fact that there were six possible dates, or 'styles', for the beginning of the year. The year could begin on 1 January (the Circumcision of Christ), 1 March (following the old Roman style), 25 March (the Feast of the Annunciation), Easter, 1 September (following the traditional Byzantine style), or 25 December (the Nativity of Christ). Once adopted, these styles could, however, change in the course of history. In England, for example, the 'Nativity' style was in use until the eleventh century, when the 'Annunciation' style was adopted. In 1752, this was replaced by the 'Circumcision' style. Although, when the Gregorian calendar was introduced into England, the beginning of the year was fixed to 1 January, the financial (tax) year was – and still is – left at the old March date. Obviously, with the 'loss' of eleven days in the calendar there were quite substantial financial implications: a loss of 11 days rent, tax *etc*. It was therefore decided to add eleven days to the traditional date, which explains why the tax year in England starts on 6 April.

Very few calendar reforms were undertaken after the Gregorian calendar was introduced – and none of them was successful. The French Revolutionary calendar, for example, introduced in 1793, was based on the ancient Egyptian calendar of 12 months with 30 days each plus 5 holidays, the so called *Sanculotides*. Being strictly decimal, it is divided into 'weeks' of ten days (nine working days and the tenth day free). No strict leap year rule applied, but there was supposed to be one leap day every fourth or fifth year. The epoch, or starting point, was fixed to 22 September 1792, the day of the proclamation of the Republic and the date of the autumnal equinox.

The old names of the months and of the saints' days, which were deemed inappropriate for the new Republic, were abolished and new names introduced, based on agricultural activities or weather conditions, minerals, plants, fruits and animals. Some people decided to stop using their Christian names and adopted the new name of the appropriate date – not always quite so flattering if a girl named Catherine in honour of the saint of 25 November would, according to the French Revolutionary calendar, have to call herself 'Pig'. In 1805 Napoleon reinstated the Gregorian calendar.

The calculation of Easter, however, which was the ultimate reason for the introduction of the Gregorian calendar, brought indirectly another major change in dating which had fundamental consequences for the whole of the Christian world and ultimately our preparations for celebrating what we believe to be the millennium. In 525 the Scythian monk Dionysius Exiguus extended an existing Easter table for another 95 years (*i.e.* 5 times the 19-year cycle) to facilitate the process of calculating the date of Easter and other moveable Christian feasts. The table he was using was, as usual, still based on the era of the Roman emperor Diocletian, starting in 284 AD, although Diocletian had by no means supported Christianity, but was notorius for having persecuted Christians. Dionysius decided that it would be much more appropriate to start counting from the birth of our Lord, *ab incarnatione Domini* (AD), a date which he calculated to a year which we now call 1 AD and which corresponds to 754 *ab urbe condita*. But there are a few problems: it can be shown that a birth of Christ in the year 1 AD does not correspond to the lifetime of King Herod (who seems to have died in 4 BC) or other historical events which all took place before 1 AD. Historians believe, therefore, that Christ's birth may have taken place some time before or in 4 BC. The evidence is too hazy to be definitive, but the consequences this 'inaccuracy' has for our calendar are obvious: it may actually be that we have missed the second millennium of the birth of Christ altogether!

There is a further problem with celebrating the Millennium on 1 January 2000. Dionysius's Christian era begins in the year AD 1. He made no allowance for a year 0. If one follows Dionysus's datings, a millennium is only complete when 1000 years have been completed. In other words, the second millennium will begin only after the year 2000 is completed, or on the night of 1 January 2001. So, perhaps we had better keep back a few of those champagne bottles and quietly ponder for another year when, exactly, we should be celebrating!

2 · THE MEASUREMENT OF TIME

Cosmogram from the Codex Féjerváry–Mayer, 15th–16th century, Liverpool, National Museums on Merseyside, The Liverpool Museum [12014 Mayer]

Mesoamerican and Andean Timekeeping and Calendars

Anthony Aveni

Maya 'deep time': ultra-long cycles of time displayed in a stucco tablet from the ruins of Palenque, 8th century AD (drawing by Linda Schele)

The Maya, the Aztecs and the Inca – these three high cultures of the Americas, free from outside contact, remained hermetically sealed by two oceans until the sixteenth-century European invasion. We can rest assured, therefore, that they modelled their calendars independently, without contact or borrowing from the West.

All three cultures were bureaucratically organized and highly specialized societies with economies, artistic styles and rulerships that exerted influence far and wide. The Aztecs, for example, controlled the Valley of Mexico and the surrounding highlands all the way to the Pacific Coast from their capital of Tenochtitlan (today Mexico City). During its peak, the Inca empire of western South America stretched 2000 miles from Quito (its northern capital) through Cuzco (the capital of the south) to central Chile.

European records of the conquests of these people are both misleading and disheartening. "We found a large number of books in these characters [hieroglyphs] and, as they contained ... superstition and lies of the devil, we burned them all ...", wrote one friar who followed the conquerors to Yucatan on a Christianizing mission. Spanish chroniclers of the Inca were little different in their attitude towards the culture they encountered.

Had the intruders from the east not plundered the material record so effectively we might be in a far better position to understand native timekeeping systems as they were first encountered. We might know something of their origin and how they developed. As it is, scarcely two dozen Mesoamerican fold-screen books (so called codices) escaped destruction. Plied by diligent scholars, their contents have much to say about indigenous temporal concepts. Some of these books are royal and civic histories, others repetitive almanacs that incorporate a few astronomical (or, better, astrological) ephemerides – predictive tables based on careful skywatching laced with omens for the future.

In one of these codices, there is what might be called 'an exercise in temporal completion'. This 'cosmogram' combines time with space, nature and social concerns. It consists of a quadripartite glyph in the shape of a Maltese cross. Carefully positioned within the symmetric floral design are all the things that belong in each of the four directions of space: gods, plants, trees, birds, even parts of the body. The four directions are colour-coded. But time is also spatially divided – each region of the world is assigned its share of the 20 days of the Aztec week. The so called 'year bearers', the names of successive New Year's Days, are placed one at each of the tips of the

Uniquely in the world, the Maya ritual calendar is

Hierarchically organized societies bureaucratize time in order to extend their power

cross. Reflecting environmental reality, the tips of the cross also connote the Sun's extreme northern and southern migratory points in its annual cycle. The same concept is embodied in the glyph for the Maya word *kin* – a word which translates as Sun, day and time. Finally, circumscribing the world is the ultimate Mesoamerican number for time: 260 dots, one to each day, arrayed in 20 units of 13. These 260 days make up the Maya *tzolkin* (called by the Aztecs *tonalpohualli*), a ritual calender known as the 'count of the days'.

Uniquely in the world, the number 260 served as the base co-efficient of practically every Mesoamerican calendar that has come down to us. Whence this temporal beat known nowhere else in the world? It seems most likely that one of its factors, the number 20, was derived from the number of fingers and toes on the body once used to tally trade goods. The notation found in Maya inscriptions dating from 500 BC employs dots for ones and bars for fives. The former looks like the graphic representation of the tips of the fingers, the latter an extended hand with fingers held together to signify a gesture of completion. The other factor, the number 13, represents the number of layers in the Maya heaven. Beyond this, however, it seems that the human body can be further implicated in the origin of the *tzolkin*. The average duration between human conception and birth is close to 260 days (on the average 266). Today, Maya women in highland Guatemala still associate this sacred count with the term of their pregnancy. The *tzolkin* also turns out to be a convenient approximation to the length of the basic agricultural season in many areas of southern Mexico, where it probably originated.

Celestial phenomena are also implicated in establishing Mesoamerica's fundamental time pillar. Nine Moons (about 265 days) represent the nine 'bloods' taken away by the Moon from pregnant women to give lives to their newborn. Moreover, lunar and solar eclipses occur at seasonal intervals commensurate with the *tzolkin* in the ratio of 2:3 (three times the 'eclipse year' of 173½ days nearly equals two times 260 days). Thus, the ancient astrologer could easily warn of certain days vulnerable to the occurrence of an eclipse. The planet Venus, the patron star of war in Teotihuacan (the ancient city of highland Mexico built around 100 BC), was also revered by the Maya at a time when the New World's most precise calendar was being developed. The duration of its appearance as Morning Star averages 263 days – again close to a *tzolkin*. And if all these harmonies were not enough, in southernmost Mesoamerican latitudes the year is divisible into periods of 260 and 105 days by the (two) days when the Sun passes overhead. There is no evidence,

however, to suggest that the 'count of the days' was ever a fixed entity in the seasonal year. Like time itself, the *tzolkin* rolled on endlessly – like a temporal odometer clicking off the days without interlude or interruption.

New World peoples were also cognizant of the seasonal year. Abhorring fractions, the Maya measured their year or *haab* (Aztec = *xiuhmolpilli*) at 365 days. They divided the year into 18 months, each of which was 20 days in length, with a concluding 5-day month (an unlucky period thought to reside outside the year). Eschewing leap years, ancient Mesoamericans easily kept track of the anniversary of the tropical year within the *haab*.

Clearly, a central theme of Mesoamerican timekeeping is cycle building. They accumulated little cycles to make bigger and bigger ones. One of the larger cycles was the calendar round – a period of 52 years consisting of 18,980 days, the lowest common multiple of the *tzolkin* and the *haab* ($52 \times 365 = 73 \times 260$). This time-loop thus records the interval over which name and number combinations in both cycles repeat themselves. Perhaps not coincidentally, it is also about equal to the length of a full life. The completion of a great cycle was quite a momentous occasion – Aztec priests timed this 'year binding' event by proceeding to a special place outside ancient Mexico City called the Hill of the Star. There, they carefully watched the Pleiades to see whether they would pass the zenith. If they did, it would be a sign from the gods that a new era would be granted to humanity.

Just as it is part of human nature to cling to life, so many societies attempt to extend their power, lineage and legacy. Hierarchically organized societies are in the best position to do this. Often they bureaucratize time, giving it a deep structure that goes beyond the immediate confines of remembered generational experience. The Maya accumulated years to make scores of years. Heaping score upon score was a logical extension of their vigesimal (or 20-based) system. The 'long count' is a 5-'digit' tally found principally on carved stelae dating from the first century BC. These display the effigy of a ruler, usually in full regalia, accompanied by a hieroglyphic text that details his or her ancestral history, described in terms of the intervals between seminal events (birth, accession, conquests, marriage, death *etc*). To add depth and historical permanence, the dating of these events often seems to have been contrived to fit with repeatable cosmic time markers, such as the heliacal risings of Venus, eclipses, solstices and so on.

based on a count of 260 days (the tzolkin)

One example of an attempt to legitimize authority publicly and entrench a favoured ideology in deep time appears in the engraved stucco tablet found in the Temple of the Cross at Palenque (dated to the eighth century). The inscriptions of the tablet link genealogical time with mythical time. Events recorded include the birth of the gods at Palenque (astronomically timed by a mass conjunction of the planets), from whom the ruler commemorated is said to have descended. There are also tales of gods who existed before the beginning of the most recent era or 'creation', or prior to 3114 BC (our time). One of the glyphs names an Olmec ruler born in 993 BC. Such a luminary, if real, would be separated in time from the date of the tablet by an interval roughly equal to that between Moses and Christ. Evidently the events carved out on such inscriptions – events that extol the unbroken connection between mundane time and eternity – were intended to enhance the vision of immortality of Maya dynastic rulership.

The concept of successive creation/destruction cycles is central to understanding time in the Americas. For example, despite the terrifying effigy at its centre, the famous 'Aztec Sun Stone' provides a pictorial narrative of a cyclic cosmogony in which people play an active role. Tonatiuh, the Sun god – a flint knife depicting his lolling tongue – grips the firmament with his claws. He cries out for the blood of human sacrificial hearts that he may keep the world in motion. The four panels that surround Tonatiuh represent previous ages, or 'Suns', as the Aztecs called them. The first cosmogonic epoch (upper right) was the 'Sun of Jaguar', named after the day '4 Jaguar' in the 260-day cycle on which it terminated (the head of the jaguar is surrounded by four dots within the panel). During this epoch, the inhabitants of the Earth, the result of the gods' first try at a creation, were giants who dwelled in caves. But they did not till the soil as expected and so the gods sent jaguars to eat them. In the second Sun, the 'Sun of Wind', symbolized by the day '4 Wind' (upper left), another less than perfect human race was blown away by the wind. The gods transformed these creatures into apes, that they might better cling to the world, an act said to account for the similarity between apes and men. In the third creation, the 'Sun of Fire-rain' (the symbol of '4 Rain' is at the lower left), some men were permitted to survive by being transformed into birds to escape from the destruction of the world by volcanic eruptions. The fourth creation, the 'Sun of Water', depicted at the lower right, ended with a flood that followed torrential downpours. But this time a transformation from men into fish kept them from perishing entirely. The symbol '4 Water' marks this epoch. The Aztecs believed they existed in the 'fifth Sun', of which the symbolic date, '4 Movement', houses the effigy of Tonatiuh and the other four ages. (The four large dots of this sign's coefficient are easily recognizable on the periphery of the four panels that denote the previous 'suns'.) According to Aztec cosmogony, the universe was destroyed and re-

created anew, each age providing an explanatory temporal framework in which to categorize different forms of life and to relate them to the present human condition.

Two distinct points about Mexican time are worth underlining in this story of creation: first, the oscillating, repetitive nature of the events taking place. Previous 'Suns' were thought to have been creative ventures that failed to achieve the necessary delicate balance between gods and people. Creation time repeats itself, but it is punctuated by periods of destruction. Secondly, each present contains a piece of the past. Each attempt at creation tries to account for the present state of humankind by referring to what remains in the world. Fish and birds are really our kin, the failed children from archaic creations. We were not destined to dominate them, as Old Testament Genesis states; rather we must revere them, for nature is part of us.

Fish and birds are the failed children of previous

According to the Aztec chronicles, the gods made sacrifices in order to bring about the world in its present condition. They performed these sacrifices at the ancient pyramids of Teotihuacan, when, in the aftermath of a struggle amongst themselves, one of their number sacrificed himself to the ceremonial fire, thus promising to become the first rising Sun. Such stories portray life like Darwinian evolution, as a struggle filled with key transitory moments. But, unlike the Western view, theirs was a cosmology with a purpose. Human action – in this case, blood sacrifice to the gods – was necessary to extend the fifth or present epoch. It mediated the balance of violent forces that might erupt – as they still do today in the fragile highland environment. After all, since the gods sacrificed themselves for us, it is only reasonable that we should offer sacrifice as payment of our debt to them.

The Inca of South America developed just as sophisticated a timekeeping system as the Maya and the Aztecs. The chronicler Bernabe Cobo tells us they organized the city of Cuzco on a system of radial *ceques* (or 'visual lines') marked out by *huacas* ('sacred places') in the manner of knots on a series of strings, resembling the Andean knotted *quipus* ('cords') on which they are said to have kept their records. Considerations of water rights and kinship lay at the foundation of the subdivision of the land into 41 *ceques*, centered on the Coricancha or Temple of the Sun (Ancestors). Timekeeping was tightly interwoven into the *ceque* system. For example, Cobo describes a significant number of the 328 *huacas* as horizon pillars that marked the annual course of the sun. One set of four pillars signalled the setting sun at intervals pivoted about one of the days (18 August), the point in the calendar opposite the day of the year when the Sun passed overhead. Evidently the passage of the Sun by successive horizontal pillars timed the planting of crops in various levels of the largely vertical environment of the Inca, where the times from planting to harvesting differed.

Serving both as a calendar and an orientation device, the 328 *huacas* of the *ceque* system measured the length of the agricultural season from planting through harvest in the Valley of Cuzco. The remaining 37 days that made up the solar cycle were not counted, for the fields lay fallow and active time did not exist. But the seasonal clock was reset (on 9 June) when the Pleiades (also marked by a *ceque*) rose simultaneously with the Sun. This concept of a 'year' less than 365 days in duration may seem peculiar to us but it arises in timekeeping systems all over the world. The Trobriand calendar is 10 lunar months long and is restarted by the appearance of the Palolo worm, a spawning marine annelid which comes to the surface of the sea at

creations. *We are not destined to dominate them*

the time of the full moon in late November. Also, the calendar of the Nuer of southern Sudan and indeed the archaic Roman calendar are both based on the 307-day gestation period of cattle.

While all the details of the Inca calendar have yet to be worked out (unlike Maya hieroglyphic writing, Inca *quipus* have yet to be deciphered), the ethnohistoric and archaeological records taken together indicate that timekeeping in the Inca capital was a co-operative social enterprise. Horizon observations made from the northern (upriver) half of the moiety of Cuzco followed the Sun on its northern course, while skywatchers in the southern (downriver) segment were more attuned to the Sun on its southerly course. The intention, like every element in the unifying *ceque* system, was to promote social harmony.

So many differences ring clear when we compare these selected Native American timekeeping systems with those of the Old World. Civilizations on the western side of the Atlantic were devoid of mechanical clocks and exhibited not a hint of a mechanical revolution, each harbouring only a minimal technology. They possessed no sundials, though the ancient Mexicans did devise 'solar observatories' to capture the light of the sun on its all-important zenith passages. That none of these high cultures developed either an interest in or a method for measuring the divisions of the day into hours, minutes *etc* is a sign that precision timekeeping mattered little to them. We do find a rare exception in the Maya codices, where astronomers seem preoccupied by narrowing the warning periods for eclipses and heliacal risings of Venus, but, even then, it is only to the nearest day. Until the archaeological record yields a single New

World instrument we must be prepared to conclude that all astronomical measurements on which native time systems rest were performed with the unaided eye.

Location and ecology condition all calendars. The sky is different at low latitudes, and the zenithal pivot emerges as seminal in the astronomies of many tropical cultures. It replaces the polar pivot of the celestial systems of Europe, the Middle East and China. Studying the Maya, Aztecs and Inca offers us an opportunity to see our own reflection in time's mirror, for, while there are differences in calendar building, there are some stark similarities with the West that make us wonder just how profound human nature might be in affecting so basic a task as meting out duration from the micro- to the macroscopic. Two suggestions along this line will suffice.

As we have seen, the Maya counted all things in units of 20, building order upon order from ones to twenties to four hundreds to eight thousands *etc*. Time counting offers the only exception to this rule. When tallying days the Maya substituted the number 360 for 400 in the third place. This they called the *tun* (year), which they reckon as 18×20 days. The next highest time unit in the long count (the *katun*) becomes $18 \times 20 \times 20 = 7200$, the next (the *baktun*) $18 \times 20 \times 20 \times 20$ and so on. The Babylonians did no less, uniquely singling out the cyclic count of time – also based on the 360-day year (later parallelled in the 360° circle) – in a special sexagesimal system, whence our 24-hour day, 60-minute hour and 60-second minute.

And, finally, there are the cyclic creations exhibited by all three cultures. They are reminiscent of the repetitive cycles in Hindu and Greek cosmologies – and perhaps even in our own modern story of creation. We may live under the aegis of Judeo-Christian linear time but some versions of the 'big bang' cosmology harbour previous creation cycles, thus viewing the present expansion as a rebound from a previous contraction. One wonders whether this insistence upon fabricating eternal repetitive time, giving the universe another chance as it were, might not run deep in the veins of all members of *homo sapiens* regardless of cultural influence.

'The Aztec Sun Stone', 1427 AD (drawing by Emily Umberger; original in Mexico City, Museo Nacional de Antropología)

The Sacred Mosque, Mecca, Saudi Arabia
photograph © Peter Sanders

Time and Space in Islam

David A. King

World map centred on Mecca, engraved brass,
late 17th century, private collection

In Islam, time is sacred time and space is sacred space. The prescriptions are, in their essence, laid down in the *Quran*, which for Muslims is God's last revelation, delivered to mankind through the last of His Prophets, Muhammad, in Mecca and Medina in the early seventh century. Further elaboration is found in the *hadith*, the sayings attributed to the Prophet. The consequences of these prescriptions constitute some of the most distinctive aspects of Muslim ritual and daily life.

The Muslim year is strictly lunar, the beginnings of the months being defined by the first visibility of the lunar crescent. There are two months that are especially sacred. The day begins at sunset, as do the months. The day and night are divided by times set aside for five prayers, the times being defined according to prescriptions revealed to the Prophet Muhammad.

The Muslim world is centred on the Kaaba in Mecca, an edifice founded, according to Muslim tradition, by Abraham. It is revered as a symbol of the presence of God. Prayer and other devotional acts are to be performed in the sacred direction towards the Kaaba,

In Islam, time is organized with respect to

which is called *qibla* in all languages of the Islamic commonwealth. Muslim religious architecture is aligned in the *qibla* and sometimes, when a significant religious edifice dominates the plan of a city, the entire city may be *qibla*-oriented. In addition to facing the Kaaba every day in prayer, Muslims are obliged to visit Mecca once in a lifetime, performing the ritual pilgrimage (*hajj*).

All this is well known to any Muslim and to any non-Muslim who has an inkling about Islam. What is not well known to either group is precisely how these religious obligations have been carried out over the centuries. Only in the past few decades have the vast historical sources – mainly unpublished manuscripts – available for the investigation of this religio-cultural phenomenon been studied. The information contained in these sources has enabled us to write one of the most colourful chapters in the history of Islamic civilization, for we are dealing with the interaction of religion and science – mainly astronomy and geography – and the effects of this interaction on daily life and on the architectural environment in a civilization that has endured for close to 1400 years. This is no monolithic civilization, but one coloured by vastly different cultures and many races of mankind, each with a rich and chequered history. In the nineteenth century, this civilization spanned the region between West Africa and the Far East, the Balkans and South Africa. Now, at the end of the twentieth century, the Muslim community also includes numerous other groups, immigrants or local converts

The sighting of the crescent at the beginning and end of Ramadan has always been important

all over the world. The unity of Muslim practice – at least with regard to calendar, prayer-times and sacred direction – has been maintained, and the same obligations which confronted the first Muslim settlers in, say, Egypt or India in the seventh and eighth centuries still confront a British or American or Chinese Muslim in the year 2000.

It is important to realise at the outset that the scholars of the religious law (who regulated Muslim society) and the scholars of astronomy and mathematics (who constituted a substantial but far less influential group) addressed these problems from different standpoints and with different methods. The legal scholars favoured what we now call 'folk science', a discipline based on pure observation of celestial phenomena, without recourse to any theory or to any mathematics beyond simple arithmetic. They also developed a sacred cosmography based on the *Quran* and the *hadith*, totally independent of the cosmography the Muslims inherited from Greek sources. The astronomers, on the other hand, addressed the

worship, and space with respect to a sacred shrine

same problems in terms of highly sophisticated mathematical procedures. Inevitably, the results proposed by the two groups were different. The historical sources, however, contain no mention of any serious conflict between them; medieval Muslim society was remarkably tolerant in this regard.

"The number of months in the sight of God is twelve (in a year) – so ordained by God the day He created the Heavens and the Earth ..." [*Quran*, 9:36].
"They ask you concerning the new moons. Say: they are but signs to mark fixed periods of time in (the affairs of) men, and for the pilgrimage" [*Quran*, 2:189].

The months of the Muslim year begin with the sighting of the lunar crescent (following the 'invisibility' of the New Moon). Two months are especially sacred: Ramadan, the holy month of fasting, and *Dhu al-hijja*, the holy month of pilgrimage. Also, all of the various religious festivals throughout the year are defined within the lunar year. According to the *Quran*, there is to be no intercalation of months – that is, there is to be no inserting an occasional month to keep the lunar year of twelve real months in line with the solar year with its twelve artificial months – for this would render the months which God had intended as sacred to be confused with other months. The lunar months are roughly 29½ days long, and the civil calendar is based on alternating months of length 29 and 30 days.

The actual calendar is based on the sighting of the crescent, which may take place on the first or second day of the civil month.

The legal scholars organized witnesses to check for the appearance of the crescent on the evening marking the beginning of the first day of the civil month. If, for various meteorological reasons, visibility was impaired, an arithmetical scheme was followed. The astronomers, on the other hand, developed procedures for determining the possibility of visibility. Shortly after sunset on the evening after a conjunction of the Sun and Moon, the New Moon must be far enough away from the Sun that a sliver of its surface be visible. The conditions proposed by a series of Muslim astronomers over the centuries involved lower limits on the apparent distance between Sun and Moon, the difference in their setting times and the altitude of the Moon above the horizon at sunset. Their results were recorded in annual ephemerides, or tables displaying the positions of the Sun, Moon and five planets visible with the naked eye, for each day of a given year, and the calculations were reproduced and a prediction made for the evening of the first day of the civil calendar. These predictions would describe the Moon in the following manner: the crescent will be seen clearly, it will be seen faintly, it will be seen with difficulty, it will not be seen.

All of this activity regarding the regulation of the calendar took place on a local level. The crescent might not be sighted on a given evening in India but might be sighted on the same evening in the Maghrib. Then Ramadan would have been celebrated with a day's difference in the two places. Nowadays, with the globalization of the Muslim community, this is regarded as problematic, and some standardization is warranted. A certain amount of confusion reigns each year at the beginning and end of Ramadan, as a result of the unwillingness of the religious authorities to listen to the astronomers. One proposal, not yet generally adopted, is that a universal Muslim calendar should be regulated by visibility at Mecca.

"O ye who believe! Fasting is prescribed to you as it was prescribed to those before you, that you may learn self-restraint .... Ramadan is the month in which was sent down the *Quran*, a guide to mankind, also clear signs for guidance and judgment (between right and wrong). So every one of you who is present (at his home) during that month should spend it in fasting ..." [*Quran*, 2:184–85].

The sighting of the crescent at the beginning and end of Ramadan has always been of particular importance. This month is a time of severity and joy: severity because the fast lasts from shortly before daybreak until sunset and prohibits eating, drinking and sexual activity; joy because the communal 'break-fast' after sunset brings entire families together and often turns into a night vigil lasting until the small meal at the end of the night before the fast resumes. Ramadan culminates in a festival ('id al-fitr) lasting several days and involving special community prayers and public entertainment. Marking the end of the rigours of fasting, it is celebrated with particular festivity.

"And perform the prayers at the two ends of the day and at the approaches of the night ..." [Quran, 11:114].
"Perform the prayers at the sun's decline until the darkness of the night and the recitation at dawn ..." [Quran, 17:78].
"The Messenger of God – may God bless Him and grant Him salvation – said: Gabriel came to me and prayed the midday prayer with me when the sun had declined and the afternoon shadow was like the width of the thong of a sandal, and he prayed the afternoon prayer with me when the shadow of every object was the same as its length .... Then on the next day he prayed the midday prayer with me when the shadow of every object was the same as twice its length ..." [Statement attributed to the Prophet Muhammad, attested in the canonical hadith collections].

The definitions of the times of prayer which became standard are outlined in principle in the hadith, and in precise detail in law books dating from the eighth century onwards. Each prayer must be performed within a certain interval of time of which the limits are astronomically defined, with the earliest part of the interval being preferred. The time for the maghrib prayer begins at sunset and lasts until nightfall. The 'isha prayer begins at nightfall and lasts until a specific fraction of the night is past. The fajr prayer begins at daybreak and must be completed by sunrise. The zuhr prayer begins at midday and lasts until the beginning of the interval for the 'asr prayer, which in turn lasts until sunset. The beginning of the 'asr is when the shadow of any vertical object has increased over its midday minimum by the length of the object. This definition, not mentioned in either the Quran or the hadith, together with regionally variant definitions for the zuhr and the 'asr and the special prayers at the festivals, provides the key to understanding the origin of the institution of prayer in Islam.

The legal scholars discussed the phenomena of sunset and sunrise and twilight and proposed simple arithmetical shadow schemes for regulating the zuhr and the 'asr. A modern equivalent would be something like this: in January the shadow of a man seven 'feet' tall is 1 foot at the zuhr and 8 feet at the 'asr, in February is 2 feet and 9 feet etc. Such schemes, in numerous variants, are proposed in medieval texts on the religious law and on folk astronomy. Thus simple observation can be used to determine the prayer times.

The astronomers, on the other hand, solved the mathematical problems relating to astronomical timekeeping and twilight phenomena. From the eighth to the nineteenth century, they produced tables for different latitudes displaying for each degree of solar longitude (roughly corresponding to each day of the year) such functions as: the length of day and night, the duration of morning and evening twilight, the time between midday and the 'asr, the time when the sun is in the direction of the qibla, and a host of other functions besides. Astronomical timekeeping was one area in which the Muslims excelled. They also produced tables displaying the time of day as a function of solar longitude and solar altitude, as well as the time of night as a function of the altitude of specific fixed stars. These tables, each serving a specific latitude, contained tens of thousands of entries. One Egyptian astronomer writing near the year 1300 even produced a table for solving problems of timekeeping by the Sun and the stars for all latitudes, albeit at the cost of having to compute over four hundred thousand entries.

As mentioned, there is no evidence of any conflict between the legal scholar and the astronomers. The muezzin who made the call to prayer from the minaret of a mosque would either be conversant with folk astronomy and use the simple procedures advocated by the legal scholars, or would be informed of the time of prayer by an astronomer equipped with tables or an appropriate instrument, such as an astrolabe or a sundial. Indeed, sundials displaying the hours of daylight and the times of the daylight prayers were a standard feature of many medieval mosques, although few have survived in a functional condition. In Egypt, in the thirteenth century, the office of mosque astronomer (muwaqqit) was introduced. These were men especially trained in astronomical timekeeping. In the present century, the methods of the medieval legal scholars have been abandoned. Until the computer age, the times of prayer were computed by the local surveying department or any other agency approved by the religious authorities. Nowadays, electronic devices are available which can be set for any locality and beep at the prayer-times. The times of prayer for different localities are available on the Internet. Most muezzins, or their pre-recorded replacements, announce the call to prayer at times recorded in special tables or given in diaries and daily newspapers.

"Turn then thy face in the direction of the sacred mosque: wherever you are, turn your faces towards it" [Quran, 2:144].
"The Kaaba is the qibla for the Sacred Mosque, the Sacred Mosque is the qibla for the sacred precincts (of Mecca and its environs), and the sacred precincts are the qibla for the inhabitants of the whole world from the place where the sun rises to the place where it sets" [Ibn al-Qass, ca. 975].

Clearly, when one is in Mecca one knows the direction of the Kaaba. But the first Muslims to occupy various regions between Spain and India had no clear idea of the local direction of Mecca. Therefore, they adopted a simple expedient, which later formed the basis of a highly developed sacred geography. They knew that when one is standing in front of the walls of the Kaaba one is facing various astronomically significant directions. In fact, the rectangular Kaaba is aligned with its major axis towards the rising of Canopus, the brightest star in the southern sky. Its minor axis points towards summer sunrise and winter sunset. Also, the corners of the edifice roughly face the cardinal directions. Furthermore, even before the advent of Islam, these four corners were associated with distant regions, namely, Syria, Iraq, the Yemen and 'the West'. It was not unreasonable therefore to postulate, for example, a *qibla* of due south for Syria. The problem of facing the Kaaba from other regions

and astronomical treatises. And, possibly as early as the ninth century, a cartographic solution to the *qibla* problem was devised in the form of a grid centred on Mecca, on which both the *qibla* and the distance to the centre were correctly represented.

In modern works on Islamic religious architecture, there reigns a great deal of confusion regarding orientations. However, astronomers were not always consulted in the laying out of mosques. Rather, the legal scholars often had the final say – proposing *qiblas* defined in terms of the cardinal directions or of astronomical risings and settings. Inevitably, sometimes the astronomers *were* consulted, but they then proposed different directions. The result is that for each major urban centre of the Islamic world, the religious architecture is oriented in a palette of different directions. The nature of the problem was appreciated by several medieval scholars – from

Astronomical timekeeping was one area in which the Muslims excelled

was solved by recourse to the corresponding feature of the edifice. Thus, the *qibla* in Andalusia was taken as the direction that one faces if one stands in front of the north-east wall of the Kaaba, namely, the rising of Canopus. Already in the first centuries of Islam, different directions were advocated by different authorities – but they were all astronomically defined. From the ninth to the sixteenth century, Muslim scholars (albeit not those versed in the mathematical sciences) proposed schemes of the world arranged in sectors around the Kaaba, with each sector associated with a *qibla* defined in terms of astronomical risings and settings.

In the ninth century, if not already in the eighth, Muslim astronomers turned their attention to the determination of the direction of Mecca from any locality as a problem of mathematical geography. They had inherited the necessary geographical data from Greek sources, which they immediately began to improve and expand upon. They had also inherited the necessary mathematical methods, including plane and spherical trigonometry, from Greek and Indian sources. Here, too, they wasted no time to prepare new trigonometric tables and in developing more streamlined mathematical techniques. Already by the early ninth century they had solved the problem – extremely complicated by medieval standards – of determining the direction of one locality on the surface of the terrestrial globe from another. Different, but essentially equivalent procedures were proposed over the centuries and, again from the ninth century onwards, tables were compiled displaying the *qibla* as an angle in degrees and minutes for each degree of longitude and latitude difference from Mecca. Sometimes, the *qibla* and the distance to Mecca were added to the lists of localities with their longitudes and latitudes found in geographical

as far afield as Andalusia, the Maghrib, Egypt, Iran and Central Asia – who have left us with discussions of the different *qiblas* used in their own areas. In short, the orientations of medieval Islamic religious architecture, some of which have long perplexed historians, are now to a large extent explained. We are, in fact, dealing with the only chapter in the new discipline of archaeoastronomy for which there is an associated textual tradition.

Both time and space are sacred in Islam. Each year, each month, each day is regulated by God's signs – the first visibility of the lunar crescent is a beautiful phenomenon, and the charming call of the *muezzin* a summons for man to worship his Creator. And the world is centred on a sacred shrine – the Kaaba is an edifice of mystery and innate power and the pilgrimage a symbol of man's search for God.

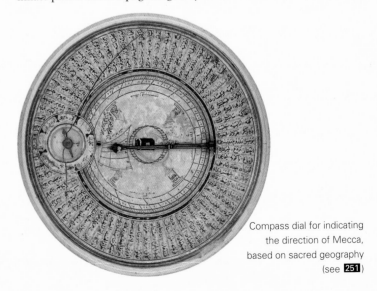

Compass dial for indicating the direction of Mecca, based on sacred geography (see **251**)

Calendars

040 Turkey (Miletus)
Parapegma stone
Marble, 108 BC
Berlin, PKB, Antikensammlung
[SK 1606 (MII)]

Parapegmata were astronomical
calendars, usually engraved in wood
or stone. They generally contained
astronomical and meteorological
predictions for every day of the year.
As the Greek civil calendar fluctuated
in relation to the months of its lunar
year, so the date for a particular
prognosis would change every year.
In the *parapegmata* (meaning 'things
fixed beside'), each prognosis was
flanked by a small hole, into which a
wooden peg was inserted. Each day,
the peg would be moved on and
'fixed beside' the next line of
information, generally consisting of
the day of the month, one or more
bits of astronomical information and a
weather forecast. This 2nd-century
calendar from Miletus offers, for
example: "Day 6: the Pleiades set in
the morning; it is winter and rainy"
and "Day 26: the summer solstice;
Orion rises in the morning; a wind
blows from the south". Not only do
these texts provide information on
the probable weather for each day of
the month, they often cite the
divergent opinions held by different
authorities. It also places the date of
the summer solstice as Skiphorian 14
or 26 June for the year 108 BC.

041 Attr. Jean Matal
Drawings after the *Menologium
Rusticum Colatianum*
Pen and ink with light-brown wash,
late 16th century
London, The British Museum, Dept.
of Greek and Roman Antiquities,
[new drawings nos. 30 and 31]

These two sheets depict the four
faces of the calendar block, which
dates from the first quarter of the 1st
century AD. For each month, there is
the depiction of the appropriate
zodiacal sign and information
concerning how many days are in
that month, when the Nones fall,
what are the relative hours of light
and darkness, which sign the Sun is
passing through, who the tutelary
god for the month is, which
agricultural duties need to be carried
out in that month and, finally, what
holidays (*feriae sacraque*) fall within
the month.

The structure of the year is often defined by those climatic
changes that signal of survival or extinction. For example, a
society that depends on agriculture will often be especially
attuned to the patterns of the rains or of heat and cold. Hunting
societies may base their calendar on the migrations of the herds.
Seafaring nations might need to construct a year in which the times it
was safe to set out to sea were differentiated from those that were not.

The earliest calendars tend to be defined by two different
concerns. Remarkably, both of these appear in one of the earliest
literary texts from ancient Greece, Hesiod's *Works and Days*,
written during the last third of the eighth century BC. Hesiod's poem
provides the basis for what we would recognize as the 'Labours of
the Months' – the cycle of agricultural occupations that fills the
year. Even though Hesiod provides a number of homilies explaining
how the behaviour of insects, birds and animals can point to
changes in the weather, for the most part the structure of his
calendar is astronomical. The cycle of the year in eighth-century
Greece was marked by the rising and setting of particular
constellations. The heliacal rising of the Pleiades, for example,
marks the time for harvest; their setting signals the time to till the
soil. Sixty days after the winter solstice, the star Arcturus (α Boo),
rises during the hours of dusk, telling you it is time to prune the
vines. When Orion and Sirius rise to the mid-heaven, it is time to
harvest the grapes.

The tradition of astronomical almanacs was extremely popular in
ancient Greece. Moreover, their authority was such that
monumental stone or bronze calendars based on these texts were
often erected in temples, or sometimes in the *agora* of a large town,
as a focal point for townsfolk and farmers to organize their year.
Only a few examples of these public calendars, or *parapegmata*,
have survived (**040**), but the tradition passed to Rome, where the
calendars became known as *menologia*. A typical Roman
menologium or *menologium rusticum* – literally, a 'farmer's
almanac' – would provide information such as the number of days
in a month, the tutelary deity and the major festivals that might be
celebrated during that month (**041**).

As unlikely as it might sound, the early Greek sources initiate a
tradition of astronomical and meteorological calendars that runs
relatively unhindered from the fifth century BC right up to this
year's edition of *Moore's Almanack*. The idea that each month has
an appropriate labour reappears in virtually all media from the end
of the Roman Empire up to the nineteenth century. Numerous
church portals, floor mosaics, manuscript pages (best known in so
called Books of Hours) and stained glass windows use the theme as
the basis for their decoration.

042 French
Book of Hours
Gold and colour on vellum, 15th
century
London, The British Library, Dept. of
Western Manuscripts [Additional Ms
11866]

The Book of Hours was the standard
Christian devotional text used during
the later Middle Ages and
Renaissance. The core of the book
was the Little Office of the Blessed
Virgin Mary, which consisted of eight
short services and prayers that were
meant to be said at specific times of
day and night – at Matins (sunrise),
Lauds, Prime (the first hour), Terce
(the third hour, ending at *ca.* 9 AM),
Sext (midday), Nones (the ninth hour,
ca. 3 PM), Vespers (early evening) and
Compline (sunset). Most Books of
Hours are prefaced with an
ecclesiastical calendar, which
provides information about the order
of the feasts and various saints' days
which should be celebrated
throughout the year. These calendars
are most often divided into months,
each page decorated with scenes of
the appropriate zodiac sign and
Labour of the Month. In this French
calendar, the decorative scheme is
quite developed, depicting scenes
from Genesis, planet gods and
Virtues and Liberal Arts.

**043 Kitao Shigemasa, Katsukawa
Shunshō and Utagawa Toyoharu**
*Day and night views of events in the
twelve months*
Colour woodblock album, early 1770s
London, The British Museum, Dept.
of Japanese Antiquities
[1945.11-10.2]

This unusual set of prints, executed
by three different artists, depicts the
twelve months of the year in terms
of day/night sequences. The artists
used this format to experiment with
ideas of light and dark, heat and cool,
bustle and quiet. For example,
Toyoharu depicts the 'Gion Festival'
and 'Cooling by the Sumida River' for
the sixth month.

044 Persian (possibly Azerbaijan)
*Candlestick with depictions of the
twelve months*
Bronze inlaid with silver, late 13th
century
Berlin, PKB, Museum für Islamische
Kunst [IB.499]

Whereas no large-scale Islamic
depictions of the Labours of the
Months appear to have survived, a
number of images of the months can
be found on objects connected with
the decorative arts, such as
metalwork and ivories. The theme
seems to have been particularly
popular as a decorative scheme for
candlesticks, of which over three
dozen examples are known. They are
generally cast in bronze, with the
decoration inlaid in silver. Some
contain generic hunting scenes and
others have small roundels of the
zodiacal signs as well. The
iconography of the months is
surprisingly close to that which
appears in the Latin West. Indeed, it
appears that these Persian cycles

may have been influenced by a type
of Limoges enamelwork (*à fonds
vermiculés*) which travelled to the
area via Byzantium. January is
shown as a man warming himself by
the fire. February has a man digging
with a spade. March shows a man
carrying a lamb over his shoulders,
like a 'good shepherd'. April and May
have men with long-handled sticks –
possibly threshing implements or
scythes. June shows a man cutting
down a plant with a sickle and July
has a man with a flat instrument
which may be used to beat grain.
September's figure carries an axe
and appears to be drinking from a
cup – perhaps in celebration of the
grape harvest. October depicts a
warrior. November's man has a cleft
stick which may be used for
gathering late fruit or nuts. And
December carries what seems to be
a bird's cage.

The second type of calendar to have survived from antiquity is organized by days that were deemed appropriate or inappropriate for the execution of certain tasks. The third part of Hesiod's *Works and Days* provides a glimpse into how these days were codified in ancient Greece. For example, the ninth and the last two days of the waxing Moon are good for the works of man. Women should weave on the eleventh. The sixth is favourable for the birth of boys, but not girls. The eighth is the day upon which you should geld your bull and your boar. Indeed, the amount of quite specific detail in this list suggests that its contents have a very ancient pedigree. Lists of 'good and evil days' also appear in early Babylonian and Egyptian sources. The concept underpinning these calendars is basically an astrological one. The good or evil influences of the planets were believed to be increased or decreased according to their relative placement in the heavens. If a planet were badly aspected on a particular day, it would be unwise to carry out any activities over which that planet was deemed to rule. The concept of good and evil days is extremely hardy and long-lived. It reappears in Rome, Persia and India as well as Christian Europe.

Roman civic calendars provide information about which activities should be carried out for nearly every day of the year. The series of abbreviations that one sees, for example, in the *Fasti Maffeiani* (see **046**) indicate on which days citizens are 'allowed' to conduct business. The 42 days marked with an F are the *dies fasti*. These were the days on which Roman citizens could initiate civil law suits. C stands for the 195 *dies comitales*, or days on which citizens could meet in assembly. An N indicates *dies nefasti* or days upon which business could not be conducted – a tradition harking back to the idea of 'evil days'. Days marked EN were both *endotercissus* ('cut') and *nefasti*. They were days upon which sacrifices were held. The other days that were considered to be unlucky were the *dies religiosi* ('religious days') and the *dies atri* ('the black days'). On the former, one tried to avoid undertaking any important action; on the latter – which usually fell on the day after the Kalends, Nones and Ides of each month – one was advised never to begin any new activity.

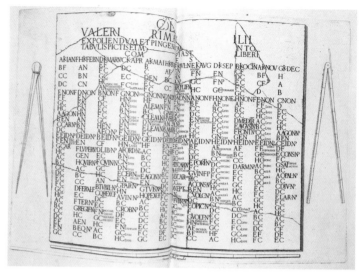

045 **Egyptian**
Calendar of good and evil days
Papyrus, 1290–1224 BC
London, The British Museum, Dept. of Egyptian Antiquities [EA 10184]

This calendar of good and evil days is one of the most complete to have survived from ancient Egypt. Missing the very opening sections of the year, it begins on 18 Thoth (18th day of 1st month) and continues well into the 9th month, missing only the last three months of the year. On the first line is the name of the month. On the second is the name of the day. There is then a marque indicating if that day carries good or bad fortune. In some cases, the fate of a person who is born on that day is added. Typical advice might be: "Do no work on this day ... slay no oxen", "Put no incense on the fire this day" or "Go not forth from thy dwelling at the time of the evening".

046 **Antonio Tempesta**
Engraving of the Fasti Maffeiani
Engraving, ca. 1575–85
London, The British Museum, Dept. of Prints and Drawings [1950.3–25.1]

The so called *Fasti Maffeiani* are named after the Maffei family, in whose house they were displayed throughout the second half of the 16th century. Fragments of the original stone calendar still survive in the Museo Capitolino in Rome. The calendar is arranged in 13 columns. The first 12 columns are each devoted to a single month, with the name of the month listed at the top of the column. At this stage, July is still called Quintilis ('the 5th month') and August is Sextilis ('the 6th month'). The final column is devoted to the intercalary days – those extra days that have to be inserted at the end of the year to bring the solar calendar back into line with the predominantly lunar cycle of the early Roman months. At the left of each column, there is a series of letters, running A–G. These signify the eight days of the Roman week. The other abbreviations in the lists indicate the 'status' of the day, such as whether it is good for business or if endeavours should be avoided. The longer inscriptions refer to special feasts and religious holidays.

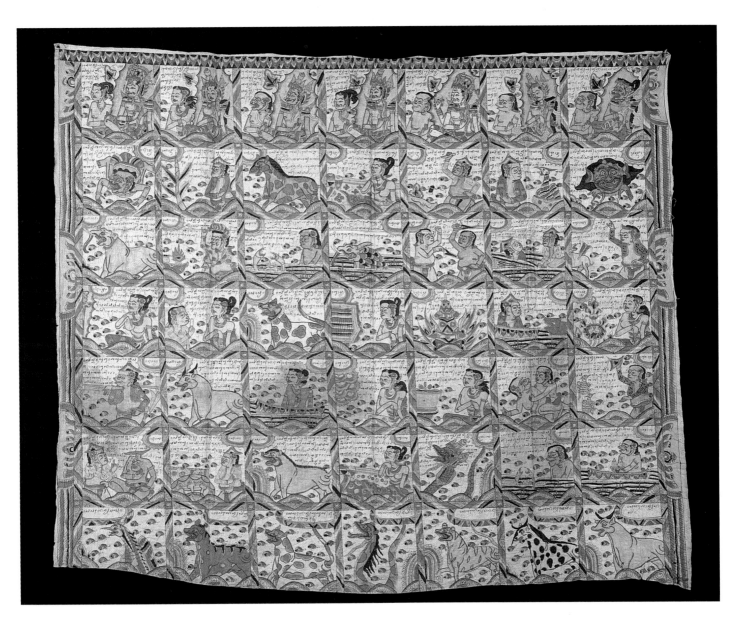

047 **Neo-Assyrian**
Calendar of good and evil days
Baked clay tablet, 705–681 BC
The Schøyen Collection [Ms 2226]

This tablet is signed by Nabû-zupuq-kěna, royal scribe to the Neo-Assyrian king Sennacherib (705–681 BC). The obverse of the tablet has a calendar with prescriptions about which months are propitious for carrying out certain activities. The first column describes the activities concerned and the remaining columns provide spaces for the names of the months for the year, beginning with Nisan. If the month is entered, it is propitious; if it is not, the activity should be avoided. For example, the first entry reads: "If he brings his wife into his house". Only the first month (Nisan) is filled in, suggesting that it is the only month in which one should bring one's wife home. The second entry, "If a baby should be born", has the second, third and fifth months filled in. There are lists of good and evil days on the reverse.

048 **Bali**
Calendar of good and evil days
Painted cloth, 20th century
Amsterdam, Koninklijk Instituut voor de Tropen, Tropenmuseum [2015–1]

Each square equals a day in the seven-day week and the gods or demons depicted indicate the day's relative benevolence or malevolence.

Also
• Borneo, *Calendar of good and evil days*, 19th century, Basle, Museum der Kulturen [II.c.1377]

049 Greek (Argos)
Selene and the zodiac
Marble, 2nd or 3rd century AD
London, The British Museum, Dept.
of Greek and Roman Antiquities [GR
1818.2–14.1/Sculpt. 2162]

From an earthly perspective, the Sun,
the Moon and all the planets appear
to travel along the same path
through the heavens. This path is
called the ecliptic because, when the
planets appear to pass in front or
behind one another, the one 'in front'
is said to eclipse the one 'behind'. In
this votive relief Selene, the goddess
of the Moon, wears a crescent moon
on her head and is surrounded by
seven stars – which may represent
the seven planets. She is also
enclosed within a niche decorated
with signs of the zodiac – those signs
through which the ecliptic passes.
The year is arranged so that the
months of spring (Aries and Taurus)
are placed at the top of the niche.
The Greek inscription is unintelligible,
possibly referring to some mysterious
Gnostic formula.

050 Babylonian
Astronomical table of New Moons
Baked clay, March 103–April 100 BC
London, The British Museum, Dept.
of Western Asiatic Antiquities [WA
34580]

As Christopher Walker and John
Britton have noted, Babylonian
astrology can be characterized by
three features: "the making of lists,
respect for tradition and extrapolation
from observed experience *ad
absurdum*". The tradition of keeping
astronomical 'diaries', as they are
called, can be dated back to the 8th
century BC. In all cases, the motive
for list-making was astrological and
focused primarily on the movements
of the planets relative to the fixed
stars. For calendrical purposes,
detailed lists of New Moons were
also kept – since the Mesopotamian
months and the year began with the
sighting of the New Moon. In
addition to this, however, lunar and
solar eclipses were believed to be
evil portents. The desire to be able
to predict them accurately and,
thereby, anticipate and alleviate their
evil effects, made the compilations
of lunar tables particularly important.

051 'Aaron, son of Losa'
Omer calendar
Wood, glass, pen and ink, 1826
London, The Jewish Museum
[JM 3636]

The Jewish calendar is defined by a
series of religious festivals. Particular
prayers are ordained for weekdays,
Sabbaths, holy days and feast days.
Omer is the name given to the
interval between the festivals of
Pessach, or Passover, and Shavuᶜot
or Pentecost ('50th day'). The *omer*
offering is made on the 16th day of
the first month, Nisan, which is the
second day of Passover. Seven full
weeks (49 days) are then counted.
On the 50th day, Pentecost begins,
commemorating the Giving of the
Law on Mount Sinai to Moses. An
announcement of which day and
which week it is in the 'counting of
the *omer*' is made every night as part
of the Maᶜariv service. A special *omer*
calendar marks the day of the count.

Also
• Simcha of Bohemia, *Prayer book*,
1745, London, The Jewish Museum
[JM 630]

053 Attr. Tshi-zun-hau-kau (Winnebago Tribe, Wisconsin)
Calendar stick
Hickory wood, 19th century (1820s?)
Bloomfield Hills MI, The Cranbrook
Institute of Science [2309]

The 19th-century anthropologists McKenney and Hall record that Tshi-zun-hau-kau, or 'He who runs with the Deer', invented a sort of 'almanac stick' that marked the divisions of time and changes of the season and of the weather. The calendar engraved on the Winnebago Stick is essentially a lunar one. The four sides of the stick are marked with notches and scratches that record the various phases of the Moon set against two half-year calendars marked in ten-day intervals. Many of the secondary markings relate to specifically lunar phenomena. For example, there is a series of half-crescent symbols that indicate the final glimpse of the waning Moon as it is seen just before sunrise and its period of invisibility. There are also recordings of the gibbous phase of the Moon, set down at periods of five days before and after the Full Moon. There also seems to be evidence that the proximity of the Moon to certain bright stars at specific times of the year has been depicted. Despite the fact that the Stick is fundamentally a lunar calendar, there is an attempt to intercalate between the lunar and solar year. The second half of the calendar year has an extra ten days to accommodate the difference between the lengths of the lunar and solar years. Moreover, there is also evidence of an awareness of the annual cycles of the planet Venus, with a series of small dots indicating the periods of its invisibility.

052 Hermann Junker
Blessing the New Moon
Oil on canvas, 1860s
London, The Jewish Museum
[JM 853]

Both Judaism and Islam inherited a number of components of the lunar calendar of the ancient Babylonians. Both place the beginning of the month as the first visible sighting of the Moon. A New Moon occurs when the Moon is positioned between the Sun and the Earth so that its unilluminated side is facing us. Immediately after this conjunction, the Moon begins to grow – or 'wax' – until it becomes a Full Moon on the fifteenth day of the month. In Judaism, the ceremony of *Kiddush Levanah* consists of a prayer of thanksgiving on the appearance of the New Moon. It is recited, usually in the synagogue courtyard, at the close of the Sabbath by a *Minyan* (or quorum of ten adult males). The Jewish year – *Rosh Hashanah* – also begins with the sighting of the New Moon in Jerusalem.

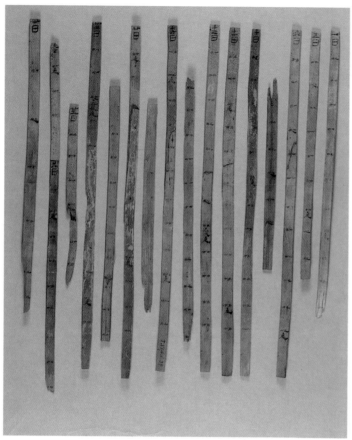

055 China (Dunhuang)

Calendar
Bamboo with ink, 63 BC
London, The British Library, Oriental
and India Office Collections
[Ms Or. 8212, nos. 208-23]

These wooden strips form about one
half of a calendar drawn up for the
year 63 BC, and they are to be read in
sequence from right to left. Each strip
bears at its head a note such as 'day
1', 'day 2', thus providing in all for the
30 or 29 days of the month in the
Chinese lunar calendar. In the
calendar in use at this time, each day
was notated not by its own number,
but by one of the 60 combinations of
the signs of the 'ten celestial stems'
and the 'twelve earthly branches',
such as *jiazi, yichou*; once 60 days
had passed, this series of 60 terms
was repeated. On each strip of the
calendar, therefore, there are
inscribed the combinations that
applied to the day, as numbered at
the head, as and when this came

round in each month of the year
(usually 12, but 13 in an intercalary
year). Thus, the sequence of terms
for the succeeding days of a given
month is set out to be read
horizontally, on the same line, from
right to left. For the calendar of this
particular year, the first day of the
first month is marked with its
notation *yiwei*, which is the 32nd
term in the series of 60; following
the passage of the 29 days as
determined for that month, the first
day of the second month is marked
with the notation *jiazi*, which is the
first term in the series.

Also

• Chinese, *Calendar for the third year
of Taiping xingguo* (978 AD), Song
Dynasty, London, The British
Library, Oriental and India Office
Collections [Stein Ms Or
8210/S.612]

054 Chinese

Mirror with cosmological decoration
Bronze, Tang dynasty (618–907 AD)
New York, The American Museum of
Natural History [70/11671]

The centre of the mirror has the four
animals of the four quarters of the
heavens (the green dragon in the
east, the scarlet phoenix in the south,
the white tiger in the west and the
black warrior, or the tortoise-cum-
snake, in the north). These are
surrounded by the twelve symbols of
the heavenly divisions, beginning with
the rat at the 'north' of the mirror and
running clockwise. The next band has
the eight trigrams associated with the
Yijing (or 'I-Ching') and beyond that is
a band containing the 28 asterisms or
'mini-constellations', associated with
the mansions of the Moon that mark
the sections of the 'zodiac'. The
penultimate border contains a poem,
in archaic seal script.

056 Aztec

Holy scriptures and census
Ink and pigments on paper and fig-bark, *ca.* 1525–50
The Schøyen Collection [Ms 1692]

This unique document combines two very different elements. On the left is a leaf from a census or legal document, written in Aztec hieroglyphs, possibly concerning a tally of labour tributes. On the right is the earliest surviving manuscript of the Scriptures written in the New World – a series of selections from Acts concerning the Conversion of St Paul translated into Nahuatl, the language of the Aztecs, and then transliterated into Roman script. The census document shows 64 heads in eight rows that are read from right to left and bottom to top. Both men and women are depicted, some of them connected to name-days. The 260 days of the *tonalpohualli* ritual was composed of two cycles, one of 20 days and one of 13. Each day of the 20-day cycle was given a name (rabbit, water, dog, monkey, grass, reed, jaguar, eagle, vulture, movement, knife, rain, flower, crocodile, wind, house, lizard, serpent, death's head or deer) and its own pictograph. The numbers in the 13-day cycle were represented as

dots. The Schøyen census represents a series of dates, which seem to be the birthdays (name-days) of the people with whom the date is connected. The dates represented include Movement 2,

Reed 4 and Scorpion 6. Movement 2 coincides with the second day of the 13-day cycle and the tenth day of the 20-day cycle. Reed 4 is the fourth day of the 13-day cycle and the sixth day of the 20-day cycle.

057 Aztec

Rattlesnake
Rhyolite porphyry, *ca.* 1200–1521 AD
Washington, D.C., Dumbarton Oaks Research Library and Collection [B-70.AS]

Both the Maya and the Aztecs observed a 260-day religious calendar, consisting of 13 'months', each of which was 20 days long. The rattlesnake, because its tail was made up of 13 separate rattles, was believed to have divine connotations and is a common motif in Aztec statuary. This snake is not a calendar *per se* but more a symbol of one of the essential ingredients of the Mesoamerican concepts of counting the time.

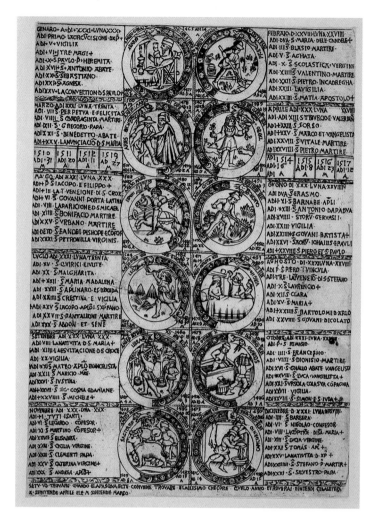

058 Italian (Florence)

Calendar for 1465–1517
Engraving, 1465
London, The British Museum, Dept.
of Prints and Drawings [V.1–74]

This single-sheet calendar of saints'
days preserves many aspects of the
classical *menologium* (see **041**). In
the small roundels running down the
inner columns of the print, there are
depictions of the Labours of the
Months. Most of these adhere to
established iconographic traditions.
January is shown feasting, February
is tilling the soil, March is pruning the
vines, and so on. The only depiction
that is unusual is for the month of
August, which has the image of a
doctor visiting a sick patient. This
may reflect the fact that Virgo (the

zodiacal sign for August) is ruled by
Mercury, the patron of the doctors.
The frame of each roundel contains a
zodiac sign for the month (each sign
shown with the rays of the Sun
streaming from its body) and a
system of shading which indicates
the relative amount of day and night
for each month. In the text of the
calendar, the cycle of lunations per
month is given as well as the feasts
and saints' days. In the spandrels
surrounding the roundels, there are
predictions of the fall of Easter for
the years 1465–1517. The date of
Easter for 1465, for example, is given
as *1465 a di 14 A* (or 14 April).

**059 German (Leipzig), Martin
Landsberg**

Ephemerides for 1487
Woodcut, 1486–87
Brunswick, Staatsbibliothek
[Broadsheet no. 7]

During the 15th century, many of the
publishers in Germany, England and
the Low Countries 'paid their rent' by
producing single-sheet almanacs or
ephemerides. The primary purpose of
the text was to provide accurate
astronomical information that could
be used for medical purposes. The
largest part of the sheet is devoted to
advice about what medical operations
might be carried out safely on which
parts of the body. Again, these
references are given relative to the
saints' days, such as "the second day

after Epiphany is good for lancing
boils". Finally, as lunar and solar
eclipses were generally considered
to be unlucky, diagrams illustrating
the astronomical conditions for the
two eclipses that occur during the
year are provided at the bottom of
the sheet.

Also

• German (Nuremberg), *Perpetual
calendar, ca.* 1750, Cambridge, The
Whipple Museum of the History of
Science [Wh.1775]

• Jehan Tabourot [Thoinet Arbeau],
Compot et Manuel Calendrier, Paris
1588, London, The British Library,
Dept. of Printed Books

By the middle of the fifteenth century, single-sheet printed calendars became the primary means for distributing information about the year. Like the Roman *menologia*, these calendars often include depictions or descriptions of the Labours of the Months, tables concerning the relative amount of light and dark for each month, illustrations of the zodiacal signs and, in some cases (such as the Florentine Calendar of 1465, **058**), instructions how to calculate the moveable feasts of the Christian year. In Northern Europe, these single-sheet calendars were often used specifically for medical purposes. These ephemerides include detailed descriptions of which planets are placed in which zodiacal signs and, most importantly, what the indications of the Moon might be to ensure the right time for the surgeon to let blood. For the illiterate who still wanted to know the structure of the ecclesiastical year, there were numerous so called pauper's almanacs, usually containing most of the information included in an ephemerides, but in pictorial form.

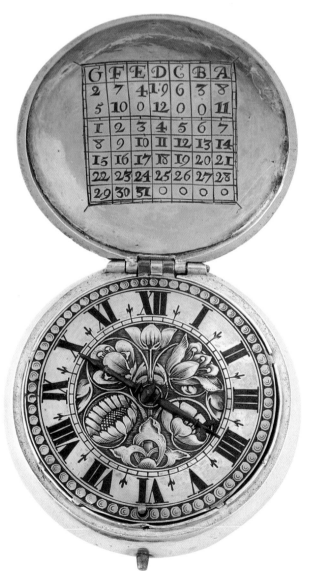

060 Thomas Alcock
Pocket watch with perpetual calendar
Mixed media, 1635
Private collection

Owing to the way the length of the months varies between 28, 30 and 31 days, there are sets of months that fall naturally into pairs because the first day of each will fall on the same day of the week. For example, if 1 January falls on a Monday, then 1 October will always fall on a Monday. The second and third lines of a perpetual calendar will tell you how these months are paired:

1	2:3	4	5	6	8	9
10	11	7	0	0	0	12

That is, 1 January will always fall on the same day as 1 October; 1 February and 1 March will always fall on the same day as 1 November, and so on. The perpetual calendar on the inside of the Alcock watch, however, looks like this:

2	7	4	1:9	6	3	8
5	10	0	12	0	0	11

The reason for the discrepancy is that, until England accepted the Gregorian Calendar Reform in 1752, the English began the year on Lady Day, 25 March. When this is the case, month 1 is March, month 2 is April, month 3 is May and so on.

061 Humfrey Cole
Astronomical compendium
Gilt brass, signed and dated 1575
London, The British Museum, Dept. of Medieval and Later Antiquities [MLA 1888. 12–2.293]

Astronomical compendia are usually 'pocket-sized' instruments containing a number of different dials. Often manufactured from gilt brass, they were intended as practical, but also decorative, objects to be used by wealthy gentlemen-scholars of the period. Humfrey Cole's compendium contains a nocturnal, a list of latitudes for principal cities in Europe, a universal equinoctial sundial, a compass and a perpetual calendar which lists 38 saints' days, Dominical Letters, Golden Numbers and epacts. The perpetual calendar forms one side of the outer casing of the instrument.

Also

• Johann Anton Linden, *Astronomical compendium*, 1596, London, The British Museum, Dept. of Medieval and Later Antiquities [MLA 1857. 11–16.1]

• John Naylor, *Astronomical clock*, 1720–25, The British Museum, Dept. of Medieval and Later Antiquities [MLA 1985. 10–15.1]

2 · THE MEASUREMENT OF TIME

063 German

Perpetual calendar

Card, 1765

London, The Science Museum

[1983.563]

062 Italian (Florence)

Easter tables

Engraving, 1461

London, The British Museum, Dept.
of Prints and Drawings

[1895.6–17.43]

The print is composed of a central
image of the Resurrected Christ,
rising from his tomb on Easter
Sunday. The roundel is surrounded by
a rota containing a series of numbers
and letters. The four corners of the
print are filled with depictions of the
Four Evangelists: John, Matthew,
Luke and Mark. As the accompanying
text makes clear, the print is intended
to help the owner 'find [the date of]
Easter forever'. The outer row of
letters indicates the name of the
month when Easter can occur (March
or April). The text says that the
calendar begins with Easter on 5 April

1461 and that has been marked with
a cross (*A* and *5* at the top of the
chart). If one counts 13 spaces to
the right in those two rows, one
discovers that the next Easter will
occur on 18 April 1462. Counting 13
more spaces to the right, the next
Easter is 10 April 1463, then 1 April
1464, 14 April 1465, 6 April 1466
(with a small 'helper label' to ensure
that you have understood the
system) and so on. The claim is that
if one always counts 13 spaces to
the right, the calendar will allow you
to find Easter forever. The two inner
rings provide the date in February or
March upon which the Sunday of the
Quinquagesima, or the Sunday
immediately preceding Lent, falls.

This German calendar is constructed
from a series of volvelles, or rotatable
discs, that can be moved to provide a
whole gamut of information about
any particular day of the year. At the
left side of the sheet, there is a
weekly calendar. The volvelle at the
top provides the name of the month,
the most important feast days, the
zodiacal sign, and relative lengths of
light and dark in the day within that
month. At the bottom, there is a
'Moon dial' showing the phase of the
Moon day by day as well as the
length of time it is visible in the sky.

Also

• Probably French, *Nocturnal with a
calendar of good and evil days*, late
16th century, London, The British
Museum, Dept. of Medieval and
Later Antiquities [MLA 1867.
3–10.1]

064 Swedish

Pauper's calendar

Ivory, *ca.* 1500

The Schøyen Collection [Ms. 1577]

In Scandinavian countries, there is a
long tradition for constructing
ecclesiastical calendars on wood or
bone. In this case, the calendar has
been drawn on pieces of stained
walrus tusk. The consistency in the
iconography of the saints throughout
the ages makes this calendar legible
to almost any 'reader'. In the calendar
for the month of November, for
example, the Feast of All Saints on 1
November is depicted with a small
bell. All Soul's Day (2 November) has
a skeleton with an arrow. St Martin's
Day (11 November) has the saint
riding a horse. For December, the
feast of the Immaculate Conception
has a mother and child (8 December)
and Christmas (25 December) has a
small manger scene.

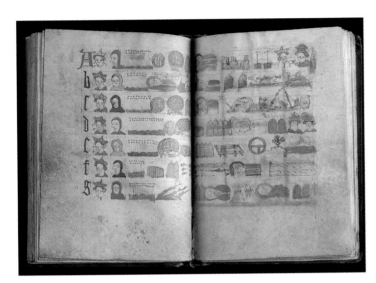

065 English (Yorkshire)

Pauper's almanac
Pen and colours on vellum, ca. 1425
The Schøyen Collection [Ms. 1581]

Almanacs were also produced for those whose ability to read was limited or lacking. Instead of words, pictures and diagrams were used to convey essential information about the ecclesiastical feasts and festivals. The 'language' of these so called 'pauper's almanacs' is remarkably consistent across the whole of Europe. One intriguing feature of this manuscript is the opening devoted to prognostications. In this case, the fortunes are determined according to the Dominical Letter of the year and they concern issues such as the abundance of crops (depicted as fields of wheat), a good or bad harvest (a sieve), honey production (a beehive), wars (kings arguing), fires and plague. During the 'A' years, for example, there will be a tolerably good harvest, but farm animals will die (evidenced by the depiction of a cow or a horse on its back with its legs straight up in the air) and there will be war (kings arguing). During the 'C' years there will be shipwrecks, and the harvests will always be extremely poor during an 'F' year .

066 Northern Dutch

Pauper's calendar
Ink and colours on parchment roll, late 14th century
The Hague, Koninklijke Bibliotheek [Ms 130 E 26]

This pauper's calendar consists of two long pieces of vellum, pasted end-on-end, that can be rolled up and stored in a small carrying case. The feast days of the different saints are conveyed by abbreviated Latin inscriptions and small pictograms. For example, in the month of February, St Peter's day is represented by his keys and St Matthew's Day by a small hatchet. Below these drawings, there is a perpetual calendar composed of Dominical Letters, which have two unusual features. First, the small feathers that are attached to the top of some of the numbers indicate that on that day the vigil (or nocturnal services held before a particular feast) should be held. Secondly, those Dominical Letters seeming to sprout long 'tails', which are covered by cross-hatches and half-circles, are ciphers that can be traced to wooden tally sticks that were widely used throughout many of the northern European peasant cultures. Each tally represents a Golden Number and relates to the numerical system by which one can determine the date of the next New Moon (in order to calculate the date of Easter). Similar 'tally ciphers' appear at the beginning of each month. The cipher in red indicates the hours of daylight and the one in black shows the hours of darkness for each month. A horizontal line crossing the vertical line equals 10, a half circle 5 and all short lines are integers. For March, there will be 12 hours of sunlight and 12 hours of darkness.

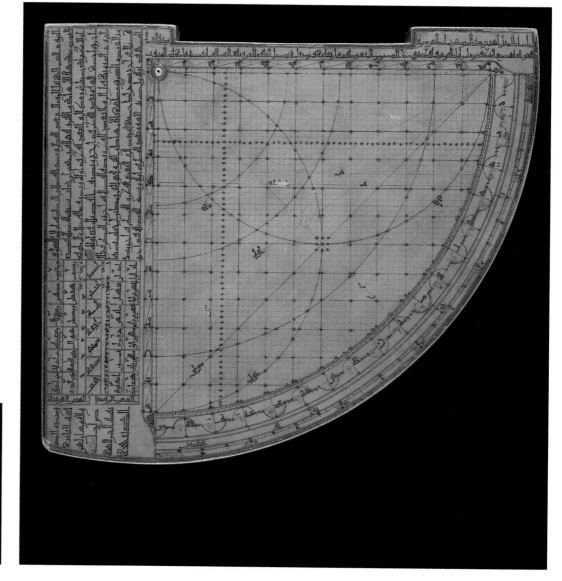

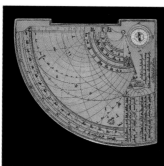

067 **Muhammad al-Sakāsī al-Jarkasī**

Quadrant with perpetual calendars
Fruitwood covered by lacquered paper, AH 1308 (1891–92 AD)
London, The British Museum,
Dept. of Oriental Antiquities
[OA 1997.2–10.1]

The information contained within the various scales and diagrams of this Turkish quadrant has only recently been deciphered by Silke Ackermann and David King. This is the sort of instrument that a merchant or official might need while trying to carry out

his business in the far-flung and multi-cultural corners of the Ottoman Empire during the 18th and 19th centuries. In the small rectangular panel on one side, there is a perpetual calendar calculated for the Islamic lunar year. Its earliest date is AH 1308 and it has the idiosyncratically Turkish feature of being calculated for cycles of eight years. Also on this side, there is a second perpetual calendar calculated for the Turkish solar year. The Turks were unique in Islam for adopting a solar year for all their financial transactions, while using the Islamic

lunar calendar for their religious festivals (this is possibly one reason why Turkish perpetual calendars are particularly abundant). The name of this financial year was *malīye*. *Malīye* calendars were also slightly odd in the fact that they marked leap years one year early. Therefore, according to the *malīye* calendar, this quadrant was made in AH 1307; according to the Islamic calendar it is AH 1308. On the side with the compass, there is a correspondence table between four different calendars: the Islamic; the Turkish *malīye* year; the Christian Coptic

calendar and a French Julian calendar, rendered into Ottoman transcriptions. Each of these is laid out month by month relative to the Arabic names of the months. If one takes the first Arabic month of Nissan, according to the *malīye* calendar, the month starts five days after Nissan begins; according to the French calendar, the first day of *mars* (March) starts twelve days after Nissan; and, calculating astronomically, the first point of the sign of Aries begins 23 days after the beginning of Nissan.

069 English

Clogg almanac
Wood, 17th century
Cambridge, The Whipple Museum of
the History of Science [Wh. 1125]

Clogg almanacs and prim staffs are
very basic ecclesiastical calendars
that appear in a number of northern
European countries as early as the
11th century. The name of the prim
staff derived from the Latin phrase,
prima luna (the New Moon).
These calendars are usually marked
with Dominical Letters and Golden
Numbers, to aid in calculating the
moveable feasts of the Christian year.
There are also small pictorial
symbols, most of which seem to
encapsulate variants on the Labours
of the Months or references to
special holidays. In this Clogg
almanac, the Dominical Letters are
omitted and the Golden Letters are
reshaped into ciphers – a simplified
version of Roman numerals,
consisting of X, V and dots for the
units. The feast days are marked by
slashes, grooves or extremely
simplified pictograms.

Also

• Danish, *A prim staff in the form of a
sword*, 17th century, Cambridge,
The Whipple Museum of the History
of Science [Wh. 1126]

068 Persian

Astronomical almanac (taqwīm)
Ink and gold on paper, 18th century
Dublin, The Chester Beatty Library
[CBL T 454]

Possibly arranged by Mustafā ben
Muhammad Kātib-zāda, this almanac
contains a series of useful calendars,
tables and even a small wooden
astrolabe. The first section is devoted
to instructions on correcting a sundial
(*i.e.* how to take account of different
latitudes). The second consists of
tables for discovering the starting
date of the Persian New Year
(*narrūz*). The next section provides a
table for determining upon which day
the first day of the month will fall.
The perpetual calendar in the
almanac (fols. 6v–12r) also includes
information about good and evil
days. Some of these prescriptions
have a historical rationale attached to
them (such as "evil day; on this day
Cain killed Abel"), while others just
give indications about the actions
that should be carried out or avoided
(such as "good day to have a shave
in the *hammām*"). There are also
tables for calculating sunrise and
sunset times throughout the year
and a table of the longest days in
principal cities.

070 **Wenceslaus Hollar**

Father Time carrying the Pope back to Rome
Engraving, ca. 1641
London, The British Museum, Dept. of Prints and Drawings

When Elizabeth I received the Papal Bull of 1582, astronomers and scholars urged her to make the change. The Protestant bishops blocked the move, however. Hollar's print captures the popular mood of the period. Father Time is shown carrying the Pope and his noxious reform back to Rome. As the poem below the image explains, it is Time's daughter, Truth, who warns him of the evils promulgated by the Pope – "This load of vaniti, this Pedler's packe/ This trunke of trash".

Also

• William Hogarth, *An Election Entertainment*, February 1755, London, The British Museum, Dept. of Prints and Drawings [CC.2–182]

• Christophorus Clavius, *Inter gravissimas*, from *Opera Mathematica* (Vol. 5), Mainz 1612, London, The British Library, Dept. of Printed Books

071 **English**

The Ladies Diary for the Year of Our Lord, 1752
Printed book, London 1751–52
London, The British Library, Dept. of Printed Books

By the time the British decided to adopt the Gregorian calendar, it was necessary to deduct eleven days from the Julian calendar before it would align properly with the stars. In 1752, Wednesday 2 September was followed by Thursday 14 September. A typical almanac for the period, *The Ladies Diary*, records the change, noting that "September hath only XIX Days this Year". The reason for the change is explained in the body of the monthly calendar itself – in the hole left by the missing 11 days. The 1427 years that had elapsed between the institution of the Julian calendar at the "Council of Nice" [sic] have meant that the present calendar "by account is forward" by 11 days.

○ The Julian calendar was established on the premise that the solar year was 365¼ days long. The solar year is actually only 365.24219 days long. This discrepancy, small as it might seem, meant that after hundreds of years astronomical phenomena, such as the spring equinox, crept backwards against the calendar itself. Indeed, one of the factors that had prompted the Venerable Bede's research into the structure of the calendar in the ninth century was the fact that the vernal equinox was falling on 21 March, rather than on 25 March as tradition had stipulated. Nevertheless, it was not until the elevation of Pope Gregory XIII in the last years of the sixteenth century that anything was done about it.

In 1578, Gregory XIII sent a proposal to all the rulers in the Latin West, asking them to consider a reformation of the calendar which would bring it back in line with the stars. The reform itself was relatively minor, focusing mainly on a slight reorganization of the manner in which the leap years were inserted. The major obstacle was the fact that in order to start the new calendar on the right footing and to adjust the major religious festivals to occur at the appropriate time, ten days of the current calendar had to be dropped. After much political manoeuvring, the Pope signed a Papal Bull on 24 February 1582 announcing the change. The Bull is known as *Inter gravissimas* after the first words of its Latin text.

Most of the Catholic countries in Europe adopted the Gregorian calendar relatively quickly, but the Protestant countries were much slower to accept the reform. For them, changing the calendar was not merely an astronomical issue, it was a political and theological one as well. Gregory was a fervent supporter of the Counter-Reformation and one needs only to recall that, as thanks for the Pope's support, Philip II of Spain had sent him the severed head of one of the leaders of the French Huguenots in the St Bartholomew's Day Massacre in 1582 to understand how Protestants might have viewed any edict from Rome with suspicion and distaste. The Protestant areas of Germany and the Netherlands, for example, did not accept the Gregorian calendar until 1700. The English, in particular, were extremely reluctant to move. They saw the reform, quite simply, as a Papist plot to bring the Protestant rebels back into the fold of the Mother Church. Adoption of the Gregorian calendar became a heated issue and, as such, ripe fodder for satirists such as William Hogarth, who included a figure holding an electioneering pamphlet on the subject in his painting of *An Election Entertainment*. One of the other issues for the English, however, was financial compensation. In particular, if a man were to lose ten or eleven days in wages or earnings, would he be liable for payment of tax which might be calculated monthly? In the end, however, astronomy prevailed and the English adopted the Gregorian calendar in 1752.

073 **France**

Republican calendar
Printed broadsheet, 1793/94 (l'An II)
Paris, Musée Carnavalet [G.4140]

Calendars hold memories and convey allegiances. One of the first tasks of the Republican government following the French Revolution of 1789 was to produce a new calendar – a calendar free from saints' days and feudal associations. The proposal was to create a decimal calendar, composed of 12 months of 30 days each. The month was divided into three 10-day weeks. The beginning of the year was set at the autumnal equinox, apparently for two reasons: first, the equinox coincided with the anniversary of the foundation of the Republic (22 September) and, secondly, as the time of year when day and night are equal, it symbolized the equality of all men. Each of the months was re-named, in order: Vendémiaire (month of the grape harvest), Brumaire (foggy month), Frimaire (frosty), Nivose (snowy), Pluviôse (rainy), Ventose (windy), Germinal (month of budding plants), Floréal (flowery), Prairial (month of the meadows), Messidor (month for reaping), Thermidor (hot month), Fructidor (month of gathering fruit). These were translated by English satirists as Wheezy, Sneezy, Freezy, Slippy, Drippy, Nippy, Showery, Flowery, Bowery, Wheaty, Heaty and Sweety. Despite the fact that all publishers and horologists were directed to promote the new system, it was tremendously difficult to maintain a decimal system when the rest of Europe were using the old duodecimal system.

Also
• France, *Republican calendar*,
  1794/95 (l'An III)
  Paris, Musée Carnavalet [G.22885]

072 **Italian (Siena)**

Pope Gregory XIII being addressed by the Commission for Calendar Reform
Tempera on wood, July 1582–June 1583
Siena, Archivio di Stato [Tavolette di Biccherna, no. 72]

The *tavolette di Biccherna* were originally the decorated wooden book covers used to bind the six-monthly financial registers kept by the office for public finance in Siena, the Biccherna. Soon, it became tradition that each cover related to events that had occurred during the previous six months. The *tavoletta* for the period of July 1582–June 1583 depicts a conflation of the events which led to the institution of the Gregorian calendar. On the left, the enthroned Pope is shown listening to the arguments of his commission. The main speaker is pointing to a fanciful celestial diagram that appears to hover in the sky outside the open window. The scale above the zodiacal signs is a calendar showing how the dates have advanced – with some hypothetical date now falling two thirds of the way through the sign of Scorpio. The bottom scale shows how, by dropping ten days from the calendar, one brings the same date backwards to the beginning degrees of the sign of Scorpio, where it should be.

Chinese, Gnomon chronometer (gnomon reconstructed),
2nd –1st century BC (see **098**)

Cyclical and Linear Concepts of Time in China

Michael Loewe

Astronomical clock tower by Su Song at Kaifeng, 1090 AD

In the early Chinese writings two concepts of time are evident, one cyclical and one linear. Motives such as a search for permanence and for reconciliation with the changes of nature that disrupt human continuity lay behind these concepts, but there was also a need to place human activity securely within the larger context of the movements witnessed in the heavens, on the Earth and among all living creatures. The hope of choosing the right and fortunate moment to ensure that an action would be successful ran alongside the will to achieve a life of harmony with the forces of nature. Hope of deathlessness in time eternal was matched by a conscious recognition of the passage of succeeding generations, seasons and years.

Time was seen as a thread or line (*ji*) that linked past and present. It provided a starting point towards which men and women could trace their ancestry and the permanent existence of their kin, stretching from one generation to another. A ruling dynasty would likewise see itself as the latest in descent within a whole series of properly constituted and acknowledged authorities that had sprung into being in the mists of the remote past and would continue for ten thousand generations to come. But time was also seen as a scale or track (*li*), along which there ran remorselessly the ever repetitive cycles of birth, death and rebirth, charted alike in the movements of the stars, the growth and decay of vegetation, the births of sons and daughters and the deaths of grandfathers or grandmothers. Recognition of the stage that had been reached in the cycle led to a life of contentment and harmony.

From at least the fifth century BC, Chinese astronomers saw the relationship of the Sun to other heavenly bodies in terms of a yearly circuit or 'zodiac' of 28 uneven divisions, but measurement of the phases of the Moon dictated acceptance of temporal divisions (the lunar month) of 29 and a fraction, which led to a year divided roughly into 12 equal segments, called 'lunar mansions'. Added to this, there were several other cycles that had philosophical or proto-scientific significance. The whole operation and movement of the universe were seen to depend on the overriding alternation of the two complementary forces of *yin*, female, mysterious and receptive, and *yang*, male, open and active. No sudden events marked the

Time was seen as a thread or line

change of dominance from *yin* to *yang*, or from *yang* to *yin*: it evolved through a series of five stages. Other schools of thought saw this universal movement taking place by degrees through a sequence of no less than 64 stages, symbolized by 64 hexagrams, which were later reduced to the well known eight trigrams.

Official orders had to be so timed that there would be no conflict with the cycle of being

Owing to these series of cycles, the numbers 5, 6, 8, 10, 12, 28, 64 and 365 each formed a significant factor in understanding the aspects and movements of the universe. As these schemes were partly in conflict with one another, a full understanding of the universe depended on the creation of a comprehensive system which could provide for their reconciliation. Of all these elements, factors or figures, it was of the greatest importance to discern and demonstrate the means whereby the rules of a cyclical 5 and a linear 6 could live harmoniously together. Solution of this problem challenged some of the best minds of early China. It also affected the practical means of measuring and dividing time and certain methods of divination (see further below).

From around 800 BC, writers began to insist that authority to rule had been granted as a sacred charge from Heaven. For an emperor to demonstrate that his power was a direct mandate from Heaven involved ensuring that the decisions of his officials would conform with Heaven's own cosmic system. Orders had to be timed in such a way that, at the moment when they were obeyed, there would be no conflict between these actions and the whole cycle of being. Officials had to phase their activities so that they would be appropriate to the year, season, month, day and hour when they were to take place. Thus, as spring was seen as the season of growth on earth, it would be improper to violate its nature by hewing down forests at that time. By the same token, winter, as the season of death before rebirth, would be the correct time for implementing the death sentence of a criminal.

A universally accepted calendar was also required for the basic tasks of government, such as the registration of the population, the conscription of males for service and the collection of taxes. Officials who were required to impose a curfew within the cities needed a precise method of measuring the hours of the day and their divisions, as did those who were responsible for the completion of routine tasks such as the delivery of official mail or the maintenance of military patrols. The records of such duties were not complete without a note of the hour when they were finished.

(ji); or as a scale or track (li)

At the times when the rule of China was divided among several self-styled kings or emperors, each ruler was anxious to establish a calendar – framed on a luni-solar basis – that would be regarded as accurate and which could be enforced as legally binding. Recognition of such a calendar could be tantamount to recognizing that the ruler of the area possessed proper authority to command

obedience. No ruler would be willing to concede that his rival's dominion was based on a correct understanding of the divisions of time.

With the establishment of the first empire in 221 BC, it became essential for a dynastic house to claim the right to issue its own calendar and to ban others from doing so. The imperial calendar must be accurate; copies must be readily available to officials in the capital; and they must be distributed to distant outposts at the borders by the imperial courier service. The emperor relied on his astronomers, working in the imperial observatory, to maintain close contact with the natural order of the seasons and to keep in consonance with the five phases whereby birth, death and rebirth succeed one another. He would command that his farmers would so time their ploughing, sowing, hoeing and reaping that they matched the corresponding activities of the heavenly world. Moreover, on the occasion of the state cults, such as the New Year, both he and his priests would present their prayers and their offerings at the precise moment when they were due.

A calendar that is based on the phases of the Moon can be maintained with reasonable accuracy by means of observation; but the succession of the months very quickly ceases to correspond with the sequences of the Sun's apparent movement and the seasons of nature. Normally, the year comprised twelve months, but, in order to reconcile the cycles of Earth, Moon and Sun, it was necessary to insert an extra, intercalary, month once every 33 months. In addition, the lengths of the months necessarily varied, as no calendar could handle a month which lasted for $29\frac{43}{81}$ days. It was necessary to determine which months would be fixed as long (30 days) and which ones were short (29 days).

A further problem was left to be solved by only the most senior of these officials and may even have required the approval of the emperor. As in European cultures, in China there was no absolute principle that determined which month of the year should be regarded as its first. Whichever month was chosen fixed the date for the timing of all the important religious ceremonies in which an emperor marked the beginning of a new year.

Perhaps as early as 1500 BC, two series of written terms were used, one of ten and one of twelve signs or characters. These are sometimes known as the 'ten celestial stems' and the 'twelve earthly branches'. Their combination together provided a cycle of sixty terms which soon featured in calendrical reckoning. Enumeration of days by the terms of this sexagenary series imposed a cyclical view of the divisions of time and their sequences which transcended other ways of counting. In addition, from at least the first century AD, twelve animals had been chosen to symbolize the twelve years of the series. This practice is still in use today. For example, the years 1998, 1999 and 2000 have been recognized as the years of the tiger, the hare and the dragon.

According to the linear as opposed to the cyclical concept, years were at first numbered sequentially from the first year of a monarch's reign. But, as early as *ca.* 116 BC, a different practice arose to satisfy an emperor's wish to proclaim his achievements or to call for popular support. Within a few years of the beginning of his reign, Han Wudi decreed that for the following year a title should be adopted which would form a starting point for a fresh enumeration. Such 'reign titles' often included an auspicious or grandiloquent term or sentiment that might express a prayer for bliss – such as *Tianbao*, the 'blessed treasures of Heaven', adopted for 742 AD, or *Qingning*, 'pure tranquillity', used from 1055. Or they might mark the completion of a project of which the government was proud. For example, *Heping* or 'the pacification of the river' was used to mark repairs to the dykes of the Yellow River in 28 BC. From the start of the Ming Dynasty in 1368, the term chosen in the year of an emperor's accession remained in force throughout his reign, as is best known in the case of two emperors of the Qing Dynasty who reigned under the titles of *Kangxi* ('joyful delight', 1662–1722) and *Qianlong* ('superior glory', 1736–95). Such terms duly appear as 'date marks' on the porcelain manufactured during those years, usually without further identification. When precise specification was necessary, for example for dating official documents, the particular year of the reign period could either be denoted by its sexagenary term or by its ordinal number in the series. For example, the year 1700 could be defined either as *Kangxi gengzhen*, using the cyclical sign, or as *Kangxi 39*, in numerical terms.

Difficulties could and did arise – particularly since the cycle of 60 was often seen as forming a natural limit to the term of human capacity or endeavour. Should an emperor's rule extend over 60 years, it might seem that a natural limit had been exceeded, with the danger of untoward events that might ensue. In fact, only twice in Chinese history has an emperor's reign lasted more than 60 years. The Qianlong Emperor actually took the step of abdicating, possibly to avoid the danger that might arise from his exceeding the limit imposed by the universal system of being. Had he not done so, official documents dated at *Qianlong bingzhen* could have meant either 1736 or 1796.

Choice of the right day, hour or moment for an action – be it religious or lay, of public or of private concern – was a matter of profound importance to China's rulers and officials as well as to her farmers and merchants. To obtain guidance for such a choice, they looked to help from occult sources; and the records of such messages provide the earliest examples of Chinese writing that we possess (from *ca.* 1500 BC). The types of question that were raised included the choice of the day to make offering to the gods or to embark on a major project, or whether activities undertaken in the forthcoming period of ten days would have a fortunate outcome or be doomed to failure.

By around 200 BC other methods of enquiry had developed. Whether or not an activity would be successful was seen to depend on the cosmic circumstances in which it was undertaken. The time

The cycle of sixty was often seen as

for which it was proposed was compared with the relative positions of the Sun and the Earth, and with the date of birth of the individual who was concerned. To check that these circumstances were not in conflict, diviners used a square board, within which there was a revolving, circular disc. Both square and circle were marked with divisions and notations of some of the cycles already noted, such as those of the 10 or the 12 series of terms, or that of the 28 segments of the zodiac. Manipulation of the two parts of the board enabled an expert diviner to ascertain whether the correct combination of times had been met.

By perhaps the eleventh century, these boards assumed a new form (see **283**) with the addition of a centrally mounted magnetic compass needle. Such instruments were the ancestors of the highly complex 'geomantic compasses' used today. The multiple circles and characters inscribed on the surfaces of these compasses allow the diviners to consult the details both of time and of space or location. They may thus advise their clients whether a proposed site would be suitable for erecting a house, bank or office, and whether it would be surrounded by benign influences or open to the onslaught of evil.

From the third century BC or earlier, the standard form of stationery that served official needs and was normally used for literary, historical, legal and medical texts consisted of vertical strips of wood measuring one foot (23 cm) in length (**055**). Cords that were laid horizontally bound the strips together, so that they formed a surface which carried successive columns of writing, to be read from right to left. They could also carry diagrams that ran from one strip to its neighbours. Almanacs drawn up in this way are known from surviving manuscripts which date from the third century BC. They provided a more direct and almost a mechanical means of determining which days would be suitable for a chosen action.

Perhaps the most important element in determining whether or not a day would be suitable was the sexagenary term by which the day was denoted. In particular, the component character in the term that denoted one of the 'twelve earthly branches' demanded careful attention.

Almanacs set out to show which of these terms pertained to each day of a calendar year. They were drawn up as a table of columns, showing the 'twelve earthly branches' in sequence, with a guide strip at the right indicating the month to which they applied. Prescriptions below the table told the enquirer on which days it would be auspicious to sacrifice, to attempt a cure of illness or to start ploughing the fields, or when to avoid such actions. In simple terms,

forming a natural limit to human endeavours

the system may be compared with calendars of good and evil days that one finds in the West.

Fragments of copies of the imperial calendar of the Han Dynasty date from 134 BC onwards. As with almanacs, calendars were framed in tabular form with a guiding strip placed at the right-hand edge that showed to what headings the entries in the table should be referred. Read horizontally, the calendar gave the terms for the successive days throughout the year for which the calendar had been drawn up. Read vertically, each of the 30 strips told the reader the sexagenary terms for each of the first, second or third days *etc* in the 12 or 13 months of the year.

A remarkable example of continuity may be seen in the production of China's calendars. In addition to the basic information just described, some of the earliest examples included a catchword that drew immediate attention to the properties of a particular day and its inbuilt fortune. Even though most modern calendars in China follow the Western (Gregorian) form, some – even those displayed in banks – still carry the same catchwords as seen on calendars hundreds of years old. Apparently, the knowledge of good and evil days is still required by the astute businessman of the twentieth century.

Life in time eternal was the unvoiced hope of those who provided the dead with material symbols to escort their souls to the realm of immortality. The bronze mirrors reverently buried in the graves of emperors, princes and officials and their consorts were intended to fulfil this purpose. One particular type of mirror, found in graves of between 50 BC and 50 AD, bears symbols and emblems that were believed to lead the soul to the most felicitous of all possible situations of the whole system of the universe.

The design of these 'TLV' mirrors replicated the universe. Their circular shape enclosed a square, in the same way as the circular heavens enclose the square earth. A roll of clouds at the outer rim showed the way to pass from Earth to Heaven. Around the square, figures of elves and other creatures represented the immortal inhabitants of the next world. The symbols within and around the square represented and reconciled the two schemes of the universe, based respectively on five and on six. Four linear devices, shaped like T, marked the centres of the four sides of the square; facing them were four linear devices shaped like L, placed at the edge of the circular surround. Between the Ts and the Ls four stylized animals represented four of the five divisions of the universe both in time and space and their stages of operation. At the centre of the mirror, a semi-circular boss or mound symbolized the point of apex at midsummer, and the element of Earth.

Set in their correct positions, these symbols portrayed the cosmos, conceived to operate in its five successive stages. But in addition to guiding the soul to eternity within this correctly arrayed microcosm of five, it was also desirable to guide it to a suitable place within the cosmos as conceived in terms of twelve. So, twelve studs or bosses were placed within the central square of the mirror, each one being identified by its term in the series of the 'twelve earthly branches' and together marking the stages whereby time was advanced.

An inscription running round the circle spelt out the meaning of these symbols. They conveyed a life which endured beyond the confines of measured time; they were intended to take the soul to a situation of perfect harmony, wherein the two concepts of the universe meshed together without conflict. The circle, with its cyclical view of time, and the square were at peace with one another. As one further piece of symbolism, seen in a few choice examples of these mirrors, tiny medallions depicted the ever present influences of *yang*, rising to power in the north-east, and *yin* beginning her turn of dominance from the south-west.

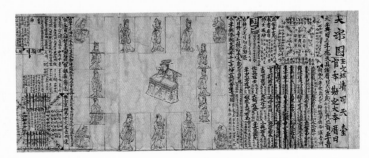

Calendar for the third year of *Taiping xingguo* (978 AD) (see **055**)

Visiting a Shinto shrine

Cycles, Seasons and Stages of Life

Time in a Japanese Context

Joy Hendry

Straw objects made in Kyushu
(see p. 82)

The Ise Shrine
(see p. 82)

The Japanese have been quite eclectic in gathering and incorporating cultural influences. Their understanding of time reflects a willingness to add to, rather than replace, ideas. Thus the calendar sent out as a gift by Japanese embassies around the world today looks identical to a Western calendar – giving away nothing of the intricacy of the calendar consulted internally in Japan to establish 'good days' to carry out rites and festivals. Neither does the precision of Japanese trains, clocks and watches give much idea of an interesting cultural propensity towards flexibility in the use of ideas of time.

Japan has assimilated ideas about time in the same way that she has assimilated religious ideas. These have come primarily from China, Korea, India and Europe, but also from her prehistoric connections with the islands of Austronesia to the south. This eclecticism makes for an interesting set of possibilities. Japanese skills of pragmatism and adaptability mean that different systems of thought may be called upon for different occasions, or even for different aspects of the same occasion. Thus, for example, the children who visit a Shinto shrine in mid-November, when, according to tradition, they are supposed to be 'aged' three, five and seven years, may have had their 'age' calculated according to one of two quite distinct systems. The older, 'Japanese' system reckons the ages of children according to the number of 'New Years' a child has lived through and it starts the counting with a baseline of being aged 'one' at birth. According to the newer 'Western' system, a child only reaches 'one' on the anniversary of their birth. This flexibility of approach allows the convenience for many of taking a group of siblings to the shrine in the same year.

In Japan, the New Year is now celebrated on 1 January but both this, and the November date when children visit shrines, are relatively new additions to the Japanese calendar. They were imposed by the government in the nineteenth century to reinforce political centralization and replace dates which varied widely from region to region. The present date for the New Year season is now the most widely recognized holiday of the year, but there are still several other dates when new beginnings are celebrated, suggesting that the idea of starting afresh – making a new beginning – is as important an aspect of Japanese possibilities for dividing up time as accurate chronology is.

The indigenous Japanese word for time (*toki*) stands very much for a point in time, a particular moment or occasion, rather than an abstract continuing entity. There is even a sense that time can be 'folded' or 'manipulated' according to ecological or social needs. The character used to write *toki* is combined with the idea of 'space' to represent an 'interval' of time. Alternatively, it can be combined with

There is a sense that time can be 'folded' or manipulated according to need

a character which can also mean 'change' or 'replace' to represent the notion of a period or generation – again suggesting the concept of renewal.

The idea of renewal associated with periods of time is reinforced in the way years are counted in Japan. They are marked with the posthumous names of emperors, starting with the number 'one' to denote the first year of their reign. The year in which an emperor dies is the end of that particular historical period and, simultaneously, the start of a new one. The year 2000 AD, for example, will be the twelfth year of *Heisei*, the name already assigned for the reign of the present emperor. This year also marks twelve years since the death of Emperor Hirohito, whose period, known as *Showa*, extended for 63 years.

Periods of history are also recorded under the names of families which held power during the time, or under the names of the cities which they chose as their capitals, as another method of renewal was to move the seat of power to a new location to mark a new era. Different parts of the country, therefore, become associated with particular periods of history and places can serve as markers of 'time'. For example, the city of Nara was the capital of Japan during the seventh and eighth centuries AD. Kyōto (literally, 'capital city') was known as *Heian-kyō* (or 'Heian-capital') from the ninth to the twelfth centuries. The city of Tokyo is, again literally, the 'eastern capital'. It became the seat of power in the seventeenth century, when it was known as Edo. The name of the city was changed in the nineteenth century to mark a new, 'modern' period.

The lack of a notion of time as a continuing entity, marked with events and periods but otherwise relatively homogeneous, is found in other Asian countries. Japan has also incorporated much of the Chinese system, where there is a stronger sense that there are good and bad times for carrying out certain ventures. This sense of there being lucky and unlucky days is based on the idea that time is measured in cycles.

Many of us have heard of the animals of the so called 'Chinese zodiac', each of which is associated with a particular year within a twelve-year cycle, known in Japan as 12 'terrestrial branches'. Years, days and hours are all assigned an animal derived from this system and, in Japan, these animals are named as rat (*ne*), ox (*ushi*), tiger (*tora*), hare or rabbit (*u*), dragon (*tatsu*), snake (*mi*), horse (*uma*), sheep (*hitsuji*), monkey (*saru*), cock or bird (*tori*), dog (*inu*) and wild

boar (*i*). Each year, day or hour is also assigned one of the five elements from which the universe is held to be composed – wood, fire, earth, metal and water – and combined, in turn, with the bipartite classification of *yin* and *yang*, forming another cycle of ten 'celestial stems'. This series of combinations means that it takes sixty years (days or hours) before the identical combination returns. This is the main reason why the number sixty and the age of sixty years is so important in Japan. When a person completes his own, personal, sixty-year cycle, he or she is usually fêted. In some parts of Japan, sixty-year-olds are even presented with red garments – such as jackets, nappies or aprons – which are said to mark a return to dependency as they hand over their responsibilities to their children.

The precise configuration of these celestial and terrestrial qualities at birth is said to affect the character of a person. Moreover, their likely happiness in marriage is related to the configuration of their partner. Experts may be consulted on this complicated subject. The expert may advise a change of name, or even a change in the characters used to write a name, in order to offset a bad combination. An advantage to the system is that an expert may be called upon to offer a negative prognosis in order to provide a diplomatic excuse on behalf of a reluctant bride or groom. One year, in particular, is said to hold disastrous consequences for men, who should avoid any woman born during *hinoeuma* – the active aspect of fire in a horse year. There is regularly a dramatic drop in the birth-rates in these years, suggesting either that parents avoid having children then or that they subsequently falsify the birth certificates.

Another cyclical system which affects decisions in Japan is a recurring six-day period which has connections with Buddhism. This cycle includes a day of great safety (*daian*), a day of great danger (*butsumetsu*), and days with more specific warnings related to the time of day. The six days are described as *sengachi* ('early victory'), *tomobiki* ('friend-pulling', when funerals should be avoided), *sakimake* ('haste loses'), *butsumetsu* ('Buddha's death' – the most inauspicious day), *daian* ('great safety'), and *shakkō* ('red mouth' – a dangerous day when high Noon is the only safe time). Weddings, funerals, journeys and business deals are often scheduled to take place on good days according to this cycle and wedding halls are particularly pressed when a *daian* falls on a Sunday, allowing most people time off work to attend the ceremony. The opening of new business ventures may also be carefully timetabled to fall auspiciously within all these cyclical systems. Also, the seven-day working week is another cyclical system to be taken into account,

Starting afresh – making a new beginning – is an important aspect of Japanese attitudes

and it is interesting to note that the Japanese, in naming the days, have added their own five elements to days for the sun and moon (the days are literally 'sun-day' (*nichiyōbi*), 'moon-day' (*getsuyōbi*), 'fire-day' (*kayōbi*), 'water-day' (*suiyōbi*), 'wood-day' (*mokuyobi*), 'metal-day' (*kinyōbi*) and 'earth-day' (*doyōbi*), with Sun-day corresponding to Sunday as used in English).

Another interesting combination of cyclical renewal and place-marking is to be found at the highly sacred imperial shrine of Ise, a seventh-century building which is carefully rebuilt every twenty years – both to keep it in good condition and to preserve the architectural style of the original period. This method of conservation was introduced by decree of the Emperor Tenmu, who ruled during that period, and the regular reconstruction is said to be a "kind of prayer for the renewal of the life of the Japanese nation, a rejuvenation of the life of Japan as a whole, and its eternal progress" (*Jingū Shikinen Sengū*).

An obvious model for cyclical renewal is found in seasonal changes which Japanese associate with their own climate (indeed, the Japanese sometimes express surprise to find the pattern of four seasons also exists elsewhere). Seasonal change is celebrated culturally in a whole variety of ways. On the specific dates that mark the beginning of each season, the Japanese will change the beautiful scrolls which hang in their homes so that their subject-matter reflects the appropriate season; according to this season they will also use a different cloth *fukusa* to wrap gifts and other objects; they will choose a different kimono for ceremonial occasions, and they hang plastic renditions of seasonal flora and foliage on lamp posts on city streets. Seasonal change also signals a change in the weight of regular clothing and the cultural importance of this pattern is particularly clear when a sharp change in temperature is steadfastly ignored until the specific date of seasonal change arrives.

As well as being associated with particular seasons, annual festivals mark points of change and renewal. New Year is undoubtedly the biggest and most important festival, preceded by an enormous effort to clear outstanding business and clean everything up so that none of the 'untidiness' of the previous year will carry over into the new one. It is marked with all manner of decorations, dress, food and customary practice which continues throughout the first half of January. Social relations are renewed with ritual greetings at this time and cards are dispatched to confirm them. Ritual visits are also made to Shinto shrines to greet the deity and purchase new protective goods for the house, car, office and so forth.

Once a girl is born into a family, they will celebrate Girls' Day on the third day of the third month. The arrival of a boy inaugurates

celebrations of Boys' Day on the fifth day of the fifth month. The special autumnal festival in September is often shifted to coincide with the appearance of the large, impressive harvest moon. November is now associated with the seven, five and three festival for children, held on the fourteenth. *Setsubun*, or the old New Year, is celebrated by throwing beans out of the house to the cry, 'Devils out, fortune in'. Images of devils can be seen all over Japan during this part of the year. In March, the opening of the first cherry blossom in Japan is announced on national television and, as the flowers burst into their short-lived bloom, parties of families, friends and workmates carry food and drink out under the trees and celebrate together. The academic new year is also held in the spring, while in the Western world it is held after a summer break.

In the middle of the hot, summer season in Japan, there is an important Buddhist festival (*Bon*) for remembering the souls of ancestors. As this is a time when many city people return to their country origins, another example of pragmatism has been to vary the dates regionally to avoid excessive congestion in the transport system. People dress in cool, cotton garments at this time and dance

All this careful counting of years for the living is

into the night. It is also an occasion for sending gifts to people to whom one is indebted and department stores sometimes turn over a whole floor to their displays, as well as offering a wrapping and delivery service.

Perhaps most closely related to the climatic seasonal changes are the festivals held throughout Japan to usher in and out the natural cycles of planting, harvesting, and in the case of rice, transplanting. In an area of Kyushu (one of the four main islands), straw objects are made to celebrate the completion of the planting out of the rice-seedlings, which are originally tended in boxes. These objects are hung over a stream to ensure that the vital water will keep flowing. Elsewhere, festivals are held in spring and autumn to herald the arrival and departure respectively of the god of rice paddies, or the 'earth spirit'. These festivals must be held on the 'earth' day in the *yang* aspect nearest to the spring and autumnal equinoxes.

One clear way of marking the passage of time in any society is by marking stages in the lives of members of that society. Japan is no exception. Indeed, several of the annual festivals are linked with the celebration of life stages, as we have already seen for counting ages, presenting children to the Shinto shrine and celebrating the birth of girls and boys into the family. On Girls' Day families set up quite spectacular displays of dolls, comprising a lord and lady sitting at the top of a tier of steps which accommodate all the personnel and regalia required for a wedding in the flamboyant Heian period. In this way, the display offers not only a history lesson, but also a reminder that marriage is an important future goal for a girl.

For boys, helmets and suits of armour from the period of samurai strength replace the wedding paraphernalia. These stand for ideal qualities to which boys are supposed to aspire, as do the large cloth carp hung out to fly over the house. Carp demonstrate perseverance and strength in adversity by swimming against the current and even leaping up waterfalls. These colourful flag-like decorations may be seen all over Japan in the period before Boys' Day on 5 May. On both days, special seasonal food is prepared, decorations reflect contemporary flowering plants and families mark the annual progress of their children.

Marking the progress of children through time also takes place at intervals after their birth, although there is again some regional variation in the precise number of days involved. A naming ceremony occurs about seven days after the birth, a visit to the local shrine around thirty days after and a first tasting of solid food at one

mirrored by a similar counting for the dead

hundred days. In some parts of the country, there is an ancient ceremony on the first birthday of the child, though this predated the clocking up of a year in age at that time and was more concerned with a child's first steps. Other ceremonies mark the first outing, first haircut, first wearing of certain garments and so forth.

Another ancient ritual marked the attainment of adulthood approximately at puberty, but this is rarely practised now. These days, children officially become adults at a national ceremony held in the January after they attain twenty years. In the latter part of life, ages again become important. After the completion of the sixtieth-year, there are celebrations at the ages of seventy-seven and eighty-eight. Other special ages are thought to be associated with particular vulnerability; the most important of these are at thirty-three for women and forty-two for men. This is a time when people again visit the Shinto shrine.

All this careful counting of years for the living is mirrored by a similar counting for the dead as they make progress in their conversion from living beings into ancestors. Mourning ceremonies, usually of a Buddhist nature, are held after seven days, one month, three months and one year, and again, at certain fixed annual intervals, until the soul is thought to have completed this journey. The total number of years again varies regionally, but there is an interesting shared idea that the individual human being then

becomes part of a larger entity of ancestral matter and some people believe that the souls of new babies come from this source.

Relative age is regarded as important in Japanese social relations and this is the first criterion used in deciding on an appropriate level of respect and politeness in the absence of other factors. Children distinguish linguistically between older and younger playmates from an early age and their classification of schoolmates always makes clear what the relative school year might be. Individuals who move through school together tend to hold reunions in later life with great nostalgia for the former times of approximate equality. In some parts of Japan, groups of age-mates will carry out social activities together regularly, also helping each other in times of need, and saving money communally for trips and special events.

In communities, these 'age-sets' (as they might be called in anthropological terminology) move together through various 'age-grades' – starting with the 'children's group', whose sport and play is partly administered by adults, but moving into the more self-sufficient 'youth group', who also dress up and carry out the various ritual activities associated with festivals. There is then the *shōbodan*, for men (a fire-fighting corps which quickly turns out to help in any disaster), and the 'young wives group' for women. Serious participation in local politics takes place in the 'prime of life'. Afterwards, people join an 'old people's group' where they, again, spend quite some time at sport, play and possibly travel.

Thus the passage of time is clearly marked through the life course, as well as by the sexagenary cycle, and it all begins again when people die and enter what could be described as the death course. It is interesting that the predominant religious influences of Shinto and Buddhism are particularly associated with life and death respectively, and they also respectively mark the renewal of New Year and the remembrance of the *Bon* midsummer festival. These older systems do seem to have evolved an internal coherence in Japan, but the adoption of foreign influences is eclectic, as we have noted. During the year 2000, the Japanese will not only celebrate the twelfth year of *Heisei* and the year of the Dragon. They will certainly also be celebrating the advent of a Christian millennium.

The north–south polar axis of the Earth is inclined at an angle of approximately 23½°. As the Earth revolves around the Sun, it maintains this constant inclination. This is the basis of seasonal change. In the northern hemisphere, our summer falls when we are leaning towards the Sun during the months between June and August. Astronomically, these periods coincide with the Sun's apparent passage through the signs of Cancer, Leo and Virgo. The winter occurs in the northern hemisphere when the Sun appears to pass through the signs of Capricorn, Aquarius and Pisces. The seasons are reversed in the southern hemisphere.

The number of seasons a civilization might recognize depends largely on its climate and its use of astronomical data. In ancient Greece, for example – a climate considered mild by many Europeans – there were originally only three recognized seasons. Winter, spring and summer were characterized as cold, wet and dry. Once man had measured the heavens, however, and discovered the cyclical progression of the two solstices and two equinoxes, which seemed to divide the year into four equal segments, the convention for recognizing four seasons was born.

The areas surrounding the Earth's equator may feel the effects of the changing seasons less strongly than those in the more northern climes, but that is not to say that equatorial cultures do not experience and mark seasonal change. Most equatorial areas will experience a dry and a wet season or, in some cases, a hurricane, typhoon or cyclone season. These changes in climate alter the patterns of activity within a society. Among the Nuer of the Upper Nile in the Sudan, the year is divided into two halves: rain and drought. During the rainy season, from March until September, the tribes retreat to the more established villages and numerous civil ceremonies take place. During the dry season (October to February), the tribes will move to temporary camps and the young men become more actively engaged in hunting, fishing and waging war. The dry period is a time for movement and migration; the wet period is a time for consolidation and ceremony.

To outsiders, the climate endured by the Inuit well north of the Arctic circle might seem to be little more than winter, winter and a flash of slightly milder winter. Nevertheless, the Inuit themselves recognise nine different seasons, marked by subtle changes and developments in the world around them. The period of March–April is known as *avunniit*, or 'the time when premature seal pups are born'. This is differentiated from the next month, *nattian*, when 'normal seal pups are born' and *tirigluit*, when 'bearded seal pups are born'. Similarly, the three months that correspond to summer and early autumn are named *saggaruut* ('caribou shed hair'), *akullirut* ('caribou hair thickens') and *amiraijaut* ('velvet peels from caribou antlers').

The Aboriginals of the Kakadu region of Australia have five seasons. In Gundjeidmi, the Maiili language, they are called *yegge* ('the cooler, but still humid season'), *wurrgeng* ('the cold season'), *gurrung* ('the hot, dry weather season'), *gunumeleng* ('the pre-monsoon storm season'), *gudjewg* ('the monsoon season') and *banggereng* ('the knock'em down storm season'). The end of *banggereng* is signalled by *yamili*, the time when 'the grasshopper calls out that the cheeky yams are ready'.

074 **John Alexander**

The Rape of Persephone
Oil on canvas, 1720
Edinburgh, National Gallery of
Scotland [NGS 1784]

In many cultures, the variation of the seasons has been given a mythological explanation. In Greek mythology, when Hades abducted Persephone, the daughter of Demeter, the goddess was so distraught that she let all the plants of the Earth wither and die. Fearing the destruction of mankind, Zeus sent his messenger, Hermes, to beg for Persephone's return. Hades agreed, on the condition that she would be allowed to leave only if she had eaten nothing during her period underground. Unfortunately, though, she had eaten six pomegranate seeds. Finally, Hades relented and allowed Persephone to live above ground with Demeter for six months of the year; but, for the other six, she had to rejoin Hades in hell. When Persephone is with her mother, the Earth enjoys the seasons of spring and summer; when she is with Hades, the autumn and winter months grip the Earth. Alexander's sketch shows Hades abducting Persephone. A small cupid precedes them – holding the monstrous dog, Cerberus, on a lead. The decorative framing of the central part of the work is carried out in *grisaille* and shows the four seasons, which owe their existence to the abduction.

075 Jasper Johns
The Four Seasons
Intaglio prints, 1987
Private collection

In each of the four images, a shadowy protagonist (based on a tracing of the artist's own body) is depicted. His experience of the weather associated with each season can be fairly easily understood – spring rains, wintry snowflakes and a trip to the seaside in the summer. On top of this, there are numerous references to the 'art world': an image of the *Mona Lisa*; illusionistic psychological puzzles (such as the 'rabbit/duck' and the 'vase/profile');

and allusions to Johns's own work (painterly cross-hatchings, wood-graining). Finally, in each image, there is a large circle which seems to fulfil a triple role. First, it recalls a clock face. Secondly, it alludes to Leonardo's drawing of the so called 'Vitruvian Man' of perfect proportions. Thirdly, as the position of the arm within the circle marks a clear progression (in Spring, the arm is poised at the beginning of a downward swing and in Winter the swing has finished), *The Four Seasons* is also an allegory of the Four Ages of Man and the passage from birth to death.

076 French
The Four Seasons
Marble, 18th century
London, The Victoria and Albert Museum [1179–1882 to 1182–1882]

Between the late 16th and late 18th centuries, allegorical subjects became a popular theme for free-standing groups of sculptures. In these cases, the 'iconography' of the subject was never intended to be too taxing to the imagination of the viewer. Instead, the main purpose was to enchant the eye and engage the sentiment. To this end, figures of women and children were often employed. This grouping of four young children as

embodiments of the four seasons is typical of the genre. Their rudimentary attributes are sufficient to allow the viewer to identify Spring with her garland of flowers; Summer bedecked with sheaves of wheat; Autumn with the grapes of the harvest and Winter with his shawl and dead bird.

  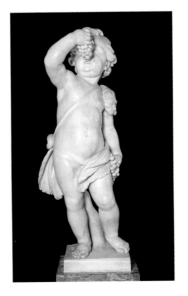 

2 · THE MEASUREMENT OF TIME

077 **Three stoats (*Mustela erminea*)**
Stuffed animals, 20th century
London, The Natural History
Museum [1977.3193, 1977.3194,
1953.457]

The change in weather during the
seasons has remarkable effects on
the natural world. A number of
species, such as stoats, experience a
change in the colouring of their coats
or their plumage, to provide better
camouflage in a differently coloured
world. As the days shorten with the
approach of winter, a hormonal signal
is triggered in the stoat that ceases
the introduction of brown pigment
into new hair growth. The whiteness
a stoat will achieve depends largely
on the latitude within which it lives.
As snow is relatively rare in southern
Britain, many of its stoats will only
attain a mottled or piebald coat.
Further north, however, the stoat will
go completely white. The stoat's
white winter coat is known as
ermine. The chromosome that
determines whitening also seems to
have a gender component. The 'go-
white' gene seems to be carried on
the 'X' chromosome and, since
females have twice the number of
'X' chromosomes, the likelihood that
they will whiten is twice that of
males.

078 **Antonio Tempesta**
'Spring', *'Summer'* and *'Winter'* from
The Four Seasons
Engraving on paper, 1592
London, The British Museum, Dept.
of Prints and Drawings [1856.1 –
12.298 to 301]

Sets of allegorical representations of
the four seasons developed from the
Labours of the Months tradition.
Particularly during the late 16th and
17th centuries, the inexpensive
medium of engraved prints provided
a growing and acquisitive middle
class with art for the home. In
Tempesta's prints, each season is
represented by a god, shown riding in
a triumphal chariot and surrounded by
an entourage who help to define the
qualities of that particular time of
year. The goddess of spring holds
wreaths and flowers in her hands.
Her chariot is drawn by birds, lambs
and smiling oxen. Ceres triumphs in
the summer and Bacchus in the
autumn. The god of winter is an old
man, shown warming his hands over
a large brazier. His companions are
rain-soaked peasants, who hold
standards proclaiming that Pisces,
Aquarius and Capricorn are the
months ruled by winter.

Also

• Jean de Court (Vigier), *Three plates
from a series of the Twelve Months,
ca.* 1565, London, The Victoria and
Albert Museum [2430–1910;
CA 2433–1910; CA 493–1912]

079 Nicolas Poussin

Phaëton asking for the chariot of Apollo
Oil on canvas, *ca.* 1630–35
Berlin, PKB, Gemäldegalerie [478 A]

The tale of Phaëton comes from Book II of Ovid's *Metamorphoses*, where the ambitious young demi-god begs his father, Helios-Apollo, to let him drive the chariot of the Sun. Apollo, knowing that Phaëton will never be up to the task, reluctantly agrees. All goes fairly well until the young man encounters the huge constellation of Scorpio, "reeking with black poisonous sweat; and threatening to sting him with its curving tail". He drops the reins, the horses bolt and Jupiter is forced to smash the chariot with one of his thunderbolts lest it destroy the Earth. The shattered chariot plunges from the sky and Phaëton is killed. The painting shows a golden-haired Apollo wearing a golden drape and golden sandals, crowned with a wreath of laurel leaves and resting his left arm on his golden lyre. He is depicted as a 'Sol-Annus' figure (see **025**), seated in the middle of the golden band of the ecliptic, on which the constellations of Taurus, Aries, Pisces, Aquarius, Scorpio and Libra are delicately drawn. The young Phaëton (looking older than his eternally youthful father) kneels before the god. Personifications of the four seasons form a loose circle around him. Spring, dressed in green and white and wearing flowers in her hair, scatters flowers. Autumn is a nude male figure who wears grape leaves in his hair and dozes through a wine-induced slumber. A chilly, grey Winter shivers between two urns of glowing embers. And Summer, wearing a pink dress, is accompanied by signs of her yearly harvest. Summer also holds the unusual attribute of a mirror in her hands – perhaps to catch and reflect the summer's heat. Father Time, or Chronos, is depicted as an old man. The attribute he is holding is not clear, but it seems as though he is biting it – an allegorical elaboration of the Ovidian saying that 'time devours all things' (*tempus edax rerum*). In the background, Apollo's golden chariot waits, the horses patiently held by personifications of the hours.

080 Japan

Kimono
Embroidered silk, 19th century
Edinburgh, National Museums of
Scotland [A.1966.31]

An appreciation of the subtleties of
the natural world has always been an
intrinsic part of Japanese culture and
aesthetics. In particular, they value
and derive pleasure from a
heightened sense of awareness
regarding the seasonal changes, with
their fleeting beauty and attendant
promises of renewal. The earliest
expressions of this sentiment can be
found in Japanese poetry, where
seasonal changes were often used to
convey the different shades of
human emotion. By setting the right
tone – the right mood – Japanese
poets were able to draw discreet
parallels between the natural world
and the personal world. By the 10th
century, poetic anthologies were
arranged by season, beginning with
spring and ending with winter.
Between the 10th and 13th
centuries, Japanese artists began to
paint folding screens and panels
depicting the landscapes as they
changed throughout the year. Typical
subjects might be a single folding
screen showing a field of bamboo
which, magically, passes through all
four seasons within one frame or
might depict an arrangement of
flowers in which individual blooms
represent a different time of the year.
It is also the tradition to wear special
garments to celebrate the arrival of
each season. This *kimono* celebrates
that precise moment when winter
finally gives way to spring. The lower
half of the garment is covered with
vibrantly coloured camellias, covered
by straw tents to protect them from
the last of the early spring snows.
Across the shoulders, the first of the
plum blossoms burst into flower,
accompanied by darting songbirds – a
sure harbinger of warmer weather.

For every culture that recognizes the cyclical rhythm of the seasons, there has to be a point at which the cycle begins. This will be the time when people come together to reaffirm old bonds, assess the events of the previous year and prepare the way for success in the coming year. Needless to say, the date that any given culture may choose to celebrate as its 'New' Year depends on which aspects of the year's cycle are deemed most important.

The Egyptians, for example, began their year with the heliacal rising of the star they called Serpet (also known as Sothis, Sirius and α CMa). The civilization of Upper Egypt depended on the annual flooding of the Nile River which brought with it rich silt deposits to feed the next year's crops. At the time in which the calendar was being developed, about 4500 BC, the annual flooding of the Nile began in late June and coincided with the sunrise appearance of Serpet.

From about the second millennium BC, the Babylonians began their year in the spring on the date of the vernal equinox. It is a tradition that survives today in the Jewish calendar. The Early Christian calendars also started with a date near the spring equinox – 25 March, or the Feast of the Annunciation of the Virgin.

Whereas the earliest recorded Roman calendars begin with the spring equinox, after the fall of the monarchy, the Romans began to date events relative to the terms in office of the consuls of the Republic. By the first century BC, the date for the installation of new consuls was set at 1 January.

The Chinese New Year lasts for two weeks. It begins with the second New Moon after the winter solstice and ends with the Feast of the Lanterns on the following Full Moon. As it is a lunar festival, its date changes, but it generally falls in the early spring, during February or early March. It marks the end of the cold winter weather and the banishment of the cruel demons of winter.

In all dates associated with the spring equinox or the passing of the winter solstice, the theme of the New Year is rebirth and renewal. Some societies, however, prefer to mark the autumnal equinox with a New Year's festival – both as a celebration of the harvest season and as a kind of spiritual 'putting the world safely to bed' in anticipation of the long winter months of cold and inactivity. In ancient Japan, an autumnal New Year (Setsubun) was celebrated. Houses were cleaned, transactions completed and the devils of the old year were purged. Similarly, many of the North American tribes, whose survival depended on sufficient stores to see them through the winter, marked their New Year with a celebration of the autumnal harvest.

In Islam, the calendar is exclusively lunar. This means that the first month of the year, Ramadan, is not fixed to the cycle of the solar year. Instead, Ramadan occurs exactly twelve lunar months after the previous Ramadan – regardless of the Sun's orientation against the ecliptic.

081 Otto van Veen
Q. Horatii Flacci emblemata imaginibus
Printed book, Amsterdam 1777
London, The British Library, Dept. of Printed Books

The association between the four seasons and the 'Four Ages of Man' can be traced back to the teachings of both Hippocrates and Pythagoras. The idea is based on the assumption that there was a direct correlation between the so called microcosm of man's life and the macrocosm of the heavens. The progress of the year – from its birth in the spring to its death in the winter – was mirrored in the Four Ages of Man. Here, an airborne putto, with butterfly wings – an image often associated with fleeting time – holds a sundial in his outstretched hands. He seems to be leading a band of four towards the distant horizon. The first is an infant, who wears a circlet of flowers around his waist (spring). The second is a youth, carrying a threshing stick and wearing sheaves of wheat in his hair (summer). The third is a heavier, more mature man, wreathed with grapes and carrying gourds underneath his arm (autumn). The fourth is a wizened old man, wrapped up against the cold and carrying a brazier (winter). The original subject of the emblem, the seventh ode in Book IV of Horace's *Carmina*, tells us: "The year warns you – the hour which forcibly carries off the kindly day warns you – Do not hope that you will last forever." In the lower right corner of the picture, there is an image of a snake biting its own tail – a well known symbol for the circuit of the year and of eternity.

082 Egyptian

New Year's flask
Green-glazed terracotta, Late period
(7th century BC)
London, The British Museum, Dept.
of Egyptian Antiquities [EA 24651]

The Egyptian solar year began with
the heliacal rising of Sirius, which
signalled the inundation of the Nile
River. During the Late period, it
became customary to mark this date
with the gift of a terracotta flask
containing fresh water from the rising
Nile. The flask was given as a good
luck token. The cow depicted on its
side represented Mehytweret, an
embodiment of 'the great flood'. The
inscriptions invoke Amun-Ra, the Sun
god, and Isis.

Also

• Japanese, *Hagoita*, 20th century,
Oxford, collection of Prof. Joy
Hendry

083 Mexico (Aztec)

The Fire Serpent, Xuihcoatl
Quartz-diorite, 1507
Washington, D.C., Dumbarton Oaks
Research Library and Collection
[B-69.AS]

In Mesoamerica, the coincidence
between the 260-day 'sacred
calendar' and the 365-day solar
calendar occurred every 52 years.
Although the completion of the solar
year was celebrated, the 'real' New
Year occurred in mid-November once
every 52 years. The Fire Serpent was
specifically associated with time and
the New Year. The glyphs on the
base of this statue denote the year
'2 Reed' and above this there is the
name of Motecuhzoma (Montezuma)
II. Together, these form a date that
coincides with our year 1507 AD and
indicate that this figure was sculpted
specifically for what turned out to be
the last of the 52-year New Year
ceremonies celebrated by the Aztecs.

084 Katsushika Hokusai (and pupils)

Album of New Year's prints
Colour woodblock prints, 1804
London, The British Museum, Dept.
of Japanese Antiquities
[1908.7-18.018]

During the New Year holidays in Edo
(modern Tokyo), it was the custom
for poets to exchange specially
commissioned prints, known as
surimono. The subject matter of
these prints varied. Sometimes they
commemorated special New Year's
activities, such as the first letter
written in the New Year or the first
bath. Or they might depict a scene
relating to the 'zodiac animal' for that
specific year. At other times, they
were decorated with images referring
to the structure of the year. In these
cases, as the Japanese used a
calendar composed of 'long' and
'short' months, the picture would
contain small hidden clues – such as
large coins for the 'long' months and
small coins for the 'short' months.
This album contains 51 small prints,
all designed for the New Year of 1804
– the Year of the Rat.

085 Rose ('Mother') Green

Corn Husk Face mask
Corn husks, 20th century
Brussels, Musées Royaux d'Art et
d'Histoire [ETAM 1489]

In many cultures, the New Year is
celebrated with the return of
ancestral spirits or helpers. The
purpose of the return is to bind the
society closer together, by reminding
each member of his or her shared
ancestry, and to offer the people
omens that might lead them towards
prosperity in the coming year. The
Husk Faces are said to represent a
race of farmers who come from the
other side of the world. They are the
people who gave the three basic
crops – corn, beans and squash – to
the earliest hunters, who later formed
the Onondaga tribes in upstate New
York. They represent supernatural
beings associated with agriculture,
the summer season and fertility. Their
appearance is always taken as a good
omen, indicative of prosperity and
bountiful crops.

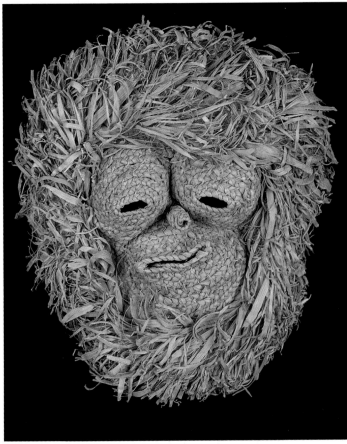

086 Shibata Zeshin

New Year subjects
Ink and colour on paper, 1848
London, The British Museum, Dept.
of Japanese Antiquities
[1982.7–1.014/Add. 699]

In Japan, the New Year ceremony is
highly orchestrated. Certain foods
must be prepared, the house must be
cleaned, debts must be settled. This
scroll by Shibata Zeshin depicts all
the necessary foods, cooking utensils
and implements, special flowers,
carefully laundered clothes and
special *sake* for the occasion. The
New Year's still life is boldly
presented, almost like a panning
camera shot, with some of the items
caught mid-scroll, but with others
disappearing off the top or bottom
edge of the piece. As one unrolls the
hand-scroll, one relives the fun and
excitement of both the preparation
and the celebration of the New Year.
As the piece is dated 'First Month
1848', it was probably given as a
New Year's present by the artist to
one of his friends.

Calendar used by an Inuk hunter, 1920s, following the introduction of Christianity to the Canadian Eastern Arctic: days of the week are represented by straight strokes, Sundays by Xs. The calendar was also used as a hunting tally, including counts of caribou, fish, seals, walrus and polar bear (from Revillon Frères, *Eskimo Life*, 1923)

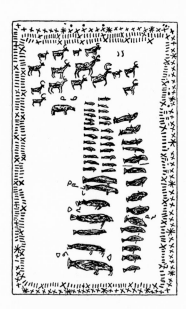

Inuit Time

John MacDonald

A generation ago, in the Arctic Quebec community of Kangiqsualujjuaq, a government development officer was explaining the virtues of hard work and efficiency to a rather polite Inuit audience. During his talk the enthusiastic official used the expression 'time is money' and his interpreter, confused but compliant, translated this tenet of capitalistic wisdom as "a watch costs a lot!".

Inuktitut, the language of the Inuit, has no word for time, not, at least, in the abstract, regimented sense commonly understood in Western industrial society. This does not mean that Inuit somehow lacked any comprehension of the links between time and so called 'economic' activity. Traditional Inuit society extolled the productive use of time, and there were consequences, beneficial or otherwise, depending on how time was used. Over-sleeping, for instance, would result in poor hunting, for it was "necessary to show the souls of the animals that one is eager to capture them", while long life (and for pregnant women an easy childbirth) was assured to those who rose early in the morning, went outside, and walked "three times round the house in the direction of the sun" (reported by Rasmussen 1929, p. 181). The habit of early rising was a social value instilled in all Inuit children. The recollections of an Inuk elder, Hubert Amarualik, interviewed in 1994, are typical: "We had to *anijaaq* ['to go outdoors'] immediately after dressing in the morning. We would observe the sky conditions, note the types of clouds, and the position of the stars".

A useful summary of the factors on which Inuit traditional perceptions of time were based, prior to European contact, is offered by Finn Gad, a Danish historian: "Formerly, [the Inuit] had been quite content with the changing of the seasons and the known migratory habits of the animals they hunted. Wind and weather might upset everything, but on the whole there was a predictable series of changes in a relatively rigid sequence, which applied also to the rising and setting of the sun, its height over the horizon, the

The Inuit calendar

Moon month	Season	Environmental markers
Siqinnaarut ('Sun is possible')	Ukiuq ('winter')	formation of new ice at floe edge
Qangattaasan ('the Sun gets higher')	Ukiuq ('winter')	walrus migrate; hunting for seals through breathing holes
Avunniit ('premature of (jar) seal pups')	Upirngaksajaaq ('towards spring')	premature jar seals born; snow faster for dog
Naittian ('seal pups')	Upirngaksaaq ('early spring')	playing *anauligaaq* (Iniuit baseball)
Tirigluit ('bearded seal pups')	Upirngaaq ('spring')	hunt basking seals; move from winter igloos to tents
Nurrait ('caribou calves')	Upirngaaq ('spring')	arrival of migratory birds, caribou calving
Manniit ('eggs')	Upirngaaq ('spring')	migratory birds nest, eider duck snaring
Saggaruut ('caribou hair sheds')	Aujaq ('Summer')	hunting seals moult, bumblebees moult
Akullirut ('caribou hair thickens')	Aujaq ('Summer')	caribou sheds skins, open-sea hunting
Amiraijaut ('velvet peels from caribou antlers')	Ukiaksajaaq ('towards autumn')	heavy morning dew, inland caribou hunting
Ukiulirut ('winter starts')	Ukiaksaaq ('autumn')	caribou mate, fish spawn, move to winter homes
Tusartuut ('hearing – news from nieghbours')	Ukiaq ('early winter')	seas freeze over, fox trapping
Tauvigjuaq ('great darkness')	Ukiuq ('winter')	*Tivajuut* (festival of renewal), seal hunting

period of darkness, the midnight sun in the north and finally the phases of the moon and tide. Besides that there was a rough system for reckoning longer periods of time, but this was more an individual measure, in relation to important events in one's own life

Accurately estimating time in the dark mornings of winter was a priority

or the life of the settlement .... The individual and his immediate family had their private calendar based on landmarks on the growing-up of their children, especially the boys" (Gad 1973, II, p. 265).

The first two measures of Inuit time, the migrations of animals and the cycle of the Sun and the Moon, can be reduced to a single category incorporating the annual solar cycle and nature's response to it, termed "ecological" or "eco-time" by the astronomer Anthony Aveni (1990). It "connects people with the environment through changes in nature to which they react" while social or structural time, including "landmark events in the life of an individual or family", "connects people with one another" and was reckoned, in a more or less linear fashion (though not numerically), over much longer periods. In the case of Iglulingmiut, for example, time was reckoned to about four or five generations. A final category, 'mythic time' – a misnomer, perhaps, given the 'timelessness' implied – might also be added. Apparently, 'mythic time' played a substantial role in Inuit consciousness, serving to link the people to their very beginnings in a distant, uncertain epoch when all was darkness and disorder. Only through the creation of the Sun, Moon, and stars, did order, light, and the predictable succession of seasons prevail.

In the past, the ecological calendar, represented by up to thirteen named lunations – or 'Moons' as they were called – was the dominant regulator of virtually all Inuit life. It was relatively flexible and, being based on recurring events in the natural world, such as the return of the Sun, the nesting of birds, the break-up of the sea ice, and so forth, usually varied according to latitude. Moreover, the sequence of lunations did not always correspond precisely with the ecological events for which they were named. Thus, in Igloolik, the 'Moon-month', *manniit*, meaning 'eggs' – a reference to the nesting of birds – could straddle the sixth or seventh lunation of the year, depending on an early or late arrival of the season.

On the astronomical front, the Inuit shared the problem common to all cultures that use a predominant Moon-based calendar system, that of trying to harmonize the lunar and solar cycles so that the sequence of lunations would, over time, keep in step with the recurring events in nature for which the Moons were named. Iglulingmiut appear to have circumvented this problem in much the

same manner as the early Romans – simply by not accounting for the time of the year when subsistence activity was at its lowest ebb. This approach is at least implicit in the explanation given for the lack of Inuktitut names for the two midwinter months: "[Avva] called them simply: *ukiup tatqe*, that is, 'the winter's moons'. These two moon periods, he said, resemble each other in that they are dark, cold, and hunting in them is difficult, so that they did not need any special designation" (Rasmussen 1930, p. 63).

While the succession of the seasons and their constituent moon-months were marked by the coincidence of various celestial and terrestrial events, the progressions of day and night were reckoned by the stars and sun alone – reckonings at which Inuit were particularly adept. An example is given by Elisha Kent Kane, an American explorer journeying in North Greenland. Late one night a number of Inuit visited his expedition's hut seeking a place to sleep. One of Kane's companions, however, a man named Petersen, apparently not given to hospitality, tried to persuade the visitors that it would soon be daylight and that they should move on: " 'No,' said one of the Inuit visitors, 'when that star gets round to that point [indicating the quarter of the heavens] and is no higher than this star [naming it], it will be the time to harness up my dogs'." Petersen was astounded; but he went out the next morning and verified the sidereal fact (Kane 1856, p. 426).

In common with other Inuit groups living in the higher northern latitudes where winter daylight is sparse, the Iglulingmiut shared the problem of timing their hunting activities to coincide with whatever light the day offered. As one elder, Martha Nasook, recalled in an interview conducted in 1990: "In those days ... everyone lived in igloos. The hunters usually had a peek hole in the igloo to observe the stars that were used to determine the time of the morning .... Everyone went to sleep early in comparison to what we do today. They would spend the early mornings getting prepared for the day's ... activities. Dogs had to be fed, traces prepared; fuel had to be readied for the *qulliq* [a soapstone lamp]; and snow or ice melted for water. On waking up there were plenty of things to be done."

Iglulingmiut use of stars for telling time is based on two principles: the revolving of the constellation *Tukturjuit* (Ursa maior) around *Nuuttuittuq* (Polaris), and the easterly rising and westerly setting of the non-circumpolar stars. Of this latter variety, the two stars of the constellation *Aagjuuk* (Altair or α Aql and Tarazed or γ Aql) were the most commonly used, not only in Igloolik but across many other regions of the Arctic. The *Aagjuuk* stars were also used to mark the winter solstice and to predict the Sun's return after the annual 'dark period'. Peter Tatigat Arnattiaq of Igloolik, interviewed in 1991, nicely sums up the dual uses of *Aagjuuk* as a keeper of both diurnal and seasonal time: "*Aagjuuk* is the morning [constellation]. The Inuit used to say *Aagjulirpuq*, meaning that the dawn is not far off .... There is another saying – *Aagjuuk agujjalirpuq*. This means that the *Aagjuuk* stars are starting to catch up with the daylight. Earlier *Aagjuuk* would appear when it was still dark but as [the season] progressed *Aagjuuk* and the daylight would coincide." Cain Iqqaqsaq, speaking in 1993, explained how the *Tukturjuit* constellation could be used for telling time: "*Tukturjuit* is useful to tell the time and direction .... [The constellation] is easily identifiable for it has the form of a caribou .... As midnight approaches it appears as if [the caribou] gets up on its hind legs and its head starts to get higher."

Accurately estimating time in the dark mornings of winter was clearly a priority for many Inuit groups across the Arctic. But the stars, as Michel Kupaaq explained in an interview held in 1996, were also used to signal the end of the waking day: "The *Ullaktut* stars [Orion's Belt] when they first appear [in the evening] are slanted; when they straighten up it's time for bed". And for the people of the Kobuk River, Alaska, the appearance of the Pleiades in a certain position would also indicate bedtime.

Inuit in the Igloolik area were treated to their first lesson in the strict regimes of European time in 1822 when Captain Edward Parry, during his search for the Northwest Passage, attempted – unsuccessfully as it turned out – to regulate the visits of Inuit to his ships on Sundays (until divine service had been performed) and, naively, to introduce them to that rather arbitrary division of time, the week. Early in the next century the explorer Vilhjalmur Stefánnson, on his second visit to the Western Canadian Arctic, thought himself similarly inconvenienced by the effects of Inuit behaviour on his schedule, but for quite the opposite reason. By then, many of his Inuit guides had embraced Christianity and were reluctant to travel on Sundays.

Stefánnson deplored the effects of Sunday observance on Inuit whale-hunting activities: "There is no regularity about the migration of the animals, and often at the height of the whaling season the crews may be encamped for a week at a time without seeing any; and then, all in one day, scores of whales may come along .... This day of opportunity is ... as likely as not to be a Sunday. When the Eskimo had learned that God had forbidden work upon the Sabbath they took the point of view that it does not profit a man that he gain the whole world if he lose his own soul, and although the catching of whales was the one thing in the world which all of them most

desired, nevertheless they agreed that the loss of one's soul was too great a price to pay even for a bow-head whale. Accordingly they would commence on Saturday afternoon to pull back their boats from the edge of the ice and get everything ready for the Sabbath observance .... It usually took them half of Monday to get everything ready for work again. In this manner they lost two days out of every seven from a harvest season of only six weeks in the year" (Stefánnson 1922, p. 91).

Spreading north from Cumberland Sound, one of Christianity's major hubs in the Canadian Eastern Arctic, news of the Christians' day of rest had reached Igloolik well ahead of the missionaries' arrival. In fact conversion to Christianity, and with it the practice of Sunday observance, apparently relieved Inuit from many of the obligations required by their own taboos. The frequent and unpredictable abstentions required by these were now exchanged for a single, predictable day. As Noah Piugaattuk recalled, speaking in 1991, "They found that it was much easier for them to observe [this] one day. So on that account they fully respected Sunday."

Christianity, however, introduced to the Igloolik area in the 1920s, also soon changed the manner in which the Inuit of the region reckoned and used time. Gradually time's references began to include the liturgical demands of the new religion. Noah Piugaattuk mentions some of the practical difficulties in adjusting to the time-keeping requirements of Christianity: "It was hard when we started to live by the new religion. We had to make sure that we did not bypass the day on which no work was to be done. It was hard as there were no such things as calendars at that time .... The women of the camp would consult each other [to determine Sunday] and, after that, they made sure to keep track of the days, making a mark on [a piece of wood] for each day that passed. Whoever was recording the days would make sure not to lose their record .... Certainly there were times when two people, from different camps, met on the trail and exchanged stories and so forth, and wondered which day it was. During periods of intense hunting no attention was paid to the days (or nights). [The hunters] would not keep track of time. This responsibility usually fell to the ones who were not engaged in doing other things .... One must realise that, at times, the hunters would not sleep for days on end when hunting conditions were favourable. Under these circumstances they would, no doubt, lose count of the days. But there were always elders that kept track of time for the rest of the camp."

Later, with increasing missionary activity in the area, the task of keeping the days in proper order was made easier: calendars became standard issue. The first calendar introduced to Igloolik by the Roman Catholic mission in the 1930s interestingly included the dates of solar eclipses along with, presumably, those of important religious festivals. As one Inuk elder, Mark Ijjangiaq, remembers, this demonstration of predictive astronomy was not lost on the Inuit parishioners: "After [the Roman Catholic priest] started to live among us he made a calendar with a small typewriter he had brought with him. This was not much of a calendar [by today's standards]. He made the letter 'I' for each weekday; for Sundays he marked an 'X'. That was the first time I had seen a calendar .... Each of the days were marked off with a pencil. Our priest knew that on certain days there would be an eclipse of the sun and he would mark these days on the calendar .... As the day of the eclipse approached we would keep a close watch on the sun, and, sure enough, the sun would start to darken."

The missionaries' calendars were soon followed by the traders' clocks which introduced the system whereby the days were arbitrarily divided into hours, the hours into minutes, and the minutes into audible seconds. Time's passage in ever-decreasing divisions could now be seen, measured, and even heard, in a context quite removed from the tempo of the natural world. Curiosities at

brought predictability and precision to the waking hours. Formerly the problem of timely waking in the morning was tackled in various ways. In addition to an 'early to bed, early to rise' routine, bladder function was mentioned by a number of hunters, who would begin the day's activities after rising for their second urination of the night. Infants, as Martha Nasook explains, could also ensure that no one overslept in the morning (or stayed up too late at night): "Once when Apak [her infant son] was not yet a year old, we were over on the mainland caribou hunting .... We did not have a clock .... We had to depend on our children to determine the time in the morning. When the morning came the children would be wide awake and it would still be dark outside. At night when they fell asleep, it was an indication that it was time [for us all] to sleep. Indeed the children in those days were instrumental in keeping time. Apak was our timepiece."

With the possible exception of Christianity – itself profoundly calendar-regulated – no other element of European culture has had a greater effect on present-day Inuit society. Western time – or as Aveni has termed it, the "imposition of order" – is both a symbol and

Clocks, like calendars before them, had a hesitant introduction to Inuit society

first, clocks, like the calendars before them, had a hesitant introduction to Inuit society. Emil Imaruittuq, speaking in 1990, remembers: "Before clocks were widely used the only way to determine time was by the stars. I suppose they found clocks more convenient than having to go outdoors to tell the time. When the clock stopped – usually when they forgot to wind it – they would say outright that it was 'dead'. It would then be reset, mainly by guess-work, according to the time of day. [In these days] they had no use for the clock's short hand at all. If, for example, the short hand was at seven and the long hand at three they would say 'it is seven o'clock'."

Inuit from the Keewatin region apparently used a method of telling the clock's time not directly involving the numerical value of the numbers on the dial. Instead, they named the hours with reference to the perceived shape of the numerals, to the orientation of hour hand, or to some regular activity associated with the hour. Martha Nasook explains: "They would give the hours names such as *ulamautinguaq* ['it looks like an axe' – 7 o'clock], *igganguaq* ['it looks like spectacles' – 8 o'clock], *qulilluanga* ['it's at the top' – 12 o'clock], *saniralluanga* ['it's at the side' – 3 o'clock] and *atilluanga* ['it's at the bottom' – 6 o'clock]. Then there was *ullurummitavik* ['time for the midday meal' – 12 o'clock] .... When the clock struck nine, they use to say *sukatirvik* ['winding time'], so at that hour the clocks would be wound."
The introduction of alarm clocks and, more recently, radio clocks

purveyor of cultural change in Inuit communities across the Arctic. The workday, the schoolday, the weekend, religious and civic holidays, scheduled airline flights, store hours, birthdays, and anniversaries – all ordered by calendar and clock – deeply affect the way in which Inuit now relate to each other and to their environment. But for all that some of the old perceptions of time tenaciously persist alongside the new. In many Arctic communities the arrival of spring triggers an irresistible urge to partake of nature's bounty, to renew the social bonds engendered by the proper coincidence of time, place and activity. A mass exodus to traditional fishing and hunting grounds ensues. Employers' clock-driven schedules fall apart as the new time temporarily gives way to the old. And recently, a young man in Igloolik, when asked for a definition of time, tellingly, and without hesitation, replied, "Time is nine to five".

This essay is excerpted from: John MacDonald, *The Arctic Sky: Inuit Astronomy, Star Lore, and Legend*, Toronto (Nunavut Research Institute and the Royal Ontario Museum) 1998.

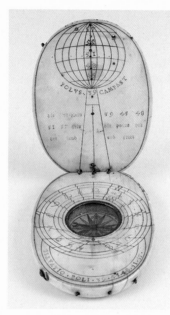

Hans Tucher, Diptych dial with nocturnal (see **091**)

Of Spheres and Shadows

by Elly Dekker

Terrestrial globe, from Gemma Frisius,
De Principiis Astronomiae et Cosmographiae, Louvain 1530

Until the seventeenth century, time was measured predominantly by observing the motion of the spherical heavens. Describing theories developed 600 years earlier, Ptolemy (*ca.* 150 AD), in his astronomical treatise known as the *Almagest*, explains in his chapter entitled 'That the heavens move like a sphere': "They [the ancients] saw that the Sun, moon and other stars were carried from the east to the west along circles which were always parallel to each other, that they began to rise up from below the Earth itself, as it were, gradually got up high, then kept on going round in a similar fashion and getting lower, until falling to earth, so to speak, they vanished completely, then, after remaining invisible for some time, again rose afresh and set; and [they] saw that the periods of these [motions], and also the places of rising and setting, were on the whole fixed and the same."

In this pattern of change, the motion of the Sun plays a major role. During the day, when the Sun moves above the horizon, it outshines all the other heavenly bodies. Only when the Sun has disappeared below the horizon at the end of the day can less bright objects such as planets and stars be seen to move across the sky. It then appears that some stars do rise and set in the same way as the Sun does during the day and that other stars never set but move in circular orbits around a fixed point – the northern pole of the universe. All the heavenly bodies – the Sun, the Moon, the planets and the stars – seem to make one complete revolution around the polar axis approximately every 24 hours.

Further observation showed that the days during winter are much shorter than during summer. Such a variation of the length of the day was believed to be the result of another motion – the annual revolution of the Sun along a great circle in the sky which was called the circle of the zodiac, or the ecliptic. This circle became divided into twelve segments, each of which became associated with the constellations through which the Sun seemed to pass. Moreover, the changes between the seasons were attributed to a specific point on that circle. Spring starts when the Sun enters the sign of Aries and autumn begins when it appears in the sign of Libra. At these dates, the periods of day and night are of equal length and, in its daily motion, the Sun can be found hugging the equator or the equinoctial (literally, the 'equal-night') line. At other times of the year, the daily track of the Sun follows a smaller circle. When the Sun enters the sign of Cancer in the summer, its daily track is along the Tropic of Cancer. Then the sun begins to turn towards the lower hemisphere, until it enters the sign of Capricorn in the winter. Then, its daily motion is along the Tropic of Capricorn, from where it begins to move towards the upper hemisphere again. The structure of the celestial sphere has always been codified by the outline of these main circles. We know that the early Greeks constructed models of the

The astrolabe was undeniably the most sophisticated invention for telling the time

heavens based on these circles – figuratively (in their poetry) and physically (as scientific instruments) – to help explain complex astronomical phenomena. The motions of the rising and setting of the Sun and the stars were the cornerstones of time measurement until the introduction of the mechanical clock. An understanding of the motions of the heavenly bodies and, especially, of the mathematics underlying them is a *conditio sine qua non* for understanding (let alone designing and making) early time-pieces, such as sundials and astrolabes.

In the so called Latin West during the early Middle Ages, the achievements of the Greek mathematicians and astronomers were virtually unknown. Yet, the rudiments of astronomy were transmitted through a number of basic, quasi-encyclopaedic works dating from the early fifth century, such as *De Nuptiis Mercurii et Philologiae* ('On the Marriage of Mercury and Philology') by Martianus Capella (*fl.* 410–39) and *De Somnio Scipionis* ('Commentary on the Dream of Scipio') by Ambrosius Theodosius Macrobius (*fl.* early fifth century). The Greek model of the universe was used by Macrobius (chap. 18) to explain how "a sign that rises with the Sun and sets with it is never visible and even nearby constellations are concealed by it". Thus, when the Sun proceeds through the zodiacal signs, a continually changing array of other signs and constellations are seen to rise and set, following a regular pattern in time.

Armed with a Greek understanding of the cosmos, medieval monks could then build upon these theories by watching the progress of the stars throughout the night. Gregory of Tours, for example, wrote his treatise *De Cursu Stellarum* ('The Tracks of the Stars') at the end of the sixth century and provided a set of rules, based on observations, to help monks to find the right time for their nocturnal prayers by recognizing certain constellations rising on the eastern horizon, the month in which they would first appear, and, sometimes, the number of hours that they would be visible during the night. The 'star clock' invented by Pacificus of Verona († 844) is similar, but slightly more sophisticated. It used the revolution of the Pole Star (α UMi) around the nearby pole of the universe as a timekeeper.

The revolutions of the Pole Star and other stars of the Greater and Lesser Bear around the pole of the world are also used in the navigational instrument known as the nocturnal. At the end of the thirteenth century, Ramon Lull, the Catalan encyclopedist, and Bernard of Verdun, best known for his treatise about an observing instrument called the *torquetum*, independently described the nocturnal (see **090** and **091**). Although not very accurate, the nocturnal has the advantage that its reliability is not dependent on one's latitude. Its simplicity served the practical needs of the navigator well into the eighteenth century.

One of the variable phenomena used for astronomical time measuring is that of the altitude of the Sun or a star above the horizon during the course of the day. Calculating the time from observation of the daily arc of the Sun or a star involves the use of complex trigonometric functions, which were unknown in the Latin West for a long time. But, even when the necessary mathematical knowledge is available, such calculations remain very time-consuming. Therefore, attempts were made to find alternative and quicker ways to solve these problems, and these led to a number of ingenious graphic solutions.

The astrolabe is undeniably the most sophisticated invention for telling time. The underlying principle of the astrolabe depends on using a stereographic projection of the celestial sphere on the plane through the equator. As a result, the sphere of the heavens and of the Earth can be, as it were, systematically flattened into a plane. Although astrolabes proper are not recorded in antiquity, planispheric projections are mentioned by Vitruvius in his description of Greek anaphoric clocks and Ptolemy provides an early description of the mathematics of the stereographic projection in his treatise, called *Planisphaerium*. The earliest treatise on the construction and use of the astrolabe is by the Byzantine astronomer, Theon of Alexandria, which was known in the Islamic world; and, through Arabic channels into Spain, knowledge of the astrolabe was transmitted into the Latin West, where it catalysed interest in a 'new astronomy' during the twelfth century.

Even though relatively few early astrolabes have survived, an amazingly large number of treatises on the construction and use of astrolabes in the Middle Ages has been conserved. This does not necessarily mean that schoolmasters, makers and users of astrolabes were familiar with the mathematics involved in the stereographic projection. Makers of astrolabes followed specific sets of instructions – as did practitioners using the instrument.

Another important graphic solution for time telling is based on the orthogonal projection of the celestial sphere on the meridian plane. In medieval times, this projection was used for an instrument known as the *organum Ptolemei*. By taking the altitude of the Sun, the exact time in equal hours can be found for any latitude. The orthogonal projection is also the basis of the universal astrolabe attributed to Joanne de Rojas (1550).

Graphic solutions of a completely different character are represented in horary quadrants and related instruments. The design of a horary quadrant, known as a *quadrans vetus* (or 'old quadrant'), is an Islamic invention attractive for the simplicity both of its construction and of its use (see **115** and **116**). The seasonal hour is determined from the intersection of a plumb line with one of a number of circular arcs drawn on the quadrant. Unfortunately, however, the *quadrans vetus* is truly accurate only for an observer standing on the equator. Away from the equator its accuracy cannot compete, for example, with that of an astrolabe. But, for domestic timekeeping, it was an easy and handy instrument that served the practical need.

The shortcomings of the 'old quadrant' may well have been at the bottom of the invention of the *quadrans novus* (or 'new quadrant', see **092**). This instrument was designed in 1288 by Jacob ben Machir ibn Tibbon (1236–1305), most commonly known by his Latin name, Profatius. In his quadrant, Profatius combined a number of features taken from the trigonometric quadrant of Islamic origin with other features from the astrolabe, which made it possible to determine the time by calculation. So that the whole circle of the heavens can be depicted in a quarter of its breadth, the instrument is constructed as if it were an astrolabe folded over twice. Therefore, there are actually four layers of the sky represented in this tiny instrument. In order to distinguish between these layers, the parts of each different layer were presented in different colours or upside down, so that, in all, four distinct parts could be identified. This *quadrans novus* has all the graphic and computational ingredients needed to obtain the correct time of the day from the measured altitude of the Sun or a star.

Since the use of the *quadrans novus* for finding the correct time was based on calculation, it demanded a high level of understanding. A simpler improvement on the *quadrans vetus* was needed and, from the fifteenth century onwards, a number of such instruments were created – such as the horary quadrant of 1438 attributed to Johannes von Gmunden (*ca.* 1384–1442).

A truly universal instrument was invented somewhere in the fifteenth century. It is now known as the rectilinear dial, an early description of which was provided by Johannes Regiomontanus (1436–1476). This instrument is an improvement of the *navicula de Venetiis* or 'little ship of Venice' (see **124**). In contrast to the circular arcs on the *quadrans vetus* for seasonal hours, there are straight lines drawn on the 'little ship of Venice' for equal hours. The idea of using straight lines may well have been borrowed from the trigonometric quadrants known from the Islamic world. Like that of the old quadrant, the graphic solution used for the 'little ship of Venice' was not perfect. Nevertheless, this medieval instrument provides a most elegant and remarkable graphic solution for finding the time in equal hours.

The most common way of finding the time during the day has been for centuries to use a shadow-casting device called a sundial. It usually consists of a style (or gnomon) for casting a shadow on a plane or curved surface, on to which the so called 'hour-lines' are marked. Diogenes Laertius tell us, in his *Lives of Eminent Philosophers* (Book II, chapter 1), that the sundial was invented by Anaximander (611–546 BC): "He was the first inventor of the gnomon and set it up for a sundial in Lacedaemon, as is stated by Favorinus in his *Miscellaneous History*, in order to mark the solstices and the equinoxes; he also constructed clocks to tell the time. He was the first to draw on a map the outline of lands and sea, and he constructed a globe as well."

The most common way of finding the time during

Nothing seems more natural than using one's own shadow (or that of a tree or pole) to determine what time of the day it is. Such schemes have been preserved in medieval manuscripts from the tenth century onwards (see **099**). Of the many obelisks erected for the special purpose of time measurement, the one in the Campus Martius in Rome, which served as the gnomon for the great sundial of Augustus, is the most famous. Yet, it cannot be denied that this most natural approach to finding time during the day is also mathematically the most complicated. The reason is that the splay of hour-lines differs considerably during the seasons: in the summer the lines are more widely arranged than in the spring or the autumn whereas in the winter there are more narrowly spaced. For this reason, the time in such dials is indicated by the shadow of the tip of the gnomon and not by the shadow of the style as a whole.

The construction of the hour-lines of dials in which the gnomon stood vertically on a horizontal plane occupied the minds of the best of the antique mathematicians. It seems that the inventor of the method of construction of these dials was Diodorus of Alexandria, but, of his original theory, only the incomplete and confusing descriptions by Vitruvius have survived. Fortunately, the method is fully described by Ptolemy in another treatise, *The Analemma*.

If there is one word connected with dialling that has been misunderstood again and again it is *analemma*. In the *Oxford English Dictionary*, there are four meanings given for it: 1) a pedestal of a sundial or a sort of sundial; 2) an orthographical projection of the celestial sphere; 3) a gnomon or astrolabe and 4) the scale of the Sun's daily declination drawn on an artificial terrestrial globe.

The last definition is the most recent one and many globes have scales engraved on to the globe gores that are labelled *The Analemma*. According to Benjamin Martin, this was invented by Joseph Harris, some time around 1740. The modern definition, however, has very little to do with the *analemma* as it was understood in antiquity. Here, the word was used exclusively to define a graphic method for working out the hour-lines and day-curves of a plane sundial with a vertical gnomon. In this sense, the *analemma* is methodologically closely related to the more familiar stereographic and orthogonal projection used in solving problems of the sphere. As described by Ptolemy, the *analemma* consists of a circular plate showing a number of lines and celestial circles in projection. As Otto Neugebauer explains in his *History of Ancient Mathematical Astronomy*, "If this plate [with its construction lines] is of metal or stone these lines are engraved, or, if it is wooden,

the day was to use a shadow-casting device

painted black and red. The surface of the plate is then covered with a layer of wax (obviously thin enough to be transparent) in which temporary lines can easily be drawn and erased".

The *analemma* has left its traces in our modern world in more than one way. Analemmatic dials are a modern development in which the disadvantages of the varying splay of the hour-lines have been met by making the vertical gnomon moveable with the season. The place of the gnomon is changed according to a rule which, in graphic form, is nowadays itself labelled *analemma* (see **106**). Indeed, this rule was part of the graphic method described by Ptolemy in his treatise. It is this very same element that is seen along the polar axis of universal equinoctial ring dials from about 1600 onwards.

The technique of constructing sundials depends on 'natural' circumstances; and makers soon realised that the vertical gnomon pointing to the zenith was not the best solution. Instead, a *polos* (that is, a style parallel to the north–south axis of the world, pointing towards the north pole), deserved to be considered the most 'natural'. The reason is not difficult to see. The orientation in space of a vertical gnomon changes with geographical latitude. A *polos* is, by definition, the same for all places on Earth.

For the construction of a dial with a *polos* this simplification is important because the shadow of its style is always in the plane running through the Sun and the axis of the world. This 'hour-plane' goes through the hour circle of the Sun and the centre of the celestial sphere and depends only on the hour of the day and not on the time of the year. To find the hour-lines for a dial with a *polos*, all one has to do is to determine the intersections of the 'hour-planes' with the

(arbitrary) plane of the dial. It is for this operation that Thomas Fale, the first English author on dialling, wrote in his *Horologiographia* of 1593: "Some teach the making of Dials by the help of the Globe as Gemma Frisius". The advantage of using the globe to aid in the construction of dials lies in the fact that one can easily read off the intersections of the various hour-lines marked on the sphere (defining the 'hour-plane') with the horizon ring. By adjusting the globe's meridian ring, one finds in a matter of minutes the data for the design of a plane dial for any latitude.

Why did such a skilled mathematician as Ptolemy not discover the advantages of dials with a style parallel to the axis of the world? The reason is that Ptolemy and his contemporaries considered a completely different problem: namely, how to indicate the time by the shadow of the sun in seasonal hours. Dials with a *polos* take advantage of the regular diurnal rotation of the celestial sphere and, therefore, indicate equal hours. The tussle between the use of equal and unequal hours is a curious one, but, intriguingly, a renewed interest in equal hours re-emerges during the fourteenth century – at the same time as the mechanical clock begins to gain a foothold.

Of course, there is a completely different and most elementary method for finding time that makes optimal use of spheres and shadows, which was published in 1530 by Gemma Frisius in his globe manual, *De Principiis Astronomiae et Cosmographiae*. With the help of a *gnomon sphaericus* ('spherical gnomon'), which was attached to the surface of the globe with some wax at the position of the Sun in the zodiac, the gnomon can always be directed towards the Sun itself by turning the globe (provided that the globe has been correctly rectified). All one has to do is turn the globe until the shadow disappears. Then the hour-circle on top of the meridian ring will show the time of day.

The determination of time has always been closely connected with the quest for understanding the structure of the world. In classical times, the construction of dials was a great challenge for mathematicians like Ptolemy. In the Renaissance, the explanation of time was, as one modern scholar has said, the final aim of the sort of cosmographical enquiry of which the work of Gemma Frisius is exemplary. Concepts of time and space have changed dramatically in recent years – mainly because our understanding of space has changed from what it used to be. What has not changed, however, is that, both in classical and modern times, the most advanced mathematical brains are needed to explore and describe them.

Non-Mechanical Time: By Night

087 Egyptian

Transit instrument (merkhet and bay)
Ebony and palm leaf, XXVI Dynasty
Berlin, PKB, Ägyptisches Museum
und Papyrussammlung [14084 and
14085]

By combining a sighting device and a scale, the Egyptians created one of the first scientific instruments which could function both as a surveying device and as a transit instrument. As such, it could be used to mark the rising and setting of particular stars and to measure the passage across the zenith (the local meridian) of the night sky. By regularly recording the transits against an equal-interval timer – such as a water clock – the ancient Egyptians were able to map the regular movements of the fixed stars and, thereby, set the basis for their timekeeping system. The L-shaped scale is called a *merkhet*, or 'indicator', and the notched palm leaf is called a *bay en imy wenut*, or the

'palm rib of the observer of the hours'. Both of these particular instruments belonged to a priest named Hor and the *bay* bears an inscription that explains it is used 'for determining the commencement of the feast and for placing every person in his hour'. In order to make an observation two priests had to sit in a north–south line (along a meridian). The slot in the *bay* was held close to the eye of the priest seated in the south. He looked towards the north and at the plumb line falling from the *merkhet*, held by an assistant priest positioned a short distance away. In this way they could mark the transit of a star as it passed across this meridian. The *merkhet* bears the inscription, 'I know the going of the Sun, the Moon and the stars, each to his place'.

The apparently regular movement of the stars across the sky can be used to measure time at night. The earliest 'star clocks' appear to have been calibrated according to horizon phenomena. The coincidence between the rising or setting of certain bright stars and the rising or setting of the Sun forms the basis of the earliest Egyptian timekeeping systems. Later, the astronomer priests became more interested in tracking the moment when these bright stars crossed the 'highest point' in the heavens, or the zenith (the north–south line of the local meridian).

Our understanding of many aspects of ancient astronomy relies on fragments – incomplete megaliths, broken bits of scientific instruments, fragmentary literary allusions. With early scientific instruments, for example, the two most basic questions are: how did the instrument work and for what purpose was it intended? Quite often, one is confronted with an instrument of which the functions no longer pertain to the needs of a given society, and the sense of what it was intended to do is often not at all clear. Indeed, this is true for many nineteenth-century instruments as well. Experts are constantly dismantling their instruments – because this is one of the few ways in which you can begin to understand what the instruments may have been used for and how they work.

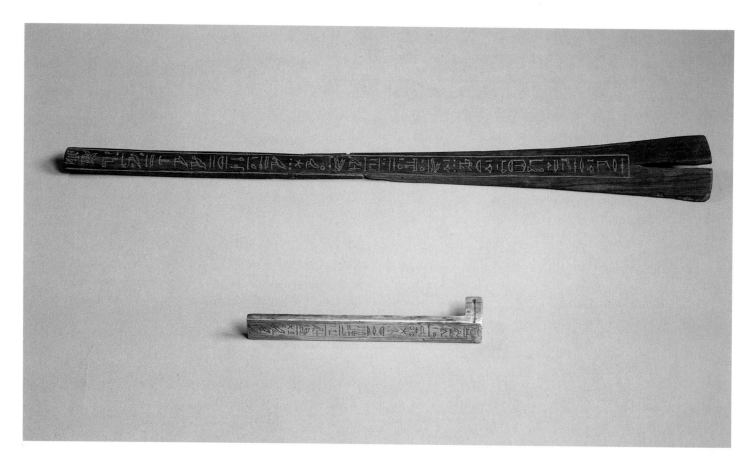

089 Chinese

Bi disc with a flanged edge
Jade, *ca.* 600–200 BC
Hamburg, Museum für Kunst und
Gewerbe [1926.85]

Circular, perforated jade discs –
referred to in literary texts from the
Western Zhou period as *bi* discs –
have been found in excavations of
general habitational debris and as part
of ritual burial goods from sites
throughout China dating from the
Late Neolithic period to the early
years of the 1st century AD. Most of
these discs are smooth-edged; a few
have three sets of serrated flanges. It
has generally been accepted that the
smooth *bi* disc played an important
ritual function connected to early
Chinese beliefs about the afterlife.
The flanged *bi* disc, however, has
been seen as a sophisticated
astronomical observing tool.
Unfortunately, there is no real
evidence to support this view.
The earliest description of a flanged
bi disc appears in the catalogue of a
19th-century collector, Wu Dacheng.
He proposed the unlikely suggestion
that these extremely expensive and

exquisite jade discs were the *xuanji*
('turning devices') mentioned in
ancient sources, and that they might
have been geared wheels that
formed part of a mechanism used to
drive an armillary sphere (this
presupposes that the armillary sphere
was in use more than 1000 years
before its invention some time during
the 1st century). Wu's thesis that the
discs might be, in some way,
'astronomical' was taken up by a
number of Western scholars.
Numerous ingenious explanations
were proffered as to how they might
have been used as a transit
instrument to map the circumpolar
stars. More recent work on early
Chinese divination, however, has
uncovered the fact that the *xuanji* is
actually a kind of divining board,
similar to a 'TLV' mirror (see 308),
and not an instrument used for
astronomical observations. This also
means that the *bi* disc is not a *xuanji*
and that it had no astronomical
purpose either.

088 Egyptian

Indicator for transit observations
(merkhet)
Bronze inlaid with electrum, Late
period (*ca.* 600 BC)
London, The Science Museum
[1929.585]

Details of this *merkhet* suggest that it
was mounted on two cylindrical
supports. There is a hole in the foot
end of the instrument through which
the thread for a plumb bob could be
inserted. The line of the plumb fell
parallel with the cylindrical supports,
ensuring that the body of the
instrument was kept level.
The inscription indicates that it was
the property of a certain priest named
Bes, the son of Khonsirdis, who was

the Observer of the Hours at the
Temple of Horus at Edfu in Upper
Egypt. The instrument is also
decorated with a small bust of the
falcon-headed Sun god.

090 Robert Yeff

Nocturnal
Boxwood, 1693
London, The National Maritime
Museum [AST 0141]

The nocturnal is a fairly simple
instrument that was used for telling
the time at night from the stars. It
could only be used in the northern
hemisphere, since it depended on
being able to see the Pole Star
(Polaris; α UMi), which is not visible
south of the equator. The outer edge
of the circular part of the instrument
has a calendar scale and the rotating
disc mounted within this circle is
divided into two periods of twelve
hours. To use the nocturnal, the 12
o'clock midnight mark was lined up
against the date. The instrument was
then held upright at arm's length and
the Pole Star was sighted through the
hole in the centre. The pointer was

then turned so that it was in line with
the 'guards' or 'guard stars' in the
constellation of the Plough or the
Great Bear (α UMa and β UMa), the
two stars furthest away from the
handle or tail of the Bear. The time
could then be read off from the scale
where the pointer cut the hour disc.
Nocturnals were in use from the 16th
to the 18th centuries.

From Petrus Apianus, *Cosmographica
per Gemmam Phrysum ... Restituta*,
Antwerp 1539

091 English

Nocturnal
Gilt brass, late 17th or early 18th
century
London, The Science Museum
[1880–41]

The use of serpents in this English
nocturnal may have been intended as
a sort of visual pun, since the
constellation of Draco, the dragon, is
regularly represented in star maps as
a long, sinuous serpent which snakes
its way between the constellations of
Ursa maior and Ursa minor. By the
late 17th century, it was fashionable
for wealthy gentlemen to have a
broad understanding of all branches
of learning, including mathematics,
the natural science and technical
subjects, as well as the arts and
literature. They often commissioned
highly decorative forms of basic
mathematical instruments, using

expensive materials such as gilt
metal, silver and ivory, indicating that
these objects were designed to
impress as well as to educate. Unlike
the simplest versions, which could be
used only with the 'guide stars' of
the Great Bear, these more elaborate
nocturnals could also be used with
the Little Bear. The bright star Kochab
(β UMi) was used as the guide star.

Also

• Hans Tucher, *Diptych dial with
  nocturnal, ca.* 1600, Cambridge, The
  Whipple Museum of the History of
  Science [Wh. 1678]

092 French?

Quadrans novus
Brass with a leather case, *ca.* 1300
Private collection

The *quadrans novus*, or 'new quadrant', is a rare and relatively complex instrument, of which only six medieval examples are known to survive. It was devised by a Jewish scholar named Jacob ben Machir ibn Tibbon, more commonly known as Profatius, who worked in Marseilles during the last half of the 13th century. The calendrical data on this instrument show that it must have been made between 1291 and 1310. The *quadrans novus* combines a number of elements taken from the trigonometric quadrant and the astrolabe. All this information is ingeniously presented within the quadrant format by 'folding' over the circular scales of the astrolabe. The introduction of red and black lettering on such a small instrument was, no doubt, prompted by the need to make it easier to determine which figures belonged to the same scale. The *quadrans novus* could be used to tell the time at night by measuring the position of one of the stars marked on the instrument. This quadrant originally had two sights along one of the straight edges and a plumb line. The scales on the curved edge allowed the position of a star or the Sun to be measured in degrees from both the equator and the north pole.

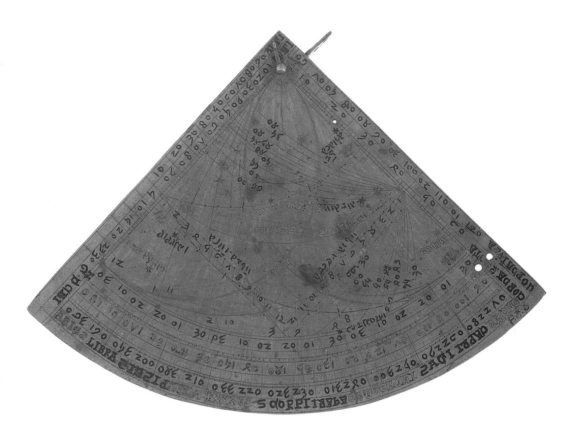

From Silvio Belli, *Libro del misurar con la vista*, Venice 1569

2 · THE MEASUREMENT OF TIME

▶ The so called Ptolemaic universe consisted of a static Earth set in the centre of a series of concentric shells. Three-dimensional models of the Ptolemaic universe, such as celestial globes and armillary spheres, seem to have been used by Greek astronomers and philosophers as a tool to explain the apparent motions of the heavens since at least the third century BC.

An astrolabe is a model of the Ptolemaic universe in which the spheres have been flattened. The front of the instrument has a pierced plate known as the *rete*. The *rete* is a planispheric map of the fixed stars with the celestial north pole set at the centre and the Tropic of Capricorn forming the outer rim of the plate. Offset to one side, there is the circle of the ecliptic, marked with the canonical twelve signs of the zodiac. The basic skeleton of the *rete* has a variable number of small pointers, each of which is set so that its tip indicates the position of one of the bright stars.

Behind the *rete*, nestled within the body of the instrument (known as the *mater*), there are a series of plates which provide a grid of co-ordinates for a specific latitude on Earth and a corresponding set of hour-lines. By setting the *rete* against the right latitude plate, one can find solutions to a wide number of astronomical problems. For example, after using the pinhole sights at each end of the *alidade* (or sighting arm) to find the altitude of the Sun or a star, one can determine the time according to either equal or unequal hours during the day or night. It is also possible to calculate the times of sunrise and sunset, find the rising or setting times for any star, or discover the exact configuration of the heavens for a date in the future or the past.

093 Mohammad ibn as-Saffār
Astrolabe
Brass, AH 420/1029–30 AD
Berlin, PKB, Staatsbibliothek
[Sprenger 2050]

Like much of the scientific learning of antiquity, knowledge of the astrolabe was lost to the Latin-speaking West for most of the early medieval period, but it was kept alive and allowed to flourish in Islam. It re-entered the West through the multicultural communities which thrived in places such as Sicily and southern Spain, where Arabic, Hebrew and Christian scholars were encouraged to work together and translate each other's texts. The inscription on this astrolabe tells us that it was made by

Mohammad ibn as-Saffār, a well known maker who lived in Toledo and Córdoba during the early years of the 11th century. It is one of the oldest surviving astrolabes in existence. The *mater* holds nine plates. The plates for Córdoba and Toledo also have Hebrew inscriptions next to the Arabic ones. And, on the back of the *mater*, there are also references to the zodiacal signs in Latin. The *rete* has pointers for 29 stars.

094 'Sergius the Persian'
Astrolabe
Bronze, July 1062
Brescia, Civici Musei d'Arte e di Storia [IC, no. 2]

There is only one known example of an astrolabe that might be called Byzantine. The text on the instrument is Greek, but the shape of the *rete* is very similar to Hispano-Moorish astrolabes, with tower-shaped star pointers and the celestial equator appearing as a band covering only one quarter of the *rete*. This supports the idea that, during the 11th century, 'Byzantine' instruments would have been heavily influenced – if not actually constructed – by Persian or Arabic craftsmen.

095 Abd al-Karim al-Misri

Astrolabe with zoomorphic pointers
Brass inlaid with silver and copper,
AH 633/1235–36 AD
London, The British Museum, Dept.
of Oriental Antiquities
[OA 1855.7-9.1]

It is not uncommon to find Arabic astrolabes in which the star pointers have become transformed into small birds or animals. In most cases, the pointer has taken on the figure of the star or constellation to which it is meant to refer – to act as a visual aid towards remembering the star. In this 13th-century Egyptian astrolabe more than a dozen zoomorphic pointers have been included. For example, depictions of birds are used as the star-pointers for Vega (α Lyr), and Altair (α Aql), a horse for *al-faras* (= Alferaz; β Peg), a goat for *wa-yuqāla* (= Alhaioh or Capella; α Aur), a scorpion for *al-gabha* (= Cabalacrab; β Sco), a snake's head for the head of the serpent held by Ophiuchus (Alhaue; α Oph) and a wonderful dancing man for *ar-ra's al-ǧāthī* (α Her), the bright double star in the head of the constellation of Hercules. A number of zoomorphic star pointers appear also in English astrolabes dating to the 14th century, such as in the *rete* of the 'Great Sloane Astrolabe' in the British Museum (see 096). This suggests that English makers of the period were directly influenced by Islamic astrolabes.

096 English

'The Great Sloane Astrolabe'
Brass, ca. 1295
London, The British Museum, Dept of Medieval and Later Antiquities
[MLA SL 54]

The Sloane astrolabe, the earliest and largest English astrolabe to have survived from the Middle Ages, shows a knowledge of Arabic astronomy and instrumentation. Of the 40 pointers, 33 have the Latinized form of the Arabic star names, such as *algenib* ('the side of Perseus'; α Per), *algomeiza* ('the fainter one'; α CMi), and *raztabin* ('the head of the dragon'; γ Dra). Also, the star pointers themselves are shaped like long, pointed tongues, which protrude from the heads of dragons, birds and dogs. We know that the astrolabe was made in England, because it preserves the name of three English saints amongst its lists of saints' days: Dunstan (19 May), Augustine of Canterbury (16 May) and Edmund (20 November). The three interior plates have arcs for unequal or temporal hours. They are designed to be used for the latitudes of 45°, 48° 30' (*Paris*), 51°, 52° (*Lundoniarium*), 53° and 55°. The London plate is also inscribed with lines demarcating the 12 astrological houses. The back of the instrument has a series of scales interspersed with decorative motifs. The first point of Aries is set at 14 March, which may provide clues as to when the instrument was made.

097 Chinese

Qinding Shu jing tushuo ('The Book of Documents')
Printed book, 1905
London, The British Library, Oriental and India Office Collections
[15520.a.1]

The Chinese use of a vertical gnomon, or pointer, to measure the summer and winter solstice has such a long history that its first practice has merged into the stuff of legends. A mythical astronomer, named Xi, is credited with being the first to use a gnomon to measure the solstices during the reign of the legendary Emperor Yao. Xi was commanded by the Emperor "to go and live at the southern borders in order to regulate the works of the South and pay respectful attention to the [summer] solstice". In this picture, Xi is shown measuring the summer solstice with a vertical gnomon.

In the northern hemisphere, every day the Sun appears to move across the sky from the eastern horizon to the western horizon. When the Sun is directly above the head of the observer, it is said to be at its zenith or its 'highest' position above the horizon. It is also the moment at which the Sun passes through the local meridian. Greenwich is located at 0° longitude. This means that when it is Noon in Greenwich, the celestial longitude of the Sun is also 0° longitude. It is Noon in New York when the Sun passes 74°W longitude; in Tokyo, when it passes 139° 45′E longitude.

We use the terms 'AM' and 'PM' to indicate whether the time is before or after Noon. The abbreviations stand for the Latins terms *ante meridiem* ('before midday') and *post meridiem* ('after midday'). All hours between midnight and midday are defined as 'AM'; hours between midday and midnight are defined as 'PM'. Noon and midnight are neither AM nor PM.

The earliest sundials were simple sticks set into the ground. As the Sun passed overhead, the shadows cast by the stick changed in both shape and direction. Longer shadows towards the west were cast by the morning Sun; short shadows facing slightly south (in the northern hemisphere) during the middle of the day; and long shadows towards the east by the late afternoon Sun. This simple phenomenon is the basis of a sundial. By dividing the patterns created by the cast shadows into equal segments, the first hour scales were established.

098 Johann Michael Vogler

Analemma dial
Gilt brass, early 18th century
London, The National Maritime Museum [AST 0368]

If one plots the line that a succession of noontime Suns appear to make in the sky along a given celestial meridian, it comes out like an elongated figure 8. This elliptical shape is known as an *analemma*. *Analemma* diagrams are often found incorporated into the paving of Italian churches. An astronomer has chipped a small hole through the roof of the building, which lets a pinpoint of light on to the floor of the church. He marks the position of the spot of sunlight every day at precisely Noon. The succession of marks create the distinctive 8-shaped figure, which in turn can be used as both a calendar and a clock. When the spot of sunlight falls on the *analemma*, it must be Noon. When it is at the

northernmost point in the 8, it must be the winter solstice; at the southernmost point, it is the summer solstice. Where the 8 crosses over itself, it is either the spring or the autumn equinox. A relatively rare type of sundial also uses the *analemma* to guide the placement of its gnomon. In this dial, the wire gnomon is attached to the hour ring at the north and south points. When the dial is correctly oriented to align with the local meridian, the bead placed at the middle point of the gnomon will cast a shadow on the correct place in the month scale exactly at Noon.

Also

• Elias Allen, *Horizontal dial*, 1606, London, The National Maritime Museum [AST0240]

• Chinese, *Gnomon chronometer*, 2nd–1st century BC, Toronto, Royal Ontario Museum [933.12.2]

099 Abbo Floriacensis

Varia Scripta Arithmetica et Chronologica
Manuscript with drawings in red, green and black ink, *ca.* 1100
Berlin, PKB, Staatsbibliothek
[Ms Phill. 1833]

Abbo was a monk working in the monastery of Fleury in France during the last half of the 10th century. Most of his writings re-examine the work of previous writers on the calendar, such as the Venerable Bede. In particular, Abbo strove to separate man's use of the reliable signs that one might derive from nature (*naturae ordo*) from the errors that were generated by an academic fondness for tradition and misinformation drawn from historical sources (*historiae fides*). In his treatise, he reconstructs the so called Roman 'foot-dial' – a very basic means, apparently used by Roman peasants during the latter years of the Roman Empire, for calculating the time by using the observer himself as the gnomon. Standing in the centre of a hand-drawn semicircle and facing north, the length of the observer's shadow could be measured by lengths of his own feet. For example, during the months of June and July, one's shadow will be exactly one foot long during the 6th hour; in the 5th and 7th hour, the shadow will be three feet long; in the 1st and 10th hour, it will be 19 feet long. As the angle between the Sun's rays and any one location on Earth will change throughout the year, different time scales have to be used for each of the months. To find the time, one inserts the information found under the headings for the different months. In January, for example, one's shadow is 9 feet long during the 6th hour. It is 29 feet during the 1st and the 10th hours.

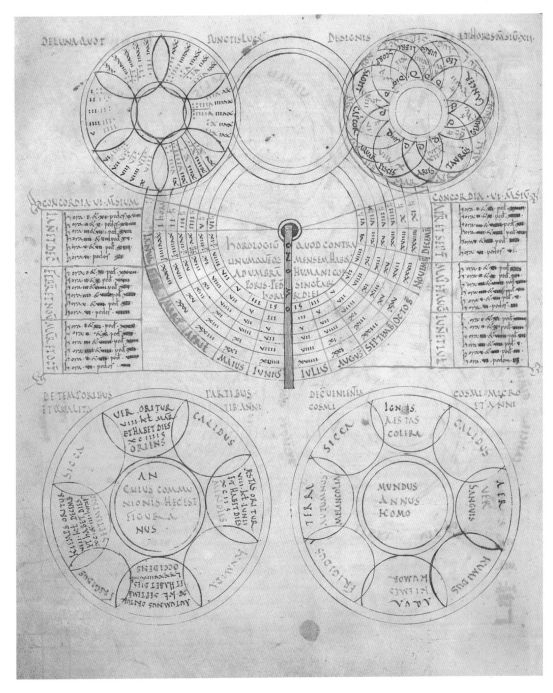

100 Egyptian

Portable gnomon block
Black granite, XXVI Dynasty
(*ca.* 550 BC)
London, The British Museum, Dept.
of Egyptian Antiquities [EA 74841]

The Egyptians also used sundials in
which the shadow cast by a fixed
pointer, or gnomon, was used to tell
the time. This small block is the
gnomon of a large horizontal sundial.
The time would have been read
according to the shadow cast by its
top edges. The vertical hieroglyphic
inscription reads: 'Plumb bob of this
sweet and glorious Sun'. It seems
that the tradition was to remove the
gnomon of a dial after use. The
plumb bob would help to ensure that
the dial was properly realigned. The
gnomon itself is shaped to resemble
a miniature shrine to the Sun god.
One side of the dial bears an image
of Ra-horakty, wearing the solar disc
on his head and carrying a sceptre
and *ankh*.

101 Egyptian

L-shaped shadow clock
Black schist, 1479–1425 BC
Berlin, PKB, Ägyptisches Museum
und Papyrussammlung [19744]

This earliest surviving Egyptian
'L-shaped shadow clock' is almost
identical with the *merkhet* (see 087),
but has an hour scale that can be
used to tell the time by the altitude
of the Sun. The time scale is marked
by five holes set at irregular intervals.
Because when the Sun is at its
zenith it appears to move more
slowly, the holes towards the foot of
the instrument are more closely
spaced than those at the flat end.

102 Roman

Scaphe dial flanked by two lions
Marble, 1st–2nd century AD
London, The British Museum, Dept.
of Greek and Roman Antiquities [GR
1821.3–1.1, Sculpt. 2545]

This Roman dial is a slightly truncated
form of the *scaphe* dial (a type known
as a *hemicyclium*). The gnomon is
placed horizontally in the U-shaped
hole at the top of the hour scale, so
that it lies parallel to the surface of
the Earth. The hour-lines are scored
so as to divide the hours of daylight
into 12 equal segments.

Also

• Egyptian (Luxor), *Vertical dial*,
  1st–2nd century AD, Berlin, PKB,
  Ägyptisches Museum und
  Papyrussammlung [20322]

• Roman (Palestrina), *Scaphe dial*, 2nd
  century AD, Liverpool, National
  Museums on Merseyside, The
  Liverpool Museum [1959.148.128]

Sundials come in all shapes and sizes. The *scaphe* dial (see 102)
was probably the most popular form for a sundial in the
Mediterranean basin during the periods of Greek and Roman
dominance. The word *scaphe* is Greek for 'anything that is dug or
scooped out' and refers to the curved, bowl-shape of these sorts of
dials. The greatest advantage of this kind of dial is that, if it is
properly set up, an equally divided hour scale can be used. This
makes both the manufacture and mass-distribution of timekeeping
devices much easier.

Most *scaphe* dials have three sets of horizontal co-ordinates
crossing their hour-lines. As the curving arc of the Sun's path
changes during the year, so the gnomon's shadow registers at a
different level. During the spring and the autumn – near the
equinoxes, when the Sun crosses the equator – the shadow would
fall near the middle line. During the winter months and near the
solstice when the Sun is lower in the sky, the top line would be
used. In the summer months, the shadow would fall towards the
bottom line. The similarity between the grid found on this sundial
and the kind of bands one finds drawn on celestial globes is not
accidental. Both record the movement of the Sun between the
Tropic of Cancer and the Tropic of Capricorn – the two solstitial
turning points – throughout the year. The middle line equals the line
of the celestial equator, the top line is the Tropic of Cancer and the
bottom line the Topic of Capricorn in both cases.

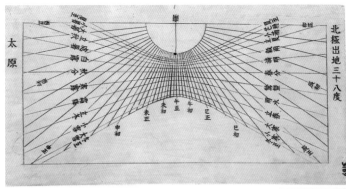

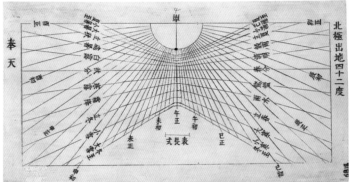

two halves of the dial. This type of indicator is known as a 'string gnomon'. The hour-lines are marked around the edge of the lower plate, from 5 AM to 7 PM. Small, self-orienting dials, such as this one, first appear in Europe during the middle years of the 15th century. They reflect an important moment when the material demands of a new and expanding middle class coincided with a sudden growth in mathematical skills in Europe. Georg Peurbach, one of the great Renaissance innovators of astronomy during the middle years of the 15th century, is credited with inventing a number of dial forms (including this one) and other measuring instruments.

104 Georg Peurbach
Pocket dial
Brass, 1455
Graz, Austria, Steiermarkisches
Landesmuseum Joanneum [4525]

Portable horizontal dials had to have a small magnetic compass so that they could be correctly aligned on a north–south line or meridian. In this very early example, the shadow is formed by a thread connecting the

103 Chinese
Horizontal dial plates
Ivory, 19th century
Cambridge, The Whipple Museum of the History of Science
[Wh.3189 a and b)]

In China, sundials derived from calendrical studies and the particular Chinese interest in using the gnomon to define the shape and length of the year for different latitudes. From the Early Han period (206 BC–8 AD), the primary use of the early sundial was to check the accuracy of water clocks. In these horizontal dial plates, the radiating lines form the hour scale and the curved horizontal lines indicate the path of the Sun during the 12 months of the year. Each plate is latitude specific.

Also
• Chinese, *Equinoctial dial*, 19th century, Cambridge, The Whipple Museum of the History of Science [Wh.192]

105 Chinese
Equatorial dial
Fire-gilt bronze, Ming dynasty
(1368–1644)
Washington, D.C., The National Museum of American History
[255037]

The main duty of Chinese astronomers was to search the heavens for 'symptoms' of cosmic imbalance. If there was an unusual celestial event it was a sign of disharmony or change elsewhere in the universe. As the regularity of measured time was one of the scales against which developments might be marked, both timekeeping devices and the formation of the calendar were an exclusively imperial concern. Large-scale sundials, such as this monumental equinoctial dial set on the back of a lion, were not erected to provide time for the people – as they might be in European cities. They were generally placed in an imperial setting to which no ordinary member of the public had access.

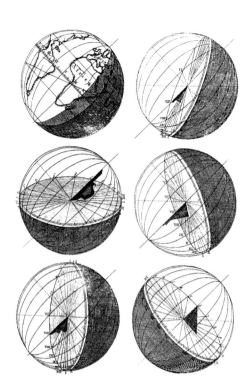

A sundial measures time by setting the length or direction of the Sun's shadow against a scale. Regardless of which time scale is being used or how the hours themselves are being counted, in order for a sundial to function, a number of components must be considered. Each of these is directly related to the fact that a sundial records the changes in shadows that have been created by the rotation of a sphere (the Earth) on an inclined axis, which revolves around a single light source (the Sun).

The three main co-ordinates that go into making a dial are the plane of the polar axis, the plane of the equator and the plane formed by the horizon of the local latitude. The three features that relate to the geometry imposed by these co-ordinates are the slant of the surface of the dial plate, the angle formed between the top surface of the gnomon and the dial plate, and the relative expanse of the splay of the hour-lines used to count the time. By examining all of these features together, it is possible to determine for which single latitude a dial has been constructed. By introducing curved dial plates, an adjustable gnomon or a multiple set of hour-lines, universal dials suitable for a number of latitudes can also be created.

From Thomas Wright, *The Use of the Globe, or the General Doctrine of the Sphere*, London 1740

106 **Johann Willebrand**
Equinoctial sundial with three scales
Silvered brass, 1726
Munich, Deutsches Museum [1707]

In an equatorial (or equinoctial) dial, the hour scale is set up so that it sits parallel to the Earth's equator. The gnomon is positioned so that it aligns with the polar axis – or exactly perpendicular to the base plate. The hour-lines converge in the centre of the dial and the divisions of the hours on the scale are of equal size. The origin of the term 'equatorial' for this type of dial is fairly obvious. The variant term for dials such as these – 'equinoctial' – refers to the position of the Sun relative to the Earth, for when the Sun is exactly over the equator at the spring and autumn equinoxes, the hours of the day and night are of equal length (*aequa* (equal) + *nox* (night)). The piece is set for a latitude close to that of Augsburg, the home of its maker.

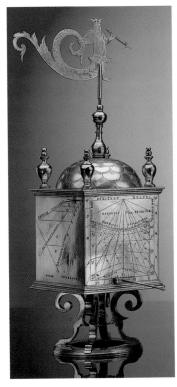

108 Czech or German

Polyhedral dial in the shape of a cube
Brass gilt, *ca.* 1600
Prague, Narodni Technike Muzeum
[24863]

Polyhedral dials are intended to
impress the viewer. This dial is
aligned by reference to the base
plate, which is marked with the four
cardinal points of the compass
(NSEW) and a circular degree scale
that is divided into two sets of 180°.
The 0° of the dial is marked to
indicate the south point. On the cube
itself, there are three vertical
sundials. A vertical dial has its
gnomon aligned with the polar axis
of the Earth. From the angle of the
gnomons one can tell that this dial
was designed for 48° 50'N – the
latitude of Prague (and Frankfurt and
Krakow). Above the cube, there is a
wind vane in the form of a crowned
mermaid with an arrow in her hand.

107 German

*Polyhedral sundial in the shape of a
tower*
Painted marble and brass, first half of
the 19th century
Munich, Deutsches Museum [2659]

In the later 18th and early 19th
centuries, there was a fashion for
novelty polyhedral dials constructed
in a variety of shapes to represent
buildings and other objects. This is a
typical example in the shape of a
tower, by an unknown maker.

Also
• South German?, *Chalice dial*,
  1596, London, The Science
  Museum [1955–149]

109 Hans Koch

Polyhedral dial
Gilt brass, 1578
Munich, Bayerisches
Nationalmuseum [Phys.283]

With 25 different dial-faces, this
polyhedral sundial demonstrates the
skill and artistry of its maker, Hans
Koch, who was a master clockmaker
in Munich. The dial was for Herzog
Ludwig von Württemberg. As with all

complex dials of this sort, the first
task is to align it along a north–south
meridian. This is made easy by the
inclusion of a small compass on the
dial's uppermost face. The other
faces carry the widest possible array
of vertical and inclining dials, each
with its own pin-gnomon. As this dial
has fixed gnomons, it has been
constructed for use at a specific
latitude: 48° 38'N – the latitude of
Württemberg.

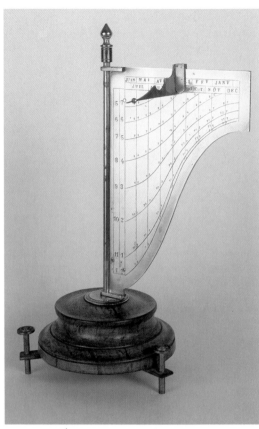

110 German (Augsburg)

Pillar dial
Gilt bronze, *ca.* 1550
London, The Science Museum
[1883–124]

Pillar dials are altitude dials. They work by measuring the altitude of the Sun above the horizon, rather than a horizontal angle from the zenith. The dial itself consists of a hollow tube. The gnomon is fixed horizontally into a slot at the top and when not in use can be stored inside the body of the dial. It is inserted so that it lines up with the correct month. The dial is then turned so that the shadow cast by the gnomon falls directly down the side of the pillar on to the appropriate place on the hour scale. The pointed tip of the gnomon's shadow indicates the time. This gilt-bronze pillar dial is mounted on a cube engraved with a variety of other types of sundials calibrated according to different hour

systems. The month scale for positioning the gnomon on the pillar dial is presented in two forms: the names of the zodiacal signs are arranged in pairs along the bottom of the pillar and, above them, there are engraved representations of the figures of the zodiacal constellations. Decorated zodiacal scales of this sort are relatively rare, one interesting example being the pillar dial depicted amongst the scientific instruments appearing in Holbein's painting of *The Ambassadors* in the National Gallery in London.

Also

- P. Gaetano di Siva P. Capuccino, *Pillar dial*, 18th century, London, The National Maritime Museum [AST 0294]

- Tibetan, *Time stele*, 19th century, London, The Science Museum [1995-0267]

111 French

Flag dial
Wood and brass, early 19th century
Liège, Musée de la Vie Wallonne
[Max Elskamp, no. 558]

The flag dial takes its name from its shape. It is an altitude dial and the pattern of its hour-lines is identical to those found on a pillar dial. Indeed, if one were to 'roll out' the surface of a pillar dial, it would look exactly like a flag dial in almost every detail. The gnomon is set to the correct date using the calendar scale along the top edge of the 'flag'. The time can then be read from the position of the tip of the gnomon's shadow. The curve of each hour-line records the relative height of the Sun throughout the day. A number of different hour-lines are needed in order to accommodate the changing path of the Sun throughout the year.

112 Bernard Daniel

Cruciform sundial
Gilt brass, 1629
London, The National Maritime Museum [AST 0477]

Until the 20th century, owning a timepiece was seen as symbolic of an elevated status within society. This fact often caused a problem for those segments of society for whom the possession of worldly goods was seen as an inappropriate personal vanity. In particular, monks and nuns, who may have had legitimate reasons for wanting to know the time of day, were dissuaded from paying too close attention to the passage of time. As the Venerable Bede himself said, a good Christian never needed to own a timekeeper that used anything but full-hour divisions. Perhaps as an attempt to skirt these issues, during the 16th and 17th centuries, small portable sundials were sometimes made in the form of a cross, so that they could act as an aid to religious devotion as well as telling the time. Some of these cruciform dials may also have been used as reliquaries and, even though the Church might advise against it, they were often elaborately decorated.

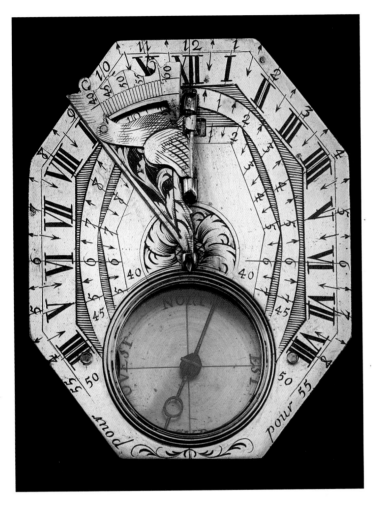

113 Pierre Sevin

Butterfield dial
Silver, late 17th century
London, The National Maritime
Museum [AST 0259]

The so called 'Butterfield dial' is a
type of portable horizontal sundial
which can be adjusted for several
different latitudes. It is named after
Michael Butterfield, an Englishman
working in Paris who popularized the
design. Even after the invention of
the pocket watch, portable sundials
remained popular. Early watches
were not good timekeepers and a dial
was needed to check the watch's
running periodically. In this dial, the
gnomon is adjustable for latitudes
between 40° and 60°N. On the base
plate there are four separate hour
scales for the latitudes of 40°, 45°,

50° and 55°. There is a compass for
aligning the dial and, on the underside
of the base, there is a list of 22
European towns with their latitudes,
including Paris, Vienna, Krakow,
Constantinople, Genoa, Rome, Lisbon
and London. For travelling, the
gnomon folds flat and the whole
instrument fitted easily into its black
fishskin carrying-case and slid into a
gentleman's pocket.

114 German

Globe dial
Gilt brass, *ca.* 1600
Bielefeld, Kunstgewerbesammlung
der Stadt Bielefeld [H-W 87]

All sundials are constructed to take
account of the anomalies created by
the fact that we live on a spherical
Earth, but the globe dial incorporates
the globe itself into the process of
determining time. Unfortunately, the
mathematics that must go into the

creation of a sundial that mirrors
natural phenomena are complex. As a
result, globe dials are extremely rare.
The globe dial can be adjusted for
use in different latitudes by means of
the degree scale on the vertical
meridian circle. The sphere must be
tilted so that its polar axis aligns with
the polar axis of the Earth. Then, as
the Sun passes overhead, the
shadow from the gnomon, or pointer,
will fall on to one of the two hour
scales on the globe.

Whereas many sundials work by charting the Sun's passage using the changing pattern of the shadows cast by a fixed gnomon, altitude dials rely on direct measurement of the height of the Sun to tell the time. For most latitudes, the apparent path that the Sun makes across the sky changes according to the time of year. In the northern hemisphere, for example, the arc will appear to be set much closer to the horizon during the winter months. If one knows the local latitude and the date, it is possible to gauge the time merely by measuring the apparent angle of the height of the Sun from a flat horizon line.

Altitude quadrants work by combining three pieces of information. With some sort of sighting mechanism, the height of the Sun is measured. A plumb bob is used to record this angle against the body of the instrument. The angle is then interpreted (or modified) with the aid of a date scale, which is usually marked with the names of the zodiacal signs to indicate different times of year. Each date provided corresponds with a different set of hour-lines, each of which mimics the path of the Sun for that particular season. These can be arranged according to either an equal or an unequal hour system and, quite often, altitude dials include both.

115 English

Horary quadrant
Gilt brass, 1398,
Dorchester, Dorset County Museum
[DORCM 1973.7]

This set of three quadrants all date from the turn of the 15th century and seem to have come from the same workshop, possibly in London. One, from the British Museum (dated 1399), bears the badge of King Richard II, a couchant white hart with a collar and chain. The Dorset quadrant is decorated with a slight variant of the white hart crest of Richard II. The white hart without a collar and chain is thought to be an emblem used by the king's half-brother, John Holland, Earl of Huntingdon and Duke of Exeter. These dials are both dated – 1398 and 1399 – and the third one can be dated to 1400 by the information contained in its perpetual calendar. The undated dial also differs in that, where the other two quadrants have a badge depicting a white hart, this one has a circular Easter table.

Each of these quadrants has scales for telling the time in equal hours as well as a degree scale for altitudes of heavenly bodies and a table of Noon altitudes of the Sun throughout the year. There are usually two sights on the upper edge and there was originally a rivet to suspend a plumb line for taking altitudes. The object was lined up through the two sights and the altitude read from the scale on the curved edge, at the point where the plumb line crossed it. The tables on the quadrants from the British Museum are for the latitude of 51° 53' or 54', a value often used for London in medieval times. The latitude on the Dorset quadrant is one degree less.

Also

• English, *Horary quadrant*, 1400 (from the calendar), London, The British Museum [MLA 1856. 6–17.155]

• English, *Horary quadrant*, 1399, London, The British Museum [MLA 1860. 5–19.1]

FIG. 169. THE ASTROLABIUM OF LANSBERG, 1635.

Philip Lansberg, *Astrolabium datis … verklaringhe van de Platte Sphaere ofte Globe van Ptolomaeus, anders Astrolabium ghenaemt …*
Middelburgh 1635

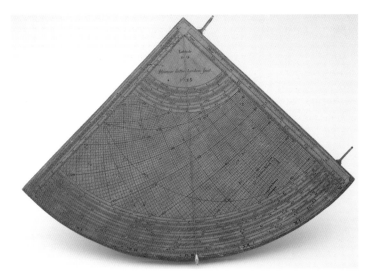

116 Henry Sutton

Horary quadrant
Printed paper on wooden board with brass sights, 1658
London, The National Maritime Museum [NAV 1042]

Quadrants can perform any number of functions – depending on what sort of material is included on them. A horary quadrant, however, is so defined because it is used mainly to tell the time. It is, essentially, an altitude dial. One uses the sights on the top edge of the instrument to find the altitude of the Sun. The string of the plumb line then falls against an hour scale, which in this case is engraved along the rim of the instrument, and the correct hour is read off. Henry Sutton was one of a number of 17th-century instrument makers working in London who produced inexpensive mathematical instruments in the form of printed sheets which could then be pasted on to board. This innovation made the instruments accessible to students, surveyors and many others who lacked sufficient wealth to buy costly brass instruments.

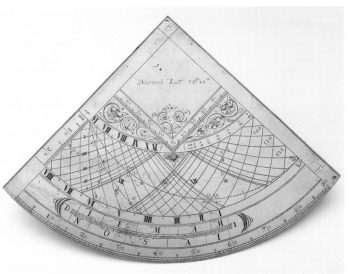

117 Attr. Walter Hayes

Gunter quadrant
Brass, third quarter of the 17th century
London, The National Maritime Museum [NAV 1033]

The Gunter quadrant is a variation on the more basic horary quadrant. It was designed by Edmund Gunter, Professor of Astronomy at Gresham College in the City of London, some time before 1623. It has a stereographic projection of the celestial co-ordinates on one side of the plate – an edited version of the sort of projections found on the latitude plates of planispheric astrolabes. There is also a date scale marked by the initials of the months, together with the usual winter and summer hour-lines and a degree scale on the curved edge. It was used to tell the time as follows: a bead was moved along the plumb line to the position of the Sun for the time of year (using the scales on the instrument). The altitude of the Sun was then measured using the sights and the plumb line. By holding the plumb line at the correct point on the degree scale the bead would cut one of the hour-lines. This dial was made for the latitude of 52° 42'N, or Norwich.

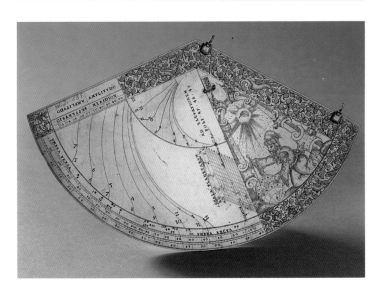

118 Christoph Schissler the Elder

Triens
Gilt brass, 1569
Nuremberg, Germanisches Nationalmuseum [WI 137]

A triens is an extended quadrant with a 120° arc forming a third of a circle. The quadrant itself has two hour scales on its obverse and can be used to tell the time for three different latitudes: 47°, 48° and 49°N. The smaller scale shows the time in unequal or planetary hours. The larger scale is for telling the time in equal hours and can be used for any time between 5 AM and 7 PM. In order to tell the time, one would set a small bead on the now missing plumb line against the zodiacal grid in the middle of the instrument. One would then take a sighting of the Sun, letting the plumb line fall free. The small bead would fall on the appropriate hour-line. On the reverse of the instrument, there are scales that can be used to calculate the times of sunrise and sunset for any date for the latitudes 47° to 49°N; and to determine the azimuth of the Sun (the horizontal angle between the points of its rising and setting and the east and west points of the compass).

119 Michael Piquer
Astrolabe
Brass with original leather case,
ca. 1550–55
London, The National Maritime
Museum [AST 0577]

Astrolabes can be used to tell the time of day by measuring the altitude of the Sun. Holding the instrument by the ring, the observer rotates the *alidade* or pointer of the instrument so that sunlight falls through both pinhole sights. The pointer intersects the date scale at the edge of the instrument, and the date given is used to set the *rete* against the appropriate latitude plate in order to tell the time. In one of the graded scales in the zodiacal ring on the front of the instrument, the date of the vernal equinox is shown as 21 March. This suggests that the astrolabe was made some time after the Gregorian Reform of 1582, when the calendar was brought back by 10 days. Other astrolabes by Piquer, however, are dated 1542 and it is unlikely that any instrument maker would have enjoyed such a long career. A closer examination shows that the graded markings that accompany the calendar have been re-engraved. Indeed, in the original engraving on the instrument, the date for the spring equinox was 10/11 March – concurring with a date during the 1550s. Obviously, what has happened is that Piquer made this instrument some time in the early 1550s. After the Gregorian Reform, and in order to keep the instrument useful, the scale was erased by burnishing and re-engraved.

Also

• Thomas Gemini, *Astrolabe, ca.* 1555, London, The National Maritime Museum [AST 0567]

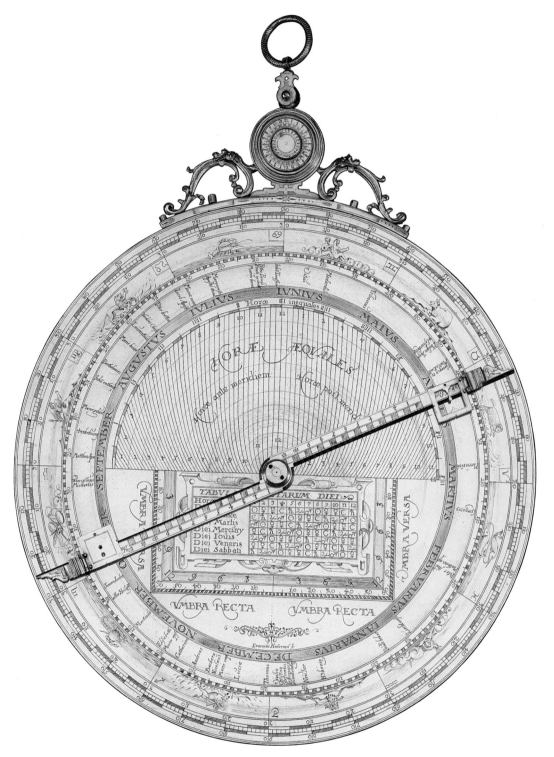

120 **Erasmus Habermel**

Astrolabe
Gilt copper, *ca.* 1585
London, The National Maritime
Museum [AST 0563]

Erasmus Habermel was one of the
finest craftsmen of the 16th century,
who held the post of astronomical
and geometrical instrument maker at
the court of Rudolf II in Prague from
1593 until his death in 1606. The
back of the astrolabe has the usual
zodiacal scales and an ecclesiastical
calendar that marks the names of
the saints' days. The central part is
occupied with a conversion scale for
translating equal (or clock) hours into
unequal (or temporal) hours. The
concentric circles are the equal
hours, counted in two sets of 12
across the diameter of the
instrument. The left side records the
hours before Noon; the right is for
the hours after Noon. The more
vertical, curving lines are the unequal
hours and are tied to the semi-
circular scale of hours written in
Roman numerals just inside the
circular band filled with the names of
the months. By using the *alidade*, or
pointer, one can see where the lines
intersect and translate one kind of a
hour into another. In the lower half of
the circle, there is a shadow square
that can be used for measuring the
heights of distant objects. The *mater*
of the astrolabe contains a single
plate for the range of 48°N to 50°N
(approximately Prague to Paris) and
the *rete* has pointers for 43 stars.
Despite the very fine level of
engraving, there is an error in the
projection of the ecliptic. This is a
feature common to many of
Habermel's astrolabes and helps to
identify this as an authentic piece.

121 John Gleave

Reconstruction of the 'Antikythera mechanism' (*ca.* 87 BC)
Mixed media, 20th century
Hebden Bridge, collection of John
Gleave (Interplanetary Models)

The so called 'Antikythera
mechanism' consists of numerous
layers of inscribed plates and toothed
gear wheels. The purpose of the
instrument was to provide direct
readings of the ecliptical and
equatorial co-ordinates of the Sun,
the Moon and four of the planets. It
had at least three dials, one of which
was a *parapegma* (see **040**),
containing calendrical information
about the rising and setting of the
major bright stars day by day for the
period of one year. There also seems
to have been a pointer for showing
the position of the Sun against the
ecliptic and a Moon dial, which may
have indicated the phases of the
Moon. Unlike the fragments of
anaphoric clocks that have survived
(see **147** and **148**), the Antikythera
mechanism shows no evidence of
having been driven by a constant
power source, such as water, air or
falling weights. Like the
parapegmata, it was probably reset
by hand at the beginning of every
day. To this end, it occupies the
slightly anomalous position of being
mechanical in its composition, but
not in its function.

122 Byzantine

Portable sundial and calendar (with
modern reconstruction)
Brass, *ca.* 520 AD
London, The Science Museum,
[1983–1393]

These fragmentary remains of a
Byzantine sundial provide valuable
information about how some of the
earlier Greek 'calendar-clocks' might
have functioned. The instrument
itself is now in four parts. The
markings on the largest, circular
plate include a scale running along
one quarter of the outer rim naming
a number of Greek towns and
provinces with their geographical
latitudes. This was used to set the
instrument to the local latitude.
There is also a quadrant scale,
marked from 0° to 90°, and a large
fan-shaped declination scale radiating
from the centre of the instrument
and covering three quarters of the
full circle, which is inscribed with the
names of the months in Greek. This
would have been used to set the
now lost U-shaped gnomon. In order
to use the instrument as a sundial,
the suspension ring would be set to
the correct latitude at the top in the
instrument; the gnomon would then
be set to the appropriate month and
the instrument turned until the
gnomon cast its shadow on the hour
scale. At the bottom of the dial,
there is a small circular dial engraved
with bust portraits of the seven
planetary gods representing the
seven days of the week. These
would have been 'clicked forward'
day by day, by a seven-toothed
wheel that has also survived. As it
moved forward, it engaged the teeth
of the accompanying 'Moon wheel',
which has 59 teeth running along its
edge, associated with two
consecutive scales, one for the 30
days of the long months and the
other for the 29 days of a short
month. As the Moon wheel was
turned, it revealed the correct phase
of the Moon for a given day through
the small aperture in the centre of
the circle of the daily planet gods.

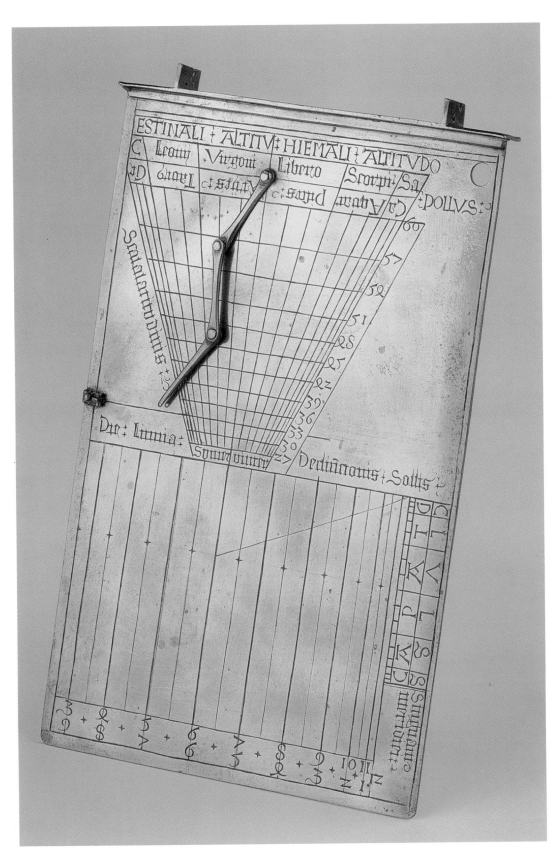

123 **German**

Universal dial with Regiomontanus projection
Brass, *ca.* 1475
Bielefeld, Kunstgewerbesammlung der Stadt Bielefeld [H-W 97]

On the front of this instrument, there is a solar quadrant. On the back, there is the earliest known example of a universal sundial using a Regiomontanus projection. The universal sundial (*quadratum horarium generale*) was invented in 1475 by the astronomer Johannes Müller (1436–1476), usually known by his Latin name of Regiomontanus after the town of his birth, Königsberg. The trapezoidal scale at the top of the instrument is divided into two sections: summer altitudes (*ESTINALE ALTITUDI*) and winter altitudes (*HIEMALI ALTITUDO*). The horizontal lines are marked according to 12 different latitudes ranging from 27°N at the bottom to 60°N at the top. The fanned vertical lines indicate the date-lines for the different signs of the zodiac (listed at the top of the scale). In the middle of the scale, there is a pointer that is divided by three joints. On the last of these joints, there was originally a plumb line with a small bead on it. The hour scale fills the bottom half of the plate (having a smaller zodiacal scale at its right side). To tell the time, one must first put the end of the pointer on the intersection between the correct latitude and date. Positioning the instrument vertically, one holds the plumb line against the zodiacal scale on the right of the hour-lines and places the small bead on the appropriate month. This establishes a specific length for the line. The instrument is then turned so that the sights at the top of the instrument are aligned towards the Sun. The plumb line can then swing free and the correct time is where the bead intersects the hour scale. The early date of this sundial suggests that it comes from the circle of Regiomontanus or even was owned by the great man himself.

124 English

Navicula sundial
Latten (leaded gunmetal), 15th century
London, The National Maritime Museum [AST 1146]

The *navicula* is essentially an altitude dial. It takes its name from the fact that it is shaped like a little Venetian galley – its full name is *navicula de Venetiis*, or 'little ship of Venice'. In order to find the time, one first sets the cursor to the correct latitude against a list of towns engraved along the mast of the ship. On this instrument the towns listed run from Exeter (sited at 51°N) to York (listed as *Eboratum* at 53° 40'N). One then tilts the mast of the ship so that its

base intersects the appropriate month in the zodiac scale at the bottom of the ship's keel. The plumb line is then laid across the same date on the other side of the instrument, with the bead positioned so that it intersects the 12 o'clock line, which runs parallel with the furthest edge of the ship's deck. The whole instrument is then tilted so that the Sun shines directly through both sets of sighting holes, located in the crenellations at the end of the ship's deck. The falling plumb line will then intersect the hour scale on the front of the instrument at the right point. The reverse of the instrument also has a scale for translating 'unequal' hours into 'equal' hours.

125 M.F. Poppel

Simple ring dial
Gilt brass, 1696
London, The National Maritime Museum [AST 0411]

Simple ring dials are very basic altitude dials. They are sometimes known as 'poke' dials because they were intended to be small enough to fit in one's 'poke' or pocket. The sliding collar on the ring can be adjusted to the correct time of year. Once aligned, the dial is suspended until a spot of sunlight falls on the graduated hour-scale on the inner side of the ring. These dials were latitude specific (this one was made for the town of Passau, 49°N), but are rarely very accurate. A development of the simple ring dial was the universal equinoctial ring dial, which is also self-orienting (so that it

does not need any scales or compasses to align it north–south); universal (it could be used at any latitude without modifications); usually quite accurate; and it could be folded flat for storage or travel and was extremely simple to use. The universal equinoctial dial was extremely popular from about 1600, the date when it is believed that it was invented by the English diallist William Oughtred.

Also

• P.I. Guymard, *Universal equinoctial ring dial*, 1740, London, The National Maritime Museum [AST 0313]

126 Humfrey Cole

Altitude sundial
Silver with tooled leather case,
ca. 1570
London, The Science Museum
[1985-100]

The sundial on the upper surface of
the dial consists of a square plate
that lies horizontally. Placed within a
quadrant, there are two sets of hour
scales: the inner one for 'unequal'
hours (labelled *the oures of the
plenet*[s]) and the outer one for
'equal' hours (labelled *the zodiac of
oures*). At one end of the plate, there
is a triangular flange of metal that is
hinged to the base plate. This is the
dial's gnomon. The decorative hole
that is cut into it would have originally
held a plumb bob. In order to tell the
time, the dial needs to be turned so
that the shadow cast by the steeper
edge of the gnomon falls exactly

parallel to one edge of the base
plate. Then, the shadow cast by the
inclined edge of the gnomon
crosses both sets of hour lines.
Along one side of the hour scales,
there is a calendar. The divisions of
the months are extended and form a
series of concentric arcs that cross
the pattern of hour-lines. The correct
time is found by noting the
intersection of the shadow-line, an
hour-line and the appropriate curving
calendar line. The Science Museum
sundial was constructed for 51° 30'N
– the latitude of London. The double
circle in one corner of the dial shows
which planets rule each hour of the
day. And, beneath the gnomon, there
is a table which lists the names of
the four winds, the four cardinal
points, the four seasons, the four
humours and the Four Ages of Man.

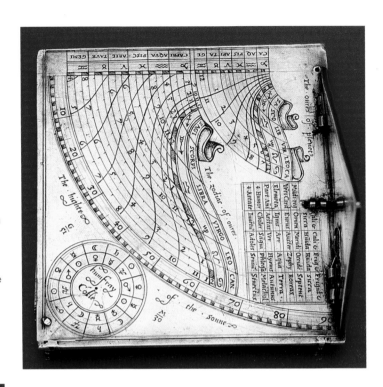

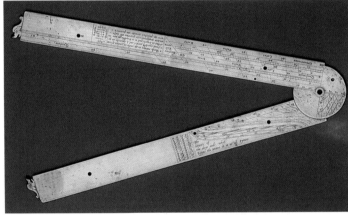

127 Humfrey Cole

Folding rule with altitude dial
Brass, 1574
London, The Science Museum
[1984–742]

For a sophisticated instrument maker,
who was used to turning his expert
hand towards the manufacture of any
number of different types of
instruments, enjoyment often lay in
trying to multiply the possible uses of
a seemingly simple instrument. This
folding rule by Cole can be used as a
regular ruler. It also has a scale for
making and reading maps, and scales

to calculate areas and volumes of
timber. When set at a 90° angle, it
can be used to measure vertical
elevations. The hour scale is drawn
for 52°N, or the latitude of London. A
table gives the dates of the Sun's
entry into each of the signs and the
first point of Aries is marked as 11
March. There are at least two ways in
which one can use the rule to
measure the time. If the ruler is
placed so that the arm containing the
hour scale is vertical and the other
arm is horizontal so that it forms an
inverted L, one can fit a pin gnomon
into the hole in the horizontal arm.

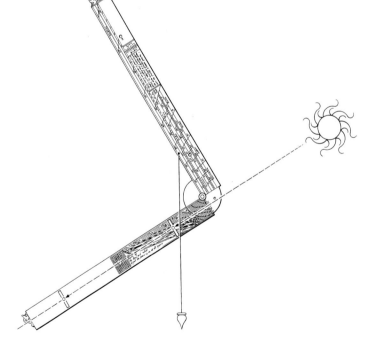

The Sun would then shine on the
upper, horizontal arm and the
gnomon would cast a shadow on the
hour scale located on the vertical
arm. Or one could hold the rule so
that the arm with the hour scale is

lined up along the same angle as the
altitude of the Sun (see illustration by
Denys Vaughan). If a plumb line were
then fitted into the hole in the other
arm, its shadow would fall across the
hour scale to mark the time.

Non-Mechanical Time: Using the Globe

Few people who appreciate antique globes recognize the fact that, during the Renaissance and certainly well into the eighteenth century, one of the primary functions of the globe was as a tool to help tell the time. As timekeeping is essentially involved with marking the motions of the heavens relative to the surface of the Earth, the globe is a perfect instrument. One turn of the globe – celestial or terrestrial – and the daily motion of the Earth is reproduced. On the well known Behaim globe of 1492, for example, there is an inscription on the horizon ring which clearly states: 'The ring is called the horizon and shows the rising and the setting of the Sun and the 12 signs'. The terrestrial globe can be used to measure time by the Sun and the celestial globe can be used to measure time against the so called fixed stars. This is one reason why all 'time telling' globes have an hour-ring placed on the meridian circle.

128 Gemma Frisius
Celestial globe
Papier-mâché core, gesso and engraved paper gores, 1537
London, The National Maritime Museum [GLB 0135]

129 *Terrestrial globe*
Papier-mâché core, gesso and engraved paper gores, 1537
Vienna, Österreichisches National Bibliothek – Globenmuseum [on loan from the Rudolf Schmidt Collection]

Although these globes are traditionally attributed to Gemma Frisius and they certainly reflect his genius, they were probably executed with the help of Gaspar van der Heyden (who is recorded primarily as a goldsmith and an artist), who was possibly the engraver of the figures; and Gerard Mercator – also an engraver, cartographer and inventor of the well known 'Mercator projection'. The figures of the constellations and the co-ordinates of the stars are based on those designed by Albrecht Dürer for his printed celestial hemispheres of 1515 (see **030**). The stand of the celestial globe is possibly 16th-century, but it is neither Flemish nor contemporary with the globe itself (the first point of Aries is marked following the Gregorian calendar, indicating that it postdates the globe by at least 50 years).

130 Erasmus Habermel

Torquetum
Firegilt copper, 1590
Hamburg, Museum für Kunst und
Gewerbe [1912.1435]

This *torquetum* is constructed so
that its three dials offer co-ordinates
according to three different
measuring systems. The bottom
plate is a horizontal dial; the middle
plate is aligned to the celestial
equator and the top dial is parallel
with the ecliptic. Many experts
believe that the *torquetum* was
primarily used to make simultaneous
readings of celestial objects against
these three different scales.
Certainly that was one of the
features that interested Erasmus
Habermel, the maker of this
instrument, for he added a second
set of sights to the equatorial plane
of the instrument. But earlier
torqueta only have sights on their
ecliptical plane. Most references to
the *torquetum* in late medieval and
early Renaissance sources indicate
that it was used for positional
astronomy and for verifying the
longitude and latitude of the stars
(for stellar mapping). The instrument
is also described as being used
specifically to track comets. This
might explain why both the
equatorial and ecliptical planes are
included. One uses ecliptical co-
ordinates for direct observation; but,
in order to 'follow' a star or a comet,
one needs to be able to adjust the
instrument so that it takes account
of the daily equatorial movement of
the skies above the observer's local
horizon. The instrument is
constructed, then, specifically so that
the equatorial plane can be turned
(hence its name – from the Latin
torquere, 'to turn') in order to
readjust the ecliptical plane for each
night's viewing. Once one knows
when a particular star is supposed to
be where, if one finds that star at the
right co-ordinate, one can use it to
tell the time.

131 Egyptian

Fragment of an Egyptian klepsydra
Basalt, *ca.* 320 BC
London, The British Museum, Dept.
of Egyptian Antiquities [EA 938]

A timekeeper that measures intervals of time against the flow of water is known as *klepsydra* from the Greek, 'water thief'. There are two basic types. An outflow *klepsydra* is a vessel in which a hole has been pierced to allow the water to escape. As it does, the decreasing levels of water left inside the vessel are measured. An inflow *klepsydra* is filled with water from an outside source. The water is measured as it rises within the vessel. The primary use of the Egyptian *klepsydra* was as a regulator against which the priests could mark the transit of the important stars.

132 Herbert Chatley

The Karnak klepsydra
Drawing, 19th century
London, The Science Museum [not numbered]

The Karnak *klepsydra*, now in the National Museum in Cairo, was discovered in the Temple of Karnak in Luxor, Upper Egypt, and dates from the reign of King Amenhotep III (1415–1380 BC). The surface is covered with hieroglyphics arranged in three rows. The top row has a series of the planet gods – Venus is depicted as the phoenix bird of Osiris and Saturn, Jupiter, the bright star Sirius and Orion are shown standing in boats. Following the gods, there are depictions of the 36 decan stars, the great celestial timekeepers of the ancient Egyptians. The middle range has depictions of the seven original celestial genii, gods of the four cardinal points, depictions of the circumpolar constellations and images of the deities of the days of the week. The bottom range has a calendar of months and month gods interspersed with prayers for the pharaoh. The large inset panel depicts Harmachis, the Sun god, and Thoth, the god of the Moon (note the large Moon as part of his crown) and of record-keeping, flanking the pharaoh. As a *klepsydra* keeps equal intervals of time and, therefore, equal hours, on the inside of the vessel there are a number of separate hour scales for each of the sets of unequal hours used for the different months during the year. The approximate flow of water for the Karnak *klepsydra* was ten drops per second.

133 Late Babylonian

Tables for setting a water clock
Baked clay, *ca.* 600–500 BC
London, The British Museum, Dept.
of Western Asiatic Antiquities [WA 29371]

One of the problems created by using a water clock was that it only marked equal intervals of time for societies that generally kept 'unequal' or seasonal hours. The Egyptians solved this by introducing different time scales against which the flow of water was measured (see 132). The Babylonians, however, varied the amount of water in the *klepsydra* – adding less water to the vessel during the shorter hours of the summer nights and more water to cover the longer period of darkness during the winter nights. As they used the *klepsydra* – or as the Babylonians called it, the *dibdibbu* – primarily for dividing the night hours into periods of three equal watches, the most important thing was to have a vessel that could run from sunset until sunrise. This tablet provides a table specifying how much water needs to be poured into the *klepsydra*.

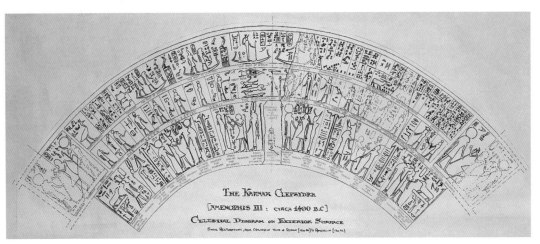

Whereas the majority of scientific instruments, such as sundials, nocturnals and astrolabes, measure time according to the relative position of the Sun, Moon and stars, the passage of time can also be measured without any reference to the heavens at all. In this instance, time is measured by interval. The interval can be of any duration – as long or short as seems appropriate – but must be measured accurately and in a consistent manner.

The ancient Babylonians, Egyptians and Chinese, for example, used the steady flow of water as the basis of their timekeeping systems. The best known interval-measuring device from the Middle Ages is the hourglass or sandglass. Both names are slight misnomers as, most often, these timekeepers neither measured an hour's worth of time, nor was the medium used to measure time sand. Usually it was powdered rock or, sometimes, crushed eggshells. In the Far East, particularly China and Japan, burning incense was often used as an interval timekeeper. Incense sticks were given a statutory length in order to help people with time measurement in the home. Later, elaborate 'incense clocks' used the reliable, slow-burning nature of incense, as well as the natural perfumes which could be incorporated within the medium, in order to create ingenious timekeepers which allowed the owner to keep time simply by using his sense of smell.

134 Vitruvius

De Architectura Libri Decem cum Commentariis Danieli Barbari ...
Printed book, Venice 1567
London, The British Library, Dept. of Printed Books

Daniele Barbaro's sumptuously illustrated edition of Vitruvius' *De Architectura* provides 'illustrations' for a number of the mechanical devices described in the book. The illustration of the geared *klepsydra* of Ctesibius the Alexandrian (IX, viii, 5–7), for example, manages to capture Vitruvius's description of how "a rack and a revolving drum, both fitted with teeth at regular intervals ... acting upon each other, induce a measured revolution and movement". Barbaro may not have fully understood the mechanics of the Greek water clock, but he does understand how a regulated driving force might be used to run a *simulacrum* of the heavens. If one compares his illustrations with modern reconstructions of Greek and Roman anaphoric dials, it is remarkable how accurate his interpretations of Vitruvius's text are.

135 Italian (Florence?)

Table klepsydra
Gilt wood, gesso and glass, *ca.* 1670
London, The National Maritime Museum [H.14]

Water clocks continued to be used well into the Renaissance and Baroque periods in Europe. Especially during the 17th century, when scientists were experimenting with natural forces, such as water pressure, steam, magnetism *etc*, in the search for mechanical innovations, it was thought that the *klepsydra* might be useful for the mensuration of long periods of time. In 1669, Domenico Martinelli designed a *klepsydra* in which a set of inflow and outflow pipes were used to provide a constant air pressure within the glass globes of the clock so that the water would flow at an even rate. A unique example of his *klepsydra* has survived. The two glass bulbs are double-graduated and connected with wax and packing cord (as Martinelli advocates). The time it measures is 30 minutes. The invention of the drum water clock has also been attributed to Martinelli. The clock has a cylindrical body made of tin or brass. The drum itself is hollow, and is divided into five or seven internal compartments, each wall of which has a small hole drilled into it. During its manufacture, water is introduced into the body of the cylinder and the cylinder is sealed. When the drum is positioned at the top of the scales, the slow flow of water between compartments causes it to turn and then drop in regular intervals down the length of a set of hour scales.

Also
• English, *Drum water clock, ca.* 1700, London, The Science Museum [1894–131]

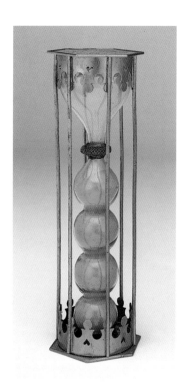

136 French

Sandglass
Brass, glass and sand, mid-18th century
London, The National Maritime Museum [AST 0054]

A sandglass works on exactly the same principle as a modern egg-timer. However, it can also be made to provide readings of fractions of a given interval by dividing the flow of sand between a number of smaller ampoules. In this glass, as each of the smaller glasses fills or empties, a quarter of an hour is recorded until the full hour is completed. Then the glass is turned over to begin the next hour. This example, with its decorative brass stand, was surely intended for domestic use.

Also

• German, *Sandglass*, 16th century, London, The National Maritime Museum [AST 0059]

• German, *Sandglass*, 17th century, The National Maritime Museum [AST 0073 and AST 0062]

137 W. & T. Gilbert

Log glass
Brass and glass filled with mercury, *ca.* 1817
London, The National Maritime Museum [NAV 0744]

Sandglasses of short duration – usually 28 seconds – were also used in navigation to assess the speed of the ship. These glasses are usually called 'log glasses' because of their relationship to another navigational instrument called a 'log'. A wooden log, attached to a rope, was thrown overboard and the rope allowed to play out for the duration of the running of the glass. Then the rope was hauled back in and measured. To make the measuring easier, the rope was knotted at regular intervals, hence the practice of referring to a ship's speed in 'knots'. Because sand could sometimes stick, the Admiralty experimented with other materials in its log glasses. For example, these experimental glasses contain mercury, and ran for 14 seconds in one direction and 28 seconds in the other. Unfortunately, mercury proved to be no more reliable than sand and this type of glass never went into general use.

138 Chinese

Incense seal (xiang yin)
Pewter with copper trim,
19th century
Washington, D.C., collection of Silvio A. Bedini

From quite early on, one comes across Chinese descriptions of periods of time as 'the time it takes to burn a stick of incense'(*Yi zhu xiang de shihou*). In some cases, they referred to something standardized – we know that standard-length sticks of incense were given to mariners to time their watches at sea and that Chinese doctors used cracked incense sticks to show their patients when medications had to be taken; in other cases, the phrase is used more poetically. The earliest incense seal, or *xiang yin*, can be traced back to

specific sects of Tantric Buddhism during the 8th century, but personal incense seals seem to have developed significantly later. This personal incense seal is constructed in three tiers. The cover grid has perforations which form the character known as 'double happiness'. The inner template that forms the pattern into which the ash will fall is shaped like the character for 'longevity' (*shou*). On its sides, there are depictions of spring flowers and the following poem:

'Cleanse the mind and smell the wondrous incense;
Official rules are impartial;
Treasure this incense burner through the generations;
The lasting fragrance of the incense resembles spring;
The fragrance of the incense goes through my red sleeves as I read my book at night;
Use with care this auspicious burner.'

A favoured form of incense seal was the *Ruyi* sceptre, a symbol derived from the shape of a sacred fungus, often called the 'plant of long life'. It was one of the Daoist emblems for longevity and often appears as the attribute of Lao Zi, the legendary founder of Daoism.

Also

• Ding Yun, *Incense seal in the shape of a Ruyi sceptre*, second half of 19th century, Washington, D.C., collection of Silvio A. Bedini

139 Japanese (Nigatsudō)

Jōkōban

Cypress wood trimmed in black
lacquer, early 19th century
Washington, D.C., collection of Silvio
A. Bedini

Introduced by the Chinese Buddhist
missionaries in the Nara period
during the later half of the 8th
century, the incense seal soon
developed into a very different form
in Japan. The base of the seal was
filled with ash, into which a series of
square-patterned furrows were
drawn, then filled with powdered
incense. Each incense square took
exactly one hour to burn. In the later
versions of the *kōbandokei*, or
incense clock, the furrows were

made by pressing the slightly
dampened ash with a template. The
jikōban (the 'time incense tray') was
used exclusively in Buddhist temples
as part of the religious rites. It was
originally designed to burn for a
specific period only, like a sandglass.
It is, perhaps, in line with this
development that *jikōban* are
sometimes called *museirō*, or
'noiseless water clocks', because
they fulfilled the same function, but

silently. In the Tokugawa period
(1600–1868), for example, the
jikōban was used to measure the
intervals between the times when
the temple bell had to be rung. But
there is evidence that the *jikōban*
was also used in temples to mark
the hours of the day. In this case,
their incense pattern was divided
into six equal segments, following
the Buddhist division of the day into
six parts – daylight being divided into

jinjō ('morning'), *nitchū* ('noon') and
nichimotsu ('evening') and the night
into *shoya* ('early night'), *chūya*
('middle night') and *goya* ('after
night'). The *jōkōban* ('constant
incense board') was used both in the
temples and in civic and domestic
settings. This *jōkōban* comes with its
original utensils, including a tray for
the ash base, a template and a
levelling bar.

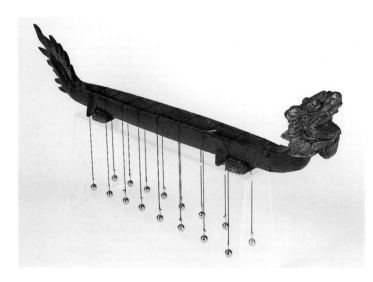

140 Chinese

Dragon-boat alarm
Wood, 19th century
London, The National Maritime
Museum [CH. 598]

One of the earliest records of
timekeeping with incense in China
describes how young messengers,
who were seldom allowed to sleep
for any more than the shortest
periods, would put a joss-stick
between their toes and light it before
they fell asleep. A similar, but more
sophisticated use of incense as part
of an 'alarm clock' can be found in
the so called 'dragon-boat alarm'.

The clock itself is shaped like the
familiar dragon boat, with a dragon's
head representing the bow and a
plumed tail indicating the stern. It
was fitted with a pewter liner that
was pierced to accommodate a
number of V-shaped wires. A stick of
incense was placed on the wire
holders and a series of fine strings
were draped over it. Tied to each end
of the fine threads were a pair of
small balls. As the incense burned, it
would – at regular intervals – burn
through the thread. This would then
release the balls which would fall
with a clang into a basin positioned
below the clock.

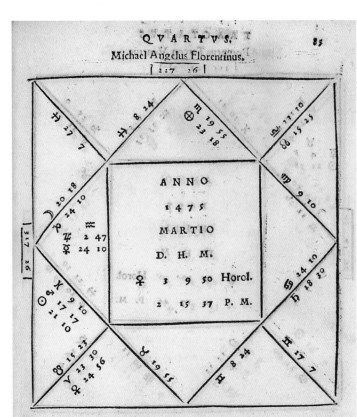

Mercurius eous a Sole 27.gradibus elongatus, in Falciferi hospitio, ab ipsa Venere irroratus exagona radiatione platica, effecerunt ipsum Michaelem Angelum sculptorem & pictorem eminentissimum, Phidia, & Praxitele clariorem cum opibus affluentissimis, quam fœlicitatem affirmare videtur Iuppiter secundæ domus hospitator in horoscopo platice supputatus, & a Venere fœliciter irrigatus. Ex sui genij doribus thesauros affluentissimos cumulauit, & a Principibus ecclesiasticis honores clarissimos.

141 **Lucas Gauricus**
Tractatus Astrologicus
Printed book, Venice 1552
London, The British Library, Dept. of
Printed Books

In 1552, the Italian astrologer Lucas Gauricus printed a series of horoscopic charts. He provided his temporal notations in two formats. In his horoscope for the birth of Michelangelo, for example, he provided two versions of the time at which the artist was born. In the central square, we are given the year (*1471*) and the month (*martio*) and three letters indicating day, hour and minute. Below this is a glyph for the planet Venus (signalling Friday). Alongside the glyph there are two

sets of days, hours and minutes: *3* (March) *9* (hours) and *50* (minutes) *Horol.* (by the clock); and *2* (March) *15* (hours) and *37* (minutes) *P.M.* (*post meridiem*). The first indicates that Michelangelo was born at 9 hours and 50 minutes 'by the clock' on 3 March. As it was Italian convention to begin the day at sunset, this notation indicates a time nearly 10 hours after sunset of what we would consider to be the previous day (2 March). The second line explains that he was born on 2 March, 15 hours and 37 minutes after midday. Counting 15 hours and 37 minutes after Noon on 2 March gives a birthtime of 3:37 AM on Friday 3 March. Both systems, then, give the same time of birth.

For those of us who live by the modern clock, it is hard to imagine that the time-telling conventions we follow are a relatively recent development. Until the middle years of the nineteenth century, however, there were very few agreed conventions for dividing the hours, their number in a day or the hour at which the day began. Not only did different systems for telling the time change from town to town, but also – depending on the purpose – one individual might employ as many as four different time systems during the course of his daily routine.

As has already been discussed, there were generally two different ways in which hours were divided: unequal and equal hours. The length of each of the unequal hours changed throughout the year as the relative percentage of hours of daylight and night varied. Equal hours were much easier to calculate, but the modern reader has to be careful when trying to interpret any temporal notation and first ascertain which of at least four systems is being used. *Horae communes* (or 'common hours') was an equal-hour system also known as 'German' or 'French' hours. Humfrey Cole refers to them as "the zodiac of oures" (see **126**). These were divided into two groups of twelve: one beginning at midday and the other at midnight. This system is the ancestor of our current system of 'AM' and 'PM'. Italian, Bohemian and 'Welsch' (foreign) hours were also equal hours. In Latin, they were described as being *horae ab occasu solis* ('hours calculated from sunset'). Zero hour was calculated at sunset (or, in certain towns, such as Siena, from a half an hour before sunset) and the day was divided into 24 equal hours. The major disadvantage of this system was that it had to be recalibrated throughout the year. So called Babylonian or Greek hours were calculated *ab ortu solis*, or 'from the moment of sunrise'. In instruments such as sundials, where only the hours of daylight are counted, a typical hour scale for Babylonian hours would run from 0 or 1 to 16. Finally, Nuremberg hours are a bewildering combination of both Babylonian and Italian hours. Daylight runs *ab ortu solis* from 1 until sunset; and night-time runs *ab occasu solis*, from 1 until the following sunrise.

As the only clearly measurable time of day is midday, when the Sun appears to pass across the local meridian, astronomers often used *horae communes* in their writings. But the day in which that hour might fall can be noted according to an altogether different system – the day could have been calculated *ab occasu solis* or from the previous midnight. Sometimes, we are helped by the inclusion of additional information – the notation of a feast day or some astronomical reading that could have only occurred on a particular date. But more often than not, it is wise to be extremely circumspect of the date or the time included in a letter or document. And when an Emilian is writing to a Florentine who is working in the Vatican for a Venetian cardinal – it is anyone's guess as to what the time, date or year might be!

142 German or French

Torquetum-type sundial for solar and sidereal time
Brass, late 17th century
London, The Science Museum
[1894–7]

As the *torquetum* developed, it seems to have changed into an instrument that could be used to make celestial measurements according to a number of different systems (see **130**). The structure of this sundial uses the *torquetum* as its model. The lower dial, set in the equatorial plane, can be used as an equatorial sundial. The upper dial, however, tells the time according to the stellar co-ordinates defined by the ecliptic. By lining the sights up towards the Sun, the upper plate shows where the Sun is against the constellations of the zodiac and it also indicates the sidereal time in terms of degrees, minutes and seconds of arc. Solar time is told by adjusting the dial so that the rod separating the upper and lower dials falls directly towards the centre of the instrument. Levelling screws, a compass and a plumb bob have all been added to the base plate to ensure that it is correctly aligned and level before any measurements are made.

143 Probably Dutch

Sabbath lamp
Pewter, 19th century
London, The Jewish Museum
[JM 1985–2,2]

The most sacred of Jewish domestic institutions is the Sabbath (*Shabat*) or Jewish day of rest. Since, by Jewish custom, each day begins at sunset, the Sabbath begins at sunset on Friday night and continues through until the following sunset. The Sabbath is welcomed through a ceremony of sanctification, called *kiddush*, and is symbolized by the lighting of two candles. These two candles commemorate the two passages from the Bible in which the Ten Commandments are listed and described (Exodus, 20:2–15, and Deuteronomy, 5:6–19). The main

difference between these two passages is that the first commands one to "remember" the laws; the second tells one to "observe" them. Before the invention of gas and electric lighting, it was the custom to have a Sabbath lamp with seven wicks. Practically, this provided enough light until bedtime and avoided the necessity of having to light the lamp on the day of rest.

144 Probably Persian

Pocket compass and sundial
Brass, 17th–18th century
London, The British Museum, Dept. of Oriental Antiquities [Richard Coll. 90.3–15.4]

In Islam, the sundial was used primarily to indicate the time for prayer. In addition to knowing the time, it was also important to know the direction of Mecca, or the *qibla*, towards which Muslims must face when they pray. The *qibla* of any given location is a trigonometric function of 1) the local latitude, 2) the latitude of Mecca and 3) the difference between the longitude of the location and the longitude of Mecca. Determining the exact *qibla* for different places on Earth took Islamic astronomers centuries to compile and *qibla* tables are a common feature in most astronomical treatises. Not only was the mathematics involved rather complicated, but establishing

longitudinal co-ordinates is never easy without being able to measure celestial phenomena in two places at the same time. As a result, perhaps, many cities acquired a traditional value for their longitude and latitude or even a traditional value for the *qibla*. Many mosques in the Maghrib, for example, point due east regardless of their relative latitude. *Qibla* tables were constantly being updated, but, once something has been written down as an authoritative source, its influence is often difficult to dislodge. This compass and sundial has a list of selected cities, mostly in modern-day Iraq and Iran, engraved on both sides of its cover. Beside each city is an indication of its *qibla*. By providing a compass with which to orient the dial, a sundial with which one might assess the correct time and a table of *qibla* to discover the correct direction towards Mecca, this sort of instrument would have been invaluable to any traveller.

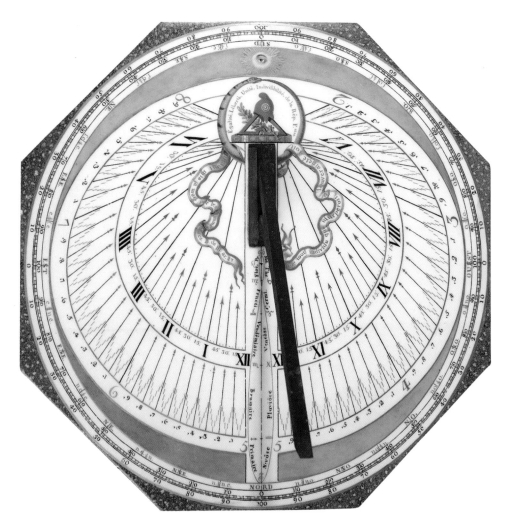

Sundial
Bronze, 1765
London, The Jewish Museum
[JM 576]

In Judaism, prayer can be private or
communal. The amount of time
spent in prayer depends on which
form of Judaism one practises. In
Hasidism, all life is centred around
the daily prayers. At the other end of
the scale, a reformed Jew – even
though there are three fixed times
for prayer services – is not obliged to
attend. The rabbis emphasize, for
example, that there must always be
a certain spontaneity in prayer: "He
who makes his prayer a fixed,
routine exercise does not make his
prayer a supplication" (*Berakhot*, IV,
4). Nevertheless, teaching one's child
to pray is an important duty. This
sundial, for example, was
constructed just for that purpose. Its
dial is marked in unequal hours and
the Hebrew inscription reads: 'Do
not forget to say *Mincha* (the
Noonday prayer) at the right time'.

145 **French (Sèvres)**
Sundial with decimal hours
Porcelain with polychrome enamel,
gold and bronze, 1793
Boston, The Museum of Fine Arts
[1980.467]

It was not only the calendar that was
'reformed' after the 1789 Revolution
in France (see **073**). There was also
an attempt to bring timekeeping into
line with a decimal system. The 24-
hour day was divided into 10 equal
hours. These hours were then
subdivided into 100 minutes each.
Decimal systems for timekeeping
had existed before, both in ancient
Egypt and in China. However,
whereas it was relatively easy to re-
christen the months, rescheduling
the hours was much more difficult.

With nearly 2000 years of training,
people had become accustomed to a
12-hour day. In almost all examples
of revolutionary time-keeping, the
10- and 12-hour dials are presented
in tandem. In this dial, for example,
the names of the Republican months
are marked on the *analemma*
without recourse to their Latinate
counterparts; but the middle dial
showing Republican, decimal hours
contains a convenient duodecimal
'crib', snugly placed slightly closer to
the gnomon.

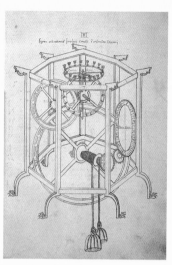

After Giovanni de' Dondi, *Manuscript drawing of the 'Astrarium'* (see **151**)

Mechanical Timekeeping in Europe: The Early Stages

From the third century BC, geared mechanisms had been used throughout the Mediterranean to create timekeepers that provided astronomical information. These simulacra of the heavens, usually driven by water-power, displayed calendrical data, showed positions of the stars relative to the horizon of the Earth, indicated the phase of the Moon and provided information about the relative hours of light and darkness (see **147** and **148**). With the disintegration of the Roman Empire, however, much of this technology was lost to the Latin West. The final classical citation of such a time-telling device can be found in a text written some time around 520 AD by the late Roman author Cassiodorus (*Institutionum*, I, xxx, 5). Even though large-scale, water-driven clocks continued to be used in Byzantium and, later, in some of the Islamic capitals of the Middle East, from the sixth century until the turn of the first millennium there is no mention of mechanical timekeepers in the Latin West.

When references to mechanical and semi-mechanical timekeeping begin to reappear around the year 1000, they are in documents relating to monks in remote monasteries needing to know what time to pray. As the times for prayer followed the Roman division of the day into temporal hours, the monks needed some sort of timekeeping device that would mark the time from one set of prayers to the next – something like a sandglass or a water clock. Prayers said during the day could be organized by consulting a sundial (the English word for midday, 'Noon', derives from the Latin name for the midday prayer, or None). In particular, though, the most pressing requirement seems to have been for a simple alarm device which would wake some unfortunate junior monk so that he could rouse the rest of the community for Matins. One of the most vexing questions for horological scholars today is whether the simple medieval alarm device developed *sui generis* from the more basic mechanisms of a sand or water clock; or whether the medieval clock was, in some manner, a revival of Greco-Roman anaphoric clocks and celestial automata. Research into the primary sources is made all the more vexing because the same word, *horologium*, can describe any means for telling the hours, including a sundial.

By the end of the eleventh century, however, large-scale water-driven mechanical clocks can be found in a number of major European centres. The clock at Bury St Edmunds, for example, was so large that, when there was a terrible fire in the monastery in 1198, the monks used the clock's water store in order to fill their buckets. The earliest evidence of weight-driven clockwork appears in the second half of the thirteenth century. The mechanics of these clocks are relatively simple. A weighted crossbar, called a foliot, is connected through a set of gears to the driving force of a set of falling weights. This force is regulated by a brake, called a verge escapement, which keeps the foliot swinging at a controlled rate.

Despite the questions about the source from which these mechanisms developed, very soon after the weight-driven clock appeared its gears were used to convey the 'time' to a display dial. In function, these early clock dials were similar to their classical forebears, often showing not only the hour, but the phase of the Moon and the positions of the Sun against the signs of the zodiac. A number of these astronomical clocks were created during the early years of the fourteenth century. For example, a large astronomical dial with automaton figures was commissioned for Norwich Cathedral in 1322, and Richard of Wallingford's well known clock for St Alban's Abbey was built between 1327 and 1336 (see **149**).

By the 1330s, the counting striking train had been introduced into the mechanics of the clock. Before this, clocks had been designed to strike either a single or a set of multiple strokes on a single bell. In 1336, the Milanese church of San Gottardo was fitted with a clock that could count the different hours by varying the pattern of its striking (*i.e.* the clock would strike one bell at 1 o'clock, two bells at 2 o'clock, and so on). This ability to, literally, 'tell' the time by the striking of bells seems to have been the key to the success of the public clock. For the first time in history, time was available to everyone within earshot.

On a smaller scale, personal timekeepers developed along similar lines. The first domestic timekeepers seem to have developed from the medieval alarm clock and would have been relatively common in Europe by the latter years of the fourteenth century. Similarly, the public astronomical automata also led the way for smaller-scaled devices. In particular, the remarkable clock created by Giovanni de' Dondi, his so called Astrarium, completed around 1375, had a set of seven dials that charted the exact position of the known planets, a 24-hour clock dial and a calendar, which not only supplied the date, but provided information on all the moveable feasts of the Christian Church, and the time for sunset and sunrise throughout the year as calculated for de' Dondi's native town of Padua. The clock itself was bought by Gian Galeazzo Visconti in 1381 and remained in his Castello at Pavia until the early years of the sixteenth century, when it disappeared without trace. Fortunately, a number of drawings and partial reconstructions of the clock were made, allowing us glimpses into the secrets of this marvel.

In the main, the mechanics of the large-scale turret clock remained unchanged for the next few centuries. Most developments during this period were concentrated within the arena of small, domestic clockwork. The most important change was the introduction of a coiled strip of metal, the mainspring, as the driving force of the timekeeper. This meant that clocks could be kept more compact and self-contained. One problem with the main spring, however, is the fact that its torque (or wound tension) is not constant. This problem was solved around 1400, when a device called the fusee was invented. The mainspring, contained within a cylindrical barrel, is connected to the gear train by a cord. This cord is, in turn, wound around a grooved cone set on the first gear of the train. When the spring is fully wound, the cord follows the thinner part of the cone; as it runs down, the cord pulls against the increasingly wider part of the cone. Once the fusee had the correct form, the driving force of the mainspring could be provided as a completely uniform power supply.

As so few clocks from this period survive in an original state, much of our understanding depends on manuscript sources. In particular, two major sources provide the bulk of information. The first is an illustrated text by the Parisian astronomer and physicist Jean Fusoris, who ran a 'horological' workshop around the turn of the sixteenth century, providing a variety of public and domestic clocks. The second is the illustrated notebook of Brother Paulus Almanus ('Paul the German'), which descibes about thirty different clocks owned by highly placed members of the clergy.

The history of watch- and clockmaking over the next few centuries is closely tied to social and economic concerns – prosperity, war and opportunities for trade. In the main, there are always two main branches of clockmaking – the Italian-French-Flemish group and the German group. During the sixteenth century, for example, the Germans tended to focus on the export market, while the Italian-French-Flemish group concentrated primarily on a home market. While the German makers expanded their trade, the French and Flemish makers were severely hampered by the revolt of the Netherlands against Spanish rule during the 1560s and 1570s. During the early seventeenth century, however, the German market was devastated by the Thirty Years' War and repeated outbreaks of plague; while the French began to recover and turned their attention to small and highly decorative watchwork. English clockmaking benefited greatly from political unrest in the Netherlands. The Clockmakers Company was founded in London in 1631.

Alongside what one might call the mainstream of clock and watch production, there has always been a small, élite band of clockmakers whose whole effort has gone into the production of precision timekeepers. Most often, these clockmakers have been, in some way, associated with astronomical observatories. At the observatory in Kassel, for example, Jost Bürgi experimented with another form of controlling mechanism in an attempt to create a more precise timekeeper. His cross-beat escapement consisted of two balances that were geared together so that they moved in opposite directions and, therefore, neutralized any outside interference (the principle of Bürgi's escapement is very close to the device utilized by John Harrison in his famous marine chronometers, 'H1' and 'H2'). He also developed the remontoire train, whereby the going train is split so that a precision train can be rewound automatically and at regular intervals by a secondary power-train. Galileo Galilei is another astronomer who was interested in creating a more precise mechanical clock. His suggestion that a swinging pendulum might be used to count the time has led many scholars to propose Galileo as the inventor of the pendulum clock (see **174**). Evidence suggests, however, that the credit for a truly practical pendulum clock must go to the Dutch scientist Christiaan Huygens (see **175**). In early 1657, Huygens commissioned the clockmaker Salomon Coster of the Hague to build the first working pendulum clock in which a complete going train with a dial and hands was controlled by the motion of the pendulum. Twenty years later, he also proposed that a spiral spring should be added to the conventional balance in order to give the balance a natural frequency (as a pendulum has). This suggestion paved the way for a timekeeper that was not only precise, but portable as well.

Beachy Head Lighthouse clock, *ca.* 1860 (see **176**)

The Growth of Modern Timekeeping

From Pendulums to Atoms

Jonathan Betts

Standard Time broadsheet, 10 November 1883
(see **165**)

By 1650, mechanical timekeepers had been in existence for nearly 400 years. Nevertheless, there had been very little fundamental change in their function throughout this period. Simple 'alarum' and striking clocks regularized the sleeping hours and daily routines in both monastery and household, while highly complex astronomical clockwork devices continued to function as simulacra of God's well machined universe. Equally, if one considers the clock strictly as a 'timekeeper', nothing really had changed.

The dramatic watershed in clockmaking technology occurred during the middle years of the seventeenth century, with Christiaan Huygens's introduction of the pendulum to clockwork in 1656. Galileo had first suggested the pendulum as a timekeeper in 1641 and the reappearance of this simple device, arguably the most important of all horological inventions, coincided with the beginning of a period of tremendous scientific advance, particularly in Holland, France and England. It became the focus of a newly discovered interest in precision timekeeping which, in turn, resulted in an extraordinarily rapid evolution in movement design. Within just twenty years, one sees the blueprint for the modern clock mechanism emerging – a mechanism that became standard (and virtually unchanged) for the following two centuries.

The catalyst for this advance in timekeeping technology was ostensibly scientific. Initially, it was thought that the pendulum might be usable in a marine chronometer in order to help find longitude at sea; but experiments with Huygens's designs soon showed this to be impracticable. Nevertheless, the pendulum's potential as a fixed timekeeper was quickly seized upon – especially for use by astronomers in order to increase the accuracy of their charting of the stars. The observatories at Greenwich, Paris and St Andrews in Scotland, amongst others, were soon using pendulum regulators to great effect. Moreover, as with most practical scientific inventions, the application of the new technology beyond the scientific arena soon followed. Even though early pendulum clocks, such as the wall clock by Edward East (*ca.* 1665, **154**) and the Haagse clock by Johannes Steffens (*ca.* 1674, **176**), were made purely for a domestic setting, they would have transformed the

Clocks were created not just to provide

existing household time standard from errors of +/- 15 minutes a day to perhaps +/- 15 seconds a day. Whereas most earlier clocks had run for just thirty hours, the majority of early pendulum clocks were made to run for a week. They were made either as spring-driven clocks like the Steffens; or they were designed, especially in England, to hang on the wall like the East. But the wall-hanging pendulum clock required very heavy weights to run for eight days –

The pendulum is arguably the most important horological invention ever made

perhaps as much as 24 lb (11 kg) in total. Indeed, the problem of fixing such heavy clocks securely to the average seventeenth-century plaster wall may have been the catalyst behind another invention, for it was just at this stage that floor-standing cases were introduced and the longcase clock – or, as most might know it, the 'grandfather clock' – was born.

One result directly derived from the increased accuracy of the new pendulum clocks was that astronomers were finally able to settle a long-standing question about time itself. An unresolved issue that had plagued astronomers for millennia was the question of the equation of time. These days, most people are unaware of the fact that the Sun is not a very regular timekeeper. Time indicated by a sundial, for example, can differ from a truly accurate clock by as much as 16 minutes fast (on 4 November) and nearly 15 minutes slow (12 February). Fortunately, this phenomenon is a cyclical one and repeats itself every year. Once sufficiently accurate clocks were available, specific values were determined to provide a printed table of 'the equation' – the difference between solar time and mean time (clock time) as it changed throughout the year for every day of the year. Such a table enabled owners to set the main clock in the household to mean (average) time, using an ordinary sundial, adding or subtracting the equation of time as appropriate for that particular day. Once a good longcase clock was thus set, it would provide a trustworthy standard for several weeks, from which many others might be put right.

Although medieval Islamic astronomers also understood the equation of time (and, apparently, produced accurate values for it), it was again Christiaan Huygens who was the first European to produce accurate tables (1660). In England, the first Astronomer Royal, John Flamsteed, finalized the first modern set of tables in the late 1670s, and these were printed in large numbers for the makers of pendulum clocks to provide for their customers. The natural progression from this improvement was the development of a clock which itself indicated the equation of time. The double longcase

information; they were also an expression of the clockmaker's intellectual and practical prowess. On another level, they functioned as a personal or institutional symbol for the owner. To own one of these clocks was to make a statement about one's investment or participation in the latest astronomical and scientific discoveries. A great institution, the Greenwich Hospital, which played a significant educational role within the community at the time, commissioned the clock from Quare in 1716. In contrast, the regulators made by George Graham had no symbolic function whatever. They had one clear and simple role: to be the most accurate timekeeper possible, a 'laboratory instrument' for use by practical astronomers in their daily work. Regulators like this (with a 'deadbeat' escapement originally designed in the 1670s by Richard Towneley and Graham's erstwhile partner Thomas Tompion) were the epitome of the best precision clock design. Throughout the eighteenth century, astronomers from around the world beat a path to London's clockmakers to order them.

Whereas fixed precision clocks were well settled in their form by the mid-eighteenth century, the question of portable precision timekeepers and their potential for addressing the vexed problem of finding longitude at sea was far from resolved. The introduction of the balance spring to watches had improved their capability considerably, but even the minute or so gain or loss per day that they could now guarantee was not nearly good enough for practical navigation. At this period, for all practical purposes, watches were still primarily articles of jewellery – affordable only by the rich who seemed not to mind that the watch itself needed to be checked and set daily if it was to be reliable as a timekeeper.

It is not within the scope of this short chapter to tell the story of the invention of the marine chronometer and John Harrison's heroic achievement. This story has been well told in many media and a visit to the Royal Observatory at Greenwich is required for those interested in this magnificent tale of scientific endeavour. Suffice to say that with his fourth timekeeper ('H4'), Harrison created not only

information, but to express the clockmaker's intellectual and practical prowess

clock by Daniel Quare (**176**) has this feature on its dial. In fact, Quare's clock is really two entirely separate clocks in one case – one rated to sidereal time (star time) for use by astronomers, and the other rated to mean time for ordinary domestic use. Double clocks like this one are very rare, but clocks with some form of complexity or other appear frequently during this horological 'Golden Age'. Clocks such as these were not created just to provide horological

the world's first truly accurate marine chronometer, but the foundation stone of all precision watches which followed. For example, once Harrison had demonstrated that a timekeeper on the scale of a watch was the only workable size for a chronometer, makers such as Pierre LeRoy in France and John Arnold in England changed tack and began to develop precision watches along these lines.

With the availability of high-precision marine chronometers, the practice of navigation was slowly but surely revolutionized; and, as with the pendulum clock, this new watch technology was soon demanded by the public at large. The Age of Enlightenment had arrived in Europe and there was a heightened interest among 'ordinary people' in all things scientific. Lecturers in 'natural philosophy', such as James Ferguson, toured the country encouraging a whole new breed of amateur 'natural philosopher', and the precision timekeeper was just one of the various instruments which could be purchased in order to pursue these new-found interests. Similarly, clocks with complex astronomical indications – popular from the late seventeenth century but reaching their height of perfection and elegance in late eighteenth- and early nineteenth-century France – played a role in the process of enlightenment. Quite distinct from the earlier, semi-religious, astrological/astronomical objects from the late Middle Ages and Renaissance, clocks such as the Raingo orrery clock (**176**) were secular and educational. However, it must be admitted that these very expensive pieces were educational in a somewhat élitist way. They were created to enlighten the gentleman dilettante – whilst undoubtedly being an opulent and impressive furnishing piece as well. At the other end of the spectrum were the rather more showy and entertaining complex astronomical, musical and automata clocks and watches, made throughout the seventeenth and the eighteenth centuries, to satisfy demand at the luxury end of the market.

Practical need was still very much an imperative in the second half of the eighteenth century, however, and, as the Industrial Revolution made its presence felt, clocks and watches began to be designed specifically for industrial and workaday functions. The workmen's clock by Thomas Tompion (*ca.* 1680, **171**) is a very early example of such an industrial use. It is a timekeeper intended purely for regulating newly made clocks and watches, without any 'time of day' capability. A similar instrument is the 'journeyman's clock' used by astronomers for timing intervals between star transits. The operation of machines and equipment, such as factory engines, and tidal indicators, such as the one by J. Mitchell (**176**), were also given the all-important dimension of a time scale. As a result, they could now provide a continuous record of production and operation. Predictably, the daily routines of the workforce of that 'brave new industrial world' were soon to be time-controlled too. It was first suggested in 1782 that the mail coaches in Britain should be run strictly to timetable and, within a couple of years, all guards were issued with watches by the Post Office – first for mail coaches, later for mail trains (**169**). Regulating one's working hours strictly by the clock was, of course, nothing new for the sailor. To ensure the safety of the crew day and night, the ship's 'watch' has existed, in one form or another, since men have been going to sea. The late eighteenth

century saw the invention of the 'noctuary' or watchman's clock (**172**), with which the night watchman was obliged to register his presence at specific times during his nocturnal inspections of a building. A number of other clocks used to record the arrival and departure of staff at the workplace were developed from the noctuary. From as early as 1844, a patent was granted for such an apparatus. Clocks designed for specific working duties were not always just to control the workforce, however. The English dial clock, for example, was a standard design which was adapted to all manner of different uses, such as the schoolroom, the railway waiting room and the factory wall. One such dial clock was modified for use in the Beachy Head lighthouse (**176**). This was used to ensure that the keeper regulated the flashes of the lighthouse optic correctly, an essential safety requirement as it was the timing ('the character') of the flashes which enabled mariners to identify which lighthouse they were approaching.

The 'ion trap' will be accurate to 1 second during

The Industrial Revolution of the late eighteenth century was undoubtedly a spur to new forms of horology, but, in the mid-nineteenth century, the combination of the electric telegraph and the railways, in Britain, Europe and the United States, brought far more dramatic and revolutionary industrial and social changes than anything which had gone before. Both the telegraph and the railways highlighted the problem caused by the use of any number of different local times across the country and around the globe. Up until this time, keeping local time was the norm. By 1800, most ordinary men (but still a minority of women) owned some form of pocket watch. These – as well as all local public clocks – would have been set according to a sundial. Thus, townspeople in Dover in eastern England, for example, saw midday nearly half an hour before the people of Penzance in the west of the country and the public clocks displayed in these local communities would have been set at two different times – nearly half an hour apart. With the new, rapid communications and transport systems, the necessity of a single standard time for the whole country became paramount. Greenwich Observatory – as the repository of the country's most accurate clocks – was the obvious supplier of this single standard. In 1852, an electric 'master and slave' clock system invented by Charles Shepherd of London was introduced at the Observatory in order to supply Britain's businesses (and anyone else concerned) with Greenwich Mean Time. GMT was first distributed across Britain using the electric telegraph system associated with the railways, which were, of course, one of the main beneficiaries of this new system. At first, the time was transmitted only through signals provided at particular times of day (usually 10 am and 1 pm). Eventually, though, it was possible to subscribe to an hourly signal.

Despite the fact that GMT was available 'on line', as it were, it was only in 1880 that Greenwich time was officially made the legal time standard for Britain. And, in 1884, with the resolutions of the International Meridian Conference held in Washington, D.C., Greenwich Mean Time was adopted as the world standard, with the primary reference for the International Time Zone System – longitude 0° – being the Greenwich Meridian itself.

As people became able to travel and communicate over long distances with relative ease, they became increasingly conscious of 'the world' as a single entity. Commercial opportunities on 'the other side of the world', for example, made knowing the time around the globe desirable. The Edwardian clockwork globe by Smith (**166**) is just one of a plethora of such objects made from the mid-nineteenth century onwards. One of the most obvious advantages of Marconi's new wireless telegraphy, introduced at the outset of the twentieth

a period as long as the lifetime of the universe

century, was the ability to distribute and receive time signals from around the world. In 1905, the US Naval Observatory began broadcasting daily signals from Washington, D.C. Originally, such signals could only be received by specialist receivers, but, in 1922, a time service to the public arrived, with the British Broadcasting Company (later Corporation) being the first to broadcast a public time signal (remarkably, in the form of the chimes of Big Ben at Westminster played on a grand piano by the announcer before the daily news broadcasts). In 1924, this was superseded by the now famous BBC 'six pip' Greenwich time signal, which has been copied by broadcasters the world over. France was the first to introduce a telephone time service in 1933. Meanwhile the synchronous motor electric clock system had been introduced in the States in 1918. Its motor derives a time standard from the carefully controlled frequency (50 Hz in Britain) at which the alternating current is supplied from the power station. Essentially, any home or office that had its electricity supplied by a power station and had a receiver like the 'Synclock' (see **163**) could 'plug in' to a constant source of correct mean time.

As early as 1928, the American scientists Horton and Morrison had created a clock controlled by a quartz oscillator, but it was to be many years before these so called 'quartz clocks' were sufficiently compact and reliable for domestic use. Only in the 1970s were independent battery-controlled quartz clocks accurate and inexpensive enough to cause the final decline of the synchronous clock. Moreover, with the introduction of commercially available radio-controlled quartz clocks in the early 1980s (corrected by digital radio time signals), synchronous clocks are now effectively

obsolete. With a radio-controlled clock or watch, everyone can now afford to have correct time to within a fraction of a second.

With effective global time distribution, there came greater synchronization and communication across the great global economies; but it did not necessarily induce greater harmony among men. Many ingenious horological devices bear witness to the fact that precise timekeeping has been equally useful for unfriendly activities, too. Scientific timekeepers by makers such as Congreve (**180**) and Hardy (**181**) were used to develop more efficient weapons of destruction. The *Machines infernales des Anglais* (**182**) of 1810 and the precision timer from a Polaris missile of 1982 (see below) are examples taken from the weapons themselves. Scientific endeavour has always been a two-edged sword. Nevertheless, the immense progress made possible by modern precision timekeeping devices like the caesium atomic clock probably far outweighs the negative aspects of their use. The Global Positioning System and the international space programmes for example, would be impossible without them.

What future benefits the improved technology of the new and extraordinary caesium fountain clock will bring can only be guessed at. The estimated stability of this new creation – currently being developed at the Paris Observatory and, amongst others, at the National Physical Laboratory in London – is, in scientific terms, 'two parts in ten to the power of fifteen'. This boils down to the almost incomprehensible accuracy of one second in 15 million years. Whatever next? Scientists at the NPL are also working on a new form of time standard, known as an 'ion trap', which is tentatively predicted to be correct to 'one part in ten to the power of eighteen'. This surely has to be an absolute time source as it would be accurate to 1 second in 10 billion years – or accurate to 1 second during a period that is believed to be the lifetime of the universe itself.

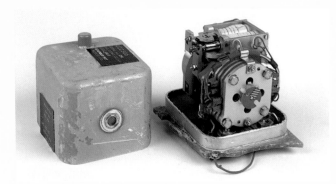

Timing device for a nuclear warhead, 1982 (see **137**)

By definition, mechanical timekeepers are those devices that utilize a constant force to drive a mechanism that can measure constant and equal intervals. The earliest mechanical timekeepers were the anaphoric water clocks of ancient Greece, which seem to have been developed some time during the third century BC. The purpose of an anaphoric clock was to provide a two-dimensional account of the daily motions of the Sun and Moon against the background of the fixed stars. The name comes from the Greek term *anaphero* ('to rise up') and refers to the rising and setting of the various bright stars that were depicted on its rotating dial. In Rome, this type of clock was called a *horologium hibernum* or 'winter clock' – a name which might refer to the fact that their primary use was for telling time at night and that, during the winter, the long hours of darkness would make such an instrument even more valuable.

In the anaphoric clock, a grid of hour-lines was superimposed on the surface of a planispheric map of the heavens. Although no complete example of an anaphoric clock has survived, the combination of celestial planisphere with a moveable grid of hour-lines has an extremely long legacy as the fundamental time-telling component of the planispheric astrolabe.

147 Roman

Fragment of an anaphoric dial
Bronze, 2nd century AD
Salzburg, Salzburger Museum [3985]

The Salzburg planisphere is the earliest surviving fragment of a mechanical timekeeper driven by a constant force. This piece was part of the main dial of a huge planispheric map of the heavens. Preserved on its front surface, one can see the images of Andromeda, Deltoton (the Triangle), Perseus and Auriga (the Charioteer). Around the edge of the planisphere, there are parts of some of the zodiacal signs: one of the fish of Pisces, Aries, Taurus and one of the Gemini. On the back surface, the names of the months are linked with the zodiac signs: [pi]*sces – Martius; Aries-Aprilis; Taurus – Maiius* [*sic*]; and *Ge*[mini]*- Iu* [nius]. The dial was part of a large water-driven clock. The flow of the water was controlled so that it drove a wheel that kept the dial turning in such a way that it was able to match the orientation of the heavens above.

148 Roman (Grand, Vosges)
Anaphoric clock dial
Copper alloy, 2nd–4th century AD
St-Germain-en-Laye, Musée des
Antiquités Nationales [inv. no. 31433]

In his description of the anaphoric clock, Vitruvius says: "For every one of the signs, there are as many holes as the corresponding months has days; and there is a boss, which seems to be holding the representation of the Sun on the dial, that is used to mark the spaces for the hours. This, as it is carried from hole to hole, completes the circuit of the full month "(*De Architectura*, IX, VIII, 9). It seems that the only thing in the anaphoric clock that was not automatic was the marking of the days. This had to be done by hand, moving small pegs day by day in the same way one did with the Greek *parapegma* calendars (see 040). Judging from the extant fragments, the anaphoric dials had a year equalling 182 or 183 holes. This suggests that, in these examples at least, each hole was used for two days. On the surface of this dial, the names of the months have been written with indications of their Kalends, Nones and Ides. It was obviously used for quite some time, as there are heavy circular scratches covering the whole of the disc.

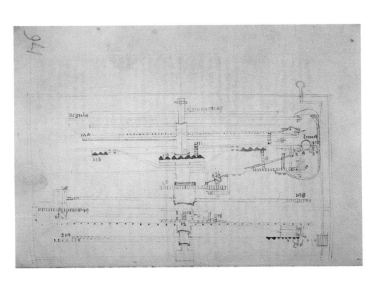

149 Peter N. Haward
Model of clock by Richard of Wallingford
Mixed media, 20th century
Needham Market, collection of Peter Haward

Richard of Wallingford
Tractatus horologii astronomici
Manuscript, mid-14th century
Oxford, Bodleian Library
[Ms Ashmole 1796]

Richard of Wallingford's clock, constructed at St Alban's Abbey between 1327 and 1336, helped to structure the daily routines of the monks. It was probably intended also as a working model to demonstrate the logical nature of the heavens and the rationality of God's work. Even though some of Richard's brethren disparaged the project as folly, it would have been an astounding and inspiring object for visitors and pilgrims, reminding them that the Abbey was a place of learning as well as spiritual refuge. This quarter-scale model of the mechanical clock has been developed from the earliest known detailed description. On the end of the frame, there is an astrolabe dial showing the positions of the fixed stars, and of the Sun and the Moon. The dial incorporates a dragon hand for indicating eclipses and has gearing to show some of the planetary motions. There was also a tidal dial and a very primitive form of hour-striking mechanism. The drawing shows an end view of the astronomical gearing of Wallingford's clock. An exceedingly early example of an engineer's working drawing, it also includes the numbers of teeth on the wheels. The dragon hand can be seen on the right-hand side of the mechanism.

Mechanical Time: Domestic Timekeepers

Most of the scholarly work on the development of the mechanical timepiece has been devoted to answering questions relating to the engineering of different aspects of the clock and its evolution. Which came first: the complicated, geared astrological *simulacra* of the heavens or the simple *alarum* device? Who first utilized the pendulum as a driving force? Why was Bürgi's cross-beat escapement not adopted more widely? And so on.

Intriguingly, very few studies of the development of clocks address the 'why' of clockmaking. Why did clocks develop in the way they did and what was the catalyst behind each new invention? The unspoken assumption is that a desire for increased precision fuelled most changes. It is much more likely, however, that the real impetus was social need. This can be seen in the way different types of clocks evolved. For example, the earliest domestic clock seems to have developed out of the monastic *alarum*. For the medieval monk, the only time you had to know was what time to pray, and for this purpose a clock's 'face' or dial was superfluous. The average Renaissance man, whose livelihood depended on commerce and communication, needed to know the time of day in order to make and meet a relatively tight schedule of appointments. Since he was a social animal, his needs directed clockmaking towards more social concerns and to the widespread development of turret clocks, travelling timepieces, night clocks and personal watches. Astronomers, in particular, demanded precision from their timekeepers and it is not surprising that, from the very beginning, most developments in what one might call 'precision' timekeeping have been associated with astronomical observatories – whether in ancient Babylon or at the Royal Observatory in Greenwich. Finally, one vital, yet often overlooked aspect of many clocks and watches is the aesthetic dimension. The changes in the shape of timekeepers prompted by the watch's function as a piece of jewellery or the clock's role as a piece of furniture should never be underestimated.

150 **French (Burgundy)**
Chamber clock
Iron, with a gilt-brass case, *ca.* 1460
London, The British Museum, Dept. of Medieval and Later Antiquities [MLA C.a. 450]

This highly important clock is a contender to be the earliest surviving spring-driven clock in existence. The clock is now weight-driven but, when made, probably in France in about 1460, it would have had springs to drive both the timekeeping and hour-striking parts (all of which would have been made of iron). What remains of the brass

case appears to be substantially original, fabricated by brazing the pieces together and then gilding. Such an early spring-driven chamber clock would have been highly prized when new, in spite of its very poor timekeeping. Undoubtedly, the motive behind the commission of such a piece was to impress and to symbolize the rank and wealth of the owner, with the domestic timekeeping role of regulating the functions of the household given a secondary consideration.

Also

- Peter Haward, *Model of de' Dondi's Astrarium* (1348–64), 1973, London, The Science Museum [1974-386]

- Giovanni de' Dondi, *De Confiendi Horologis Omnium Planetarum ...*, 15th century, Eton, Eton College Library [Ms.172]

151 Caspar Werner

Watch
Iron, with a gilt brass case, 1548
Wuppertal, Wupperthaler
Uhrenmuseum [DI6]

It was only a matter of time before the Renaissance clockmakers in southern Germany had reduced the size of spring-driven chamber and table clocks to such an extent that they could be carried on the person. Peter Henlein of Nuremberg is said to have been the first to make these tiny clocks which were "carried on the breast or in the purse". As such, they represent the first watches. Caspar Werner, a contemporary of Henlein's, made, signed and dated this example – the earliest known dated watch. The movement, which runs for 30 hours, is made entirely of iron and, as was standard practice with German watches at the time, uses a 'stackfreed' device to even out the driving force of the mainspring. The watch would probably not have kept better time than plus or minus about 20 minutes a day. The gilt brass case has an openwork cover over the dial so the single hour hand can be read against it. For night time, the dial has little 'touch pieces' by the numerals and the position of the hand can be felt in relation to them.

152 Erhard Liechti

Domestic weight-driven chamber clock
Iron, polychrome dial, 1572
Winterthur, Uhrensammlung
Kellenberger [141]

Like the majority of German and Swiss clock mechanisms up to this period, this fine chamber clock is made almost entirely of iron. Erhard Liechti was a member of the well known dynasty of Swiss clockmakers who worked in the Winterthur area for nearly three centuries. The construction of their clocks is particularly ingenious. The frame locates and interlocks together with the minimum use of pegs or pins. The clock strikes the hours and quarters and has the all-important alarm to ensure that the family woke at the required hour. Modelled on such weight-driven chamber clocks, the lantern clock first appeared in London in the late 16th century. By 1650, it had reached the height of its popularity, heralding the Golden Age of clockmaking which England enjoyed throughout the following century.

Also

• William Bowyer, *Lantern clock, ca.* 1650, London, The British Museum, Dept. of Medieval and Later Antiquities [MLA CAI.2094]

153 Abraham Louis Breguet

Pocket watch No. 1416 with
equation of time
Gold, 1812
Faversham, Harris (Belmont) Charity
[H.285]

Abraham Louis Breguet is arguably
France's most famous horologist and
is considered by some to be the
father of the modern watch. This
gold watch is a splendid and typical
example of his work, having a
delicate yet complex dial with an
annual calendar and an equation of
time indicator. There are two minute
hands, one of plain blued steel for
mean time and a wavy hand with a
gold sunburst for apparent solar
time. The watch has a lever
escapement and, according to
Breguet's records, which still survive
in Paris, was made in 1812.
Breguet's work was always
exclusive, even when he attempted
to make a design, such as the
souscription watch, that was more
affordable by ordinary people. His
styling was so avantgarde and the
quality of design and manufacture so
good that he invariably found
customers who could afford his high
prices.

154 Edward East

Pendulum wall clock
Mixed media, *ca.* 1665
Faversham, Harris (Belmont) Charity
[H.053]

Dating from before the Great Plague
of 1665, this is one of the earliest
English pendulum clocks in
existence. The elegant, yet austere,
ebony veneered case in the plain
architectural style is typical of this
early period, as are the narrow
chapter ring and engraved corner
spandrels on the fire-gilt dial. The
clock, which goes for eight days,
retains original 17th-century brass
cased lead weights. Unlike earlier
clocks, pendulum clocks might stop
during winding and were not self-
starting. They were usually therefore
provided with 'bolt and shutter'
maintaining power. A spring loaded
bolt can be made to drive the wheels
of the clock for about three minutes,
keeping the clock in motion while
the going weight is being wound up.
Edward East is one of the great
names of the Golden Age of English
horology.

155 Thomas Tompion

Travelling clock
Mixed media, *ca.* 1680
London, The British Museum, Dept.
of Medieval and Later Antiquities
[MLA 1986.3–6.1]

One of the best known of all English clockmakers, Thomas Tompion has been dubbed 'the father of English clockmaking'. His clocks and watches were of uniformly high quality in both design and construction. This superb little spring clock was specifically designed for travelling and, as such, is an early predecessor of the carriage clock. Sufficiently small and solid to be transported safely, it would be easy to set up for use after travelling or could even be used during a calm sea voyage or a steady coach ride. The case is made of gilt brass with silver mounts and the movement is quarter striking with an alarm mechanism. The solid silver *champlevé* ('raised field') dial has the alarm-setting disc at the centre and hours and minutes on the main dial. The eight-day movement has a very special feature whereby there are two options for the controller: a pendulum for the best possible timekeeping in a static position, or a balance for use when the clock is being moved or is running during a journey.

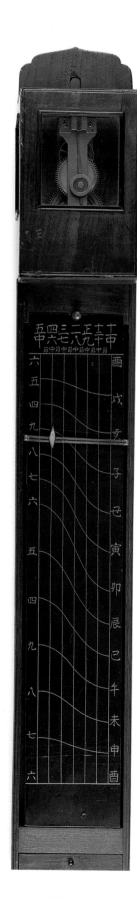

156 Japanese

Thirty-hour pillar clock
Mixed media, *ca.* 1860
Faversham, Harris (Belmont) Charity
[H.238]

The Japanese employed temporal or unequal hours to divide up their day until remarkably recently. It was only in 1873, with the revolution which ended the Tokugawa Shogunate and restored Imperial Japan, that the European system of timekeeping and calendar were introduced. Owing to the fact that, as the seasons change, the temporal hours of night and day constantly vary in length, many Japanese clocks had moveable numbers on the dial so the hand could indicate the hour correctly. Others had two oscillating controllers, one for night and one for day, to indicate the hours correctly. This pillar clock, dating from about 1860, is weight driven and has a descending pointer indicating the time against a *nami-ita* lacquered dial plate, moving from top to bottom in one day. The clock was wound up at the beginning of the Japanese day, which began at dusk. The descending pointer is mounted on a horizontal bar and, over the course of six months, the pointer is moved sideways in small increments about once every two weeks. On the left hand side, in June, at the height of summer, the hour figures are close together at first with the short night hours and lower down with the longer daylight hours. After six months, in December, the pointer will be across on the right hand side, where the daylight hours are shorter than the night hours. The clock is controlled by a little pendulum on the front of the movement, visible in the window at the top.

157 Swiss

Illuminating bedside clock
Mixed media, *ca.* 1901
Faversham, Harris (Belmont) Charity
[H.267]

By the turn of the century the traditional French carriage clock was facing competition from a large Swiss-made eight-day travelling watch known as a 'Goliath'. The Goliath was usually set in a large nickel-plated case, mounted in a silver-fronted easel frame. Its main function was for use on a bedside or dressing table. This version of the Goliath, intended for providing time at night, was patented in Britain by H.V. Manger in February 1901. The carrying box contains two dry cells in the back and has a tiny bulb mounted at the top of the front opening to illuminate the dial.

158 English

Cyclist's wristwatch
Mixed media, 19th century
London, Collection of Peter Gosnell

Wristwatches appeared in isolated examples and as novelty items as early as the 17th century; but the real beginnings of the wristwatch occur in the late 19th century. In the mid-1880s, leather 'watch wristlets' were made as an accessory for ladies. They were, essentially, converted pocket or fob watches, but were specifically marketed as a means of enabling women to become more active, unencumbered by a fob or pocket watch attached to their clothing. Small watches fitted to metal bracelets followed and, by 1910, the wristwatch was commonplace – though chiefly used by women.

159 Germany

Black Forest cuckoo clock
Mixed media, glass, wood, metal,
ca. 1785
Furtwangen, Deutsches
Uhrenmuseum [03–2202]

The Black Forest clockmaking
industry in Southern Germany is
perhaps most famous for the 'cuckoo
clock'. It was the Germans and not,
as many people believe, the Swiss
who invented this delightful idea. The
Black Forest clockmakers, who were
usually also farmers, developed their
craft as a means of occupying their
time and making money during the
snow-bound winter months, when
active farming was not possible.
Black Forest clocks were made from
the late 17th century onwards and

there are examples of early 18th-
century Black Forest clocks with
cuckoo hour-striking. Distinctively,
these clocks were made almost
entirely of wood. This was to ensure
that they were easy and quick to
produce and not because the material
was available, as the soft wood of
the Black Forest itself was unsuitable
and the wood for making these
clocks had to be brought in. Once
spring arrived, a travelling salesman
set off into the rest of Europe, and
especially to Britain, to sell the
completed clocks. They found a ready
market as they were reliable and
attractive, yet could be afforded even
by fairly poor people for timekeeping
in the home. Late 19th-century
examples had alarm mechanisms
instead of striking work.

160 Henry Jones

Night clock
Mixed media, *ca.* 1670
Faversham, Harris (Belmont) Charity
[H.082]

As Henry Jones was apprenticed to
Edward East (see **154**), much of his
work bears strong similarities to his
master's. This eight-day night clock is
similar in style to such clocks made
by East and is contemporary in date.
Its purpose was to provide time in
the bed chamber. The time is
indicated by pierced numerals on the
dial, through which the light of a lamp
or candle shone at night. The pierced
Roman numerals I–III on the upper
part of the dial are fixed and
represent the quarter-hours. The
revolving blued painted disc below

has pierced Arabic numerals to show
the hours and the position of the hour
under the quarter numerals shows
the time. Strangely, the movements
of these clocks were usually hour-
striking, often on a relatively large
bell. This was not a clock for
insomniacs. There would have been a
small shelf on the back of the rear
door for mounting the oil lamp or
candle and the inside of the case
would have been lined with thin iron
plates to form a kind of chimney to
guide the smoke and heat out. As
might be imagined, many of these
clocks caught fire, and are very rare
today.

Until the 1840s, nearly every town in the world kept local time. Local Noon was defined by the passage of the Sun across the meridian. The rest of the hours fell according to whatever timekeeping system was the local practice. For the traveller, this disparity was generally seen as a huge inconvenience. For those who travelled more widely, there seemed to be no end to the confusion. For example, how could an Englishman travelling to Venice hope to know what time it was when he was used to reckoning according to common hours (*horae communes*), while the Venetians used *horae ab occasu solis*, or counted the hours from sunset?

Until the nineteenth century, the Italian day was calculated to begin at sunset. Noon, therefore, occurred approximately 18 hours after the day began (18 hours after the previous day's sunset at around 5:00 or 6:00 PM, depending on the time of year). There was a long-established tradition in clockmaking to show Noon at the top of the dial – a practice derived from the placement of Noon in the centre of the splay of hour-lines in a sundial – and so the Venetians, for example, placed XVIII (or 18 hours after the day had begun) at the top of their clock dials (see **161**). The main inconvenience of having a clock keep 'Italian hours' is that the time of sunset changes throughout the year. In practice, this meant that Italian clocks had to be recalibrated every five to ten days. Each parish or town would publish a set of tables indicating on which days and by how much the clocks had to be reset to ensure that the church bells would ring at the right time and that the hand or hands would indicate Noon when the Sun was crossing the local meridian.

England, too, had its problems. The local time difference between London and Bristol, for example, is nearly twenty minutes. This might not matter if one made the journey by horse or coach, but once the railways were introduced, when following a London-based timetable one could arrive in Bristol nearly twenty minutes early according to local time.

From 1847, the railways used Greenwich Mean Time (GMT) as the standard for all railway timetables in England and in Scotland. But the single event which caused the most concerted push towards a single, standard time was the Great Exhibition of 1851, when over six million people travelled – most of them by train – to see the displays. For many of these visitors, it was their first experience of a standardized time system. By 1855, 98% of the public clocks in Britain were set to GMT, but it was not until August 1880 that GMT was adopted as the single, legal time standard throughout Great Britain.

161 **Antonio Canale (Canaletto)**
View of the Piazza of San Giacometto di Rialto
Oil on canvas, 1726–30
Dresden, Gemäldegalerie Alte Meister [583]

The 'modern' turret clock is derived from the late medieval mechanical clocks used in monasteries. Its primary purpose is to provide time for the community. Its characteristic feature is a large and clearly marked dial. One suspects that very few of the tourists visiting Venice have noticed that almost all of the clock dials of the city are marked according to 'Italian hours', including the well known dial on the façade of the Torre dell'Orologio in the Piazza San Marco. Each of these dials is marked for 24 hours, rather than the more conventional 12 hours. The first hour of the day is located at a position that would normally indicate 3:00 – at the right-hand side of the dial. At the top of the dial (where 12:00 would be normally), there is the Roman numeral XVIII.

162 Italian

Table clock with six-hour dial
Mixed media, *ca.* 1730
London, The British Museum, Dept.
of Medieval and Later Antiquities
[CA1–2191]

Developed from the Roman system
of dividing the day into four periods,
clocks and watches with six-hour
dials enjoyed a certain popularity in
Italy during the 17th and 18th
centuries and a number of public dials
also employed the six-hour system.
Hours were of the normal length, but
the hand circuited the dial four times
instead of twice in 24 hours. Such
dials were considered to be easier to
read and, given only an hour hand,
made a more precise estimate of the
time possible. As the striking system
employed in these clocks also
followed a six-hour system (1–6 four
times) it saved a considerable amount
of energy. This anonymous two-day
going table-clock movement, which
no longer has its case, also strikes
the quarters (on a higher bell), and
has a repeating function, operated by
pressing a button at VI on the dial, to
strike the previous hour and quarter.

163 George Elgar Hicks

*The General Post Office: One Minute
to Six*
Oil on canvas, 1859–60
London, The Museum of London
[90.276]

This painting perfectly epitomizes the
degree to which mid-Victorian society
had become more conscious, and in
turn fascinated by the precise timekeeping that increasingly governed
daily life. The scene represents the
Grand Public Hall at the General Post
Office in St Martin's le Grand,
London. It shows a wide spectrum of
London society crowding to catch the
last post at six o'clock, which,
apparently, had become something of
a tourist attraction. Since 1855,
clocks at the main London post
offices had been corrected every day
by electric telegraph signals from the
Royal Observatory at Greenwich.

Also

• Standard Time Company, *Lund
Patent Synchronized Clock, ca.
1880*, London, The National
Maritime Museum [ZBA0673]

164 Benjamin Lewis Vulliamy

*Pocket watch, with table of local
times and two minute hands*
Gold case, 1847
London, The National Maritime
Museum [CH 616]

This 18-carat gold pocket watch was
made in 1847–48 – a period when a
serious conflict had developed
between the use of local times
across the British Isles and the
adoption of new rapid communication
and transport systems, in the form of
the electric telegraph and the
railways. By 1840, the Great Western
Railway was using Greenwich Mean
Time rather than local times for all its
timetables. Local communities across
the country, however, were still
setting their watches to their own
local time. Thus, it was useful to have
a watch with two minute hands,
which could be set independently,
one to GMT and the other to local
time. Both hands then moved
together round the dial. Inside the
back of the watch case, there is an
engraved table of local time
differences between Greenwich and

a number of different towns and
ports, both in the British Isles and on
the Continent. GMT is shown as a
central dividing line with towns to
the east (such as Dover, Boulogne,
Paris, Brussels and Hamburg) shown
to the right and towns to the west
(including Liverpool, Oxford, Land's
End and Dublin) to the left. Vulliamy,
who retailed the watch, was
clockmaker to the Queen.

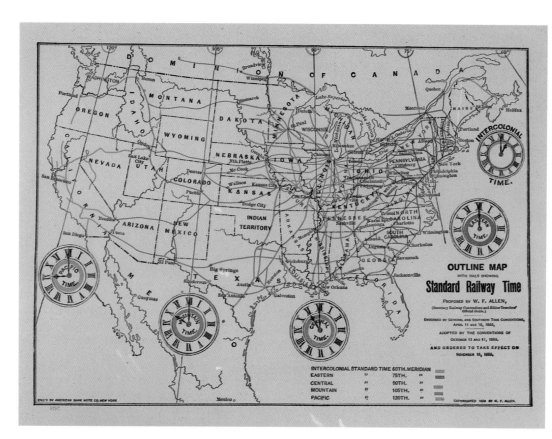

165 After Charles F. Dowd

Standard map of North American time zones
Paper, 10 November 1883
New York, New York Public Library,
Manuscripts and Rare Books
Division, William F. Allen Papers

This map was published in 1883 to show the exact location of the United States time zones as proposed by Charles F. Dowd. Dowd had first proposed a national standardized time zone system for the United States in 1870. The fact that the United States was already successfully operating a similar national system was surely decisive in the adoption of an international time zone system in 1884.

Also

• The United States, *Standard Time broadsheet*, 10 November 1883, New York, New York Public Library, Manuscripts and Rare Books Division, William F. Allen Papers [P. 88975]

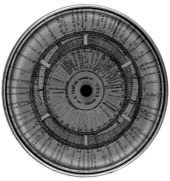

166 Cesare Pascal

Universal time-finder
Wood, glass and metal, *ca.* 1900
London, The National Maritime
Museum [AST0700]

Designed and patented by Cesare Pascal, the universal time-finder enables the user to find the relative local times for any place on the globe. In essence, it is a simple revolving table of world cities with their longitudes, printed on a paper disc, with a separate rotatable 12-hour paper disc beneath. The cities are separated into east and west hemispheres (the inner and outer parts of the disc respectively) which have been defined according to the Paris meridian, despite international acceptance in 1884 of the Greenwich Meridian as the Prime Meridian of the world. In fact, France did not officially recognize the Greenwich Meridian until 1911, preferring to keep *l'heure nationale*, which was 9 minutes and 21 seconds fast on GMT. In 1896, *l'heure nationale* was diminished by 9 minutes and 21 seconds (making it equal to GMT), but it was a further fifteen years before the word 'Greenwich' was actually included in the French legislation.

Also

• S. Smith & Son Ltd., *Globe clock*, *ca.* 1910, London, The National Maritime Museum [GLB0062]

In October 1884, the International Meridian Conference was convened in order to establish a body of timekeeping conventions for the world. Forty-one delegates from twenty-five different nations assembled in Washington, D.C. The majority were professional diplomats, though some members were invited to attend the conference by virtue of their renown as scientists or technical experts. The first act agreed the desirability of a single, prime meridian. The second act – after much heated debate – decreed that the governments represented would adopt the meridian "passing through the centre of the transit instrument of the Observatory of Greenwich as the initial meridian for longitude".

The fourth act established the desirability of a universal day, which was defined – in the fifth act – as a mean solar day, which could begin "for all the world at the moment of mean midnight of the initial meridian" (*i.e.* Greenwich) and which would be counted from 0 to 24 hours. The sixth act voiced a desire that these conventions would be extended to the measurement of astronomical and nautical days. And, finally, as a patently bizarre addition (possibly included as some form of appeasement to the French, who were clearly irritated at having lost the prime meridian), the last act expresses the hope that "all technical studies designed to regulate and extend the application of the decimal system shall be resumed". These applications have yet to make themselves apparent.

167 French

Decimal time pocket watch
Gilt and enamel, *ca.* 1800
Paris, Musée Carnavalet [OM 112]

Made at the beginning of the period in which France adopted the Revolutionary decimal calendar and time system (see **073**), this silver-cased watch indicates both decimal and conventional hours and minutes. The dials on the left show the decimal time and the calendar. In common with many such dual-function Revolutionary watches, the enamel dial depicts the female figure of France wearing a Phrygian cap to symbolize Liberty and holding scales to represent Justice. The figure is always shown facing and reading the decimal dial, with her back turned on the old-style duodecimal system indicated by the right-hand dials. As the decimal system was only in force in France for thirteen years, such watches are now very rare. The movement of this example is unsigned and the decimal minute hand is now missing.

168 Bruel à Paris

Decimal time pendulum clock
Marble, gilt brass and enamel,
ca. 1795
Paris, Musée Carnavalet [MN 627]

Made just before the turn of the 18th century, this clock was designed to indicate both duodecimal time (the conventional 12-based hours and minutes) and decimal time, imposed in Republican France in 1793. The day was divided into 10 hours, each of 100 minutes, each minute having 100 seconds. A new Republican calendar was introduced at the same time, keeping 12 months in the year but dividing each month into three *decailles* of ten days each (see **073**). The decimal system was officially in use for thirteen years, but it was never popular. It was finally abandoned in 1806. This clock strikes the duodecimal hours and half hours and beats half seconds with a central second hand. The whole clock is of superb quality and, with its elegant proportions and fine enamel dials and decorative plates, is a masterpiece.

2 · THE MEASUREMENT OF TIME

169 Thwaites & Reed
Mail guard's watch
Mixed media, 1850s
London, The National Maritime
Museum [ZAA0269]

From 1840, the Great Western
Railway ran all its timetables on
Greenwich Mean Time, having
rejected any consideration of local
time owing to the great confusion
this had been causing to passengers.
Other railway companies gradually
followed suit and, by 1850, the
majority ran to GMT. In 1880, GMT
was officially introduced as the legal
time for the whole country.
To ensure that trains ran to the
correct time, particularly those trains
carrying Her Majesty's mail, the
railway companies issued their
guards with watches. They were
made to a standard plan, being
mounted on a block of mahogany
with a brass front, and lockable so
that the hands were tamper-proof.

170 William Willett
Daylight Saving pamphlets
Paper, 1907
London, The National Maritime
Museum

It was probably for industrial reasons
that William Willett first promoted
the idea of 'daylight saving'. In his
publicity material, however, he
emphasized that it would improve
the nation's health and happiness. As
the Sun rises before 4 am in
summer, most people sleep through
the first two or three hours of
sunlight – a waste of its useful and
health-giving properties. Putting the
clocks forward one hour would give
an 'extra hour of daylight' in the
evening. Willett was so convinced of
the value of his idea that he
canvassed it internationally.

171 Thomas Tompion
Workman's clock
Mixed media, *ca.* 1680
The National Trust (Llandudno,
Powys Castle) [POW.H.06]

This clock – one of only two known
to exist – was designed by Thomas
Tompion specifically "for the use of
the workmen". It was intended for
use by watchmakers, so that they
could bring watches 'to time' during
construction in the workshop. The
time of day cannot be told by the
workman's clock, which has only a
four-hour dial, but with its short, fast-
beating pendulum (beating three
times every second) it is very useful
for short-term timing exercises.
Tompion was the first maker to
produce watches in a highly
organized and structured system.

172 J.L. & W. Smith for Thomas Larard

Eight-day night watchman's clock
Mixed media, mahogany case,
ca. 1860
Faversham, Harris (Belmont) Charity
[H.096]

'Noctuary' or 'detector clocks' ensured that the night watchman made his regular tour of the building at the appointed times. This device was invented about 1790 by William Strutt and Erasmus Darwin of Derby. A revolving hour disc, visible round the edge of the main dial, has pins projecting from its periphery at half hour intervals. This disc revolves once in 12 hours (with the hour hand). Every half hour, one of the pins comes underneath a plunger mounted on top of the case. It is the watchman's job to push the plunger down to depress each pin, every half hour throughout the night – proving that he was in that part of the building when he was supposed to be. By morning, all the pins for the night-time hours should have been pushed into the disc. Normally, in a large establishment such as a factory or a prison, there would be many of these clocks fixed at various remote places around the building and they would be locked, to prevent any tampering with the dial. Just before the appropriate pins are due to be depressed, they are automatically pushed out again by a piece behind the dial in readiness for the next record.

173 The United States

International Time Recording Catalogue
Printed book, 1914
Washington, Smithsonian Institution Libraries [Uncat. C 681.118 I 572.1914 NMAH]

The logical extension of the noctuary – the night watchman's clock – was a clock to record the exact timing of employees arriving and leaving work. Although potentially a means of pressurizing the workforce to ensure close timekeeping, such clocks had an advantage for staff in that they would also record additional hours worked and, hence, overtime payments due. Earlier examples of the clocking-in clock employed a system of workers' tokens which, when dropped into a slot in the clock's front, collected in separate quarter-hour compartments, disclosing the time of arrival. But a clock, using a printed card such as this illustration depicts, was the more precise and logical development of such devices.

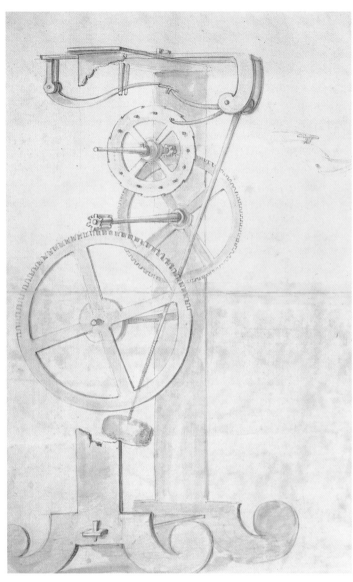

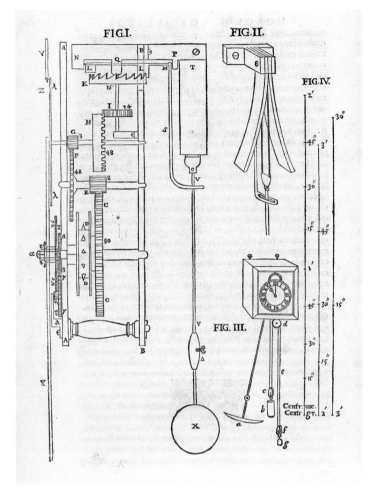

174 Vincenzo Viviani

Galileo's pinwheel escapement
Drawing on paper, *ca.* 1659
Florence, Biblioteca Nazionale
Centrale [Ms Galileiano 85, c.50]

The invention of the pendulum as the timekeeping element in clockwork is arguably the most important single advance in the whole history of horology to date. Although there is some evidence to suggest that Leonardo da Vinci · originally conceived the idea of the pendulum as a timekeeping device, there is no evidence that he ever pursued this idea and made a mechanism. The primary credit for the invention of the pendulum goes to Galileo Galilei. According to Vincenzo Viviani, one of his disciples, Galileo conceived the idea in 1641, but it was not until April 1649 that Galileo's son Vincenzio made a working model of the clock. This drawing seems to have been made by Viviani in 1659, based on an earlier incomplete design by Vincenzio and on the model Viviani had been shown. But we cannot say that Galileo made the pendulum clock a practical reality.

175 Christiaan Huygens

Horologium Oscillatorium
Printed book, Paris 1673
London, British Library, Dept. of
Printed Books

Christiaan Huygens was the man who created the first practical pendulum clock design. In an earlier work on timekeeping, the *Horologium* of 1658, Huygens described how he completed his design for a pendulum clock in late 1656 and assigned the rights to it to the Hague clockmaker Salomon Coster, in whose name the patent was taken out on 15 June 1657. Huygens's pendulum design was first illustrated in his later work, the *Horologium Oscillatorium* of 1673, the earliest modern treatise on the science of pendulum theory. In 1675, when Huygens applied the balance spring to watch work (as did Tompion for Robert Hooke in London), the improvement in timekeeping was found to be so good that comparisons were made with the application of the pendulum to clockwork. Watches were not, of course, normally fitted with pendulums as such – but there were a small number of such experimental pieces made. The Cremstorff movement, for example, is from a contemporary watch which has been altered to pendulum control.

Also

• Joachim Cremstorff, *Watch with a pendulum attachment*, 1700, Dresden, Staatliche Kunstsammlungen, Mathematisch-Physicalischer Salon [D. iv. a. 167]

176 Johannes Steffens

Haagse clock
Mixed media, 1674
Dresden, Staatliche
Kunstsammlungen, Mathematisch-
Physicalischer Salon [D.iv.b.21]

This is a fine example of an early
Dutch pendulum clock of the type
known as a 'Haagse clock' (a clock
from The Hague). The wooden case
and velvet covered dial, with the
movement attached to the back of
the dial and swinging out with it, are
typical. A number of technical
features of the movement follow on
from Renaissance designs. The single
winding hole of this clock conceals
the fact that it strikes the hour, the
going and striking parts being driven
in tandem off one spring, as often
occurs in earlier clockwork. The

increased accuracy is reflected in the
use of a large, prominent dial with a
minute hand.

Also

• Raingo, *Orrery clock, ca.* 1820,
Faversham, Harris (Belmont) Charity
[H. 201]

• Brockbank, Atkins & Moore, *No.
11,986, Beachy Head Lighthouse
clock, ca.* 1860, London, The
National Maritime Museum
[ZAA0127]

• James Mitchell, *Tidal clock/recorder,
ca.* 1740–1830, The National
Maritime Museum, [ZBA 1429]

• Everett, Edgcumbe & Co.,
Synchronous clock, ca. 1929,
Cambridge, collection of Bob Miles

177 Junghans

Radio-controlled clock
Mixed media, 1991
London, The National Maritime
Museum [CH 627]

Since the earliest days of radio
communications, it was realised that
radio waves are an ideal means of
distributing precise time. From
January 1905, the US Navy broadcast
regular time signals at Noon (Eastern
Standard Time) from Washington,
D.C. Germany and France followed
suit in 1910, the French signal being
transmitted from the Eiffel Tower at
midnight. These early time signals
required a special receiver; but in
1922, with the first public radio
broadcasts by the British
Broadcasting Company, the
opportunity was taken to provide

precise time, with the announcer
playing the Westminster chimes on a
piano. Two years later the BBC
introduced its six-pip time signal, the
brainchild of electric clock
manufacturer Frank Hope-Jones and
the Astronomer Royal, Sir Frank
Dyson. From the early 1970s,
battery-driven quartz clocks have
been made which automatically
correct themselves, on a daily basis,
using one of the digitally coded time
signals now broadcast across the
world. This example, made by the
German company Junghans, is tuned
to receive the National Physical
Laboratory's MSF 60 kHz time signal
broadcast from Rugby. Models are
also made which receive the German
signal from Frankfurt, the range of
which is about 1000 miles.

178 Daniel Quare

Double longcase clock with solar and sidereal time
Mixed media, *ca.* 1700
London, The National Maritime
Museum [ZBA 0474]

With the great strides made in astronomy, horology and the natural sciences in the 17th century, clock designs developed to show off this new-found information. The Quare double clock, for example, provided both mean solar time (clock time) and sidereal time (star time), by means of two entirely separate pendulum clocks both set within the same case. Recent advances in astronomy were represented by the inclusion of an equation of time dial, indicating the difference between apparent solar time (sundial time) and mean solar time.

Also

• George Graham, *Astronomical Regulator*, 1725, London, The National Maritime Museum [new acquisition]

179 Abraham Louis Breguet

Double pendulum regulator
Mixed media, 1825
Lent by Her Majesty the Queen
[2767]

Breguet's fertile brain was constantly searching for means of improving timekeeping in both his watches and clocks. One of his most extraordinary ideas – but one with considerable scientific justification – was to provide a regulator with two entirely independent movements, controlled by pendulums swinging in anti-phase, one in front of the other. The considerable disturbance caused by an ordinary pendulum to its mounting support during operation is virtually eliminated by having the equal and opposite actions of the two pendulums. Moreover, there is a tendency for the two pendulums to stay in anti-phase, resisting any change of phase by either pendulum and thus averaging their timekeeping.

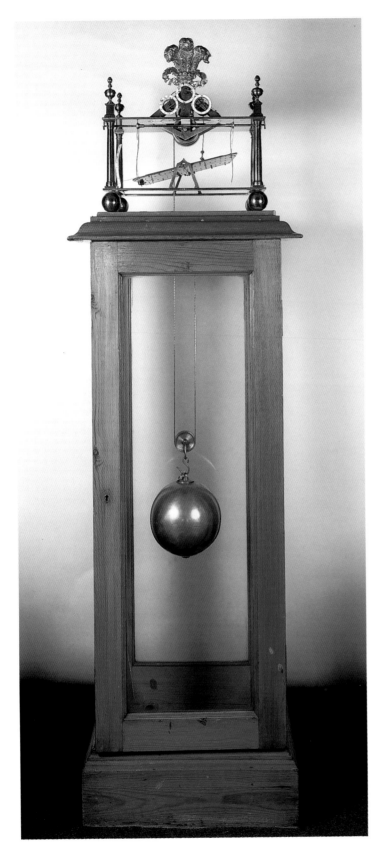

180 William Congreve
Rolling-ball clock
Mixed media, 1808
London, Royal Artillery Historical Trust
[IDRAI 9/37]

Perhaps one of the most famous of
all novelty clocks, this timekeeper
was actually intended as a serious
precision clock. William Congreve,
who enjoyed the splendid title of
'Superintendent of Military Machines'
at the Royal Laboratory at Woolwich,
specialized in developing designs for
early military rockets and was anxious
to have as accurate a timekeeper as
possible for his work. In 1808, he
patented a number of designs for
precision clocks, but the rolling-ball
clock is the most famous. This is the
very first example ever made,
Congreve's own prototype. The little
steel ball rolls down the zigzag track,
supposedly taking one second exactly
for each part, and 15 seconds to
traverse the whole track. On reaching
the end, the ball releases the
clockwork mechanism above, via a
pivoting catch, and the mechanism
tilts the tray the other way, sending
the ball rolling back in the reverse
direction. As a precision timekeeper,
the clock is hopelessly inaccurate,
owing to the friction of the ball on the
tray. Its inventor must have been very
disappointed with it.

181 William Hardy
Precision interval timer
Mixed media, ca.1825
London, The National Maritime
Museum [ZBA 0668]

There are many scientific purposes
for which very precise, short-term
timing is needed. In his description
of this invention, Hardy tells us that
it would be useful for measuring
"... the occultation of the stars and
the velocity of sound ...". This
instrument was in use at the Royal
Observatory Greenwich and was
probably purchased for very precise
timing of star transits. Pressing the
button on the front of the timer sets
it in motion and releasing it stops it
instantaneously. The timer shows
seconds on the main control dial.
The sector dial below the centre has
a hand attached to the balance itself,
the sector scale being divided into 60
equal parts each representing $1/600$ of
a second. However, Hardy has failed
to recognize that the motion of the
balance, like a pendulum, is
sinusoidal and that the divisions on
this scale should be spaced
accordingly, with those at the ends
of the scale much closer together
than those at the centre.

2 · THE MEASUREMENT OF TIME

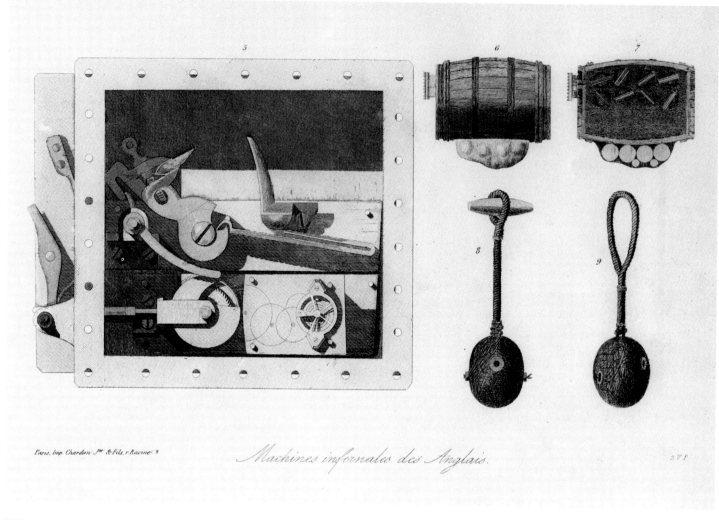

Machines infernales des Anglais.

182 French

Print of a *Timing device for a time-bomb*
Paper, *ca.* 1810
London, collection of Jonathan Betts

As a result of the success of John Harrison's H4, a much smaller scale of timekeeper was adopted and experimented on, both in France and in Britain. Expeditions of exploration were perfect opportunities to test newly developed chronometers. Captain James Cook's second and third voyages to the South Seas in the 1770s, for example, were used to test marine chronometers. Ironically, timekeeping devices in warfare are often employed to save lives, by allowing those planting weapons time to get clear of the explosion. An unexploded example of an English time-bomb must have fallen into French hands as this French print is exactly contemporary with the bomb itself. Such mechanisms were primarily used by frogmen to destroy enemy ships. They were designed to release a flintlock on to a powder charge a number of hours and minutes after the clockwork was set. The complete device was sealed in an airtight box with the powder charge and was then securely fixed to a barrel of gunpowder and metal scrap. The metal scrap was intended to disintegrate into shards, similar to a cluster bomb, causing maximum damage and fatalities. It appears that the earliest English time-bomb devices were made to the designs of the original inventor, Robert Fulton, an American engineer, in about 1804. More than a century and a half later, extremely precise timekeeping devices were needed to detonate a nuclear bomb or missile. Devices were designed to ensure the missile detonated just above ground level, achieving maximum damage and loss of life. The 'chevaline' nuclear warhead is now officially obsolete and virtually all of these devices have been destroyed.

Also

• English (possibly London), *Timing device for a time-bomb*, ca. 1810, London, The National Maritime Museum [EQS 0343]

• English, *Timing device from the 'chevaline' nuclear warhead* (fitted to the Polaris missile), 1982, private collection

• Les Frères Goyffon, *Experimental marine chronometer*, Paris hallmarks for 1775–76, Paris, Musée des Arts et Métiers [21962]

• Parkinson & Frodsham, *Pocket chronometer* (issued to Sir John Franklin for the North West Passage Expedition), 1810, London, The National Maritime Museum [AAA 2203]

183 English (London)

First World War wristwatch with protective cover
Mixed media, 1916
London, collection of Jonathan Betts

By the early 1900s, purpose-made watch bracelets had become an established fashion accessory for ladies. They were not taken up quickly by men, being seen as somewhat effeminate. The practicality of the wristwatch became apparent during the First World War. The trench warfare of 1914–18 made timing critical. If watches were not reliable and precisely synchronized, the agreed moment for 'going over the top' could be delayed, with disastrous consequences. It seems that many lives were lost during the Gallipoli campaign because watches were not correctly synchronized and there was a critical delay of several minutes between the cessation of shelling and the push forward by troops, allowing the Turkish machine guns to be remanned. In the early wristwatches, reliability was not the only problem. Breakage of the fragile glass covering the dial was another. Some, like this silver-cased example had hunter-type, spring-activated covers to protect the dials.

184 W.H. Douglas

Fire control chronograph (deck watch)
Mixed media, *ca.* 1900
Private collection

Part of Admiral Fisher's programme of improvements to the British Navy was his requirement that the naval guns should be able to "hit a thimble on the horizon". Developing such high precision shelling required the use of highly accurate chronograph watches such as this, able to continue running during the three functions of timing with the seconds hand, stopping the hand and 'flyback' to the zero position. Douglas took out several patents for inventions of this type. His deck watch, however, was never widely used, and was soon superseded by the simple stop watch.

185 Seth Thomas

Zigzag control ship's clock
Mixed media, 1942
London, The National Maritime Museum [CH 250]

To avoid submarine torpedoes in the North Atlantic during the Second World War ships followed an irregular 'zigzag' course. The zigzag control clock rang a bell when it was time to change course. A random pattern might have endangered any accompanying ships, so a number of specific zigzag patterns were stipulated in the Royal Navy's *Book of Reference*.

Mechanical Time: Sumptuous Clocks

186 Attrib. Pierre Mangot

The Royal Tudor Clock Salt
Silver-gilt, enamel, cameo and semi-precious stones, *ca.* 1530
London, The Worshipful Company of Goldsmiths [9]

Almost certainly made by one of the French court goldsmiths, this richly decorated and impressive table ornament was ostensibly a 'salt', the most important single object on the dining table. Since Roman times, salt was symbolic of high value; therefore, one's position at table 'above' or 'below' the salt denoted one's status. This salt is believed to have been a gift from King Francis I of France to Henry VIII. The piece is certainly one of three items of plate known to have survived from Henry's Jewel House. It was one of eleven clock salts in the Royal Collection in 1550. The hexagonal base originally contained the movement and the horizontal dial would have been viewed through the rock crystal cylinder in the centre of the clock. The object has undergone several alterations over the years and the present 18th-century, quarter striking movement is now contained within the crystal cylinder. The salt's finial cover is a recent restoration.

187 Probably English

Emerald containing watch, from the 'Cheapside Hoard'
Emerald, gold and enamel,
ca. 1575–1610
London, The Museum of London
[A 14162]

In 1912, an extraordinary collection of jewellery and jeweller's material was unearthed in the foundations of a house in Cheapside, near St Paul's Cathedral. Among the artefacts was this watch movement set into a solid hexagonal crystal of Colombian emerald. The dial has gold Roman numerals and is enamelled in translucent dark green, through which can be seen decorative engraved radiating lines. The movement is wound through a hole in the dial and the whole dial is covered by a hinged lid of faceted emerald set with a gold joint and catch. Unfortunately, the movement is too corroded to be removed from the case without damage, so no details of its construction are available. The watch would probably have been worn round the neck, suspended on a cord or chain from the white-enamelled gold loop on the top of the case, or it may have had a shorter attachment for holding in the hand or keeping in a pouch at the waist. The Cheapside Hoard is believed to have been part of the stock-in-trade of a 17th-century jeweller, or possibly a pawnbroker, and was hidden about 1640, perhaps because of the political uncertainty of the times.

188 **Philipp Trump**

Table clock with lion
Mixed media, *ca.* 1630–40
Munich, Bayerisches
Nationalmuseum [74 / 203]

This automaton table clock is hour
and quarter striking and indicates the
24-hour day on the rotating spherical
dial. A 12-hour dial is on the base,
along with a quarter-hour indicator
and a one to six indicator for the hour
striking. The lion is connected to the
mechanism; it moves its eyes in
synchronization with the going of the
clock and moves its jaws with the
hour striking. Automata have regularly
been a popular by-product of clock
manufacture. The simulation of the
movements of living creatures,
effects of nature and the creation of
'mysterious effects' by intricate
gearing have always found favour
with the public. One common type –
the 'mystery clock' – appears to work
and tell the time in a mysterious way.
In Houdin's clock, for example, the
single hour hand moves around the
completely clear glass dial with no
visible connection with any other part
of the clock. In fact, the base
contains the clock movement and the
dial, which appears to be one sheet
of glass with the Roman numerals on
it, is actually two sheets. The second
sheet has the hour hand attached to
it. A drive from the movement, in a
tiny shaft up the casing of the clock,
engages with a thin toothed rim on
the second glass disc and the hand is
thus made to circuit the dial
apparently without connection.

Also

• Jean-Eugène Robert Houdin,
  Mystery clock with glass dial,
  ca. 1830, Faversham, Harris
  (Belmont) Charity [H.147]

189 Adrian de Bergh (movement) and Hans Conraedt Breghtel (case)

Astronomical clock
Embossed silver, *ca.* 1600
London, The Victoria and Albert Museum [92–1870]

This enormous, richly embossed and chased solid-silver clock is technically quite simple. It has dials on three sides but all indicate a 12-hour cycle – the front dial having an hour and minute hand, the other two simply showing hours. The eight-day movement, which is signed by the maker *Adriaen van den Bergh Fecit Hague*, also strikes the hours on a bell in the top of the cupola. The case, surmounted by a figure of Father Time, is signed by the maker on an inner door to the movement: *A. Breghtel Fecit Hague*. Perhaps more than any of the other clocks displayed here, the importance of this piece rests solely in the decoration of its case. The timekeeping aspect of the clock is incidental. It was undoubtedly crafted for a wealthy patron for whom a conspicuous display of wealth was paramount.

190 South Germany

'Orpheus-type' table clock
Mixed media, *ca.* 1560–70
Munich, Bayerisches Nationalmuseum [33/125]

This is one of a group of German Renaissance table clocks, collectively known as 'the Orpheus clocks' because they all have an elaborate cast and chased gilt-brass frieze which depicts Orpheus and Eurydice at the entrance to Hades, surrounded by a number of animals (alluding to Orpheus's ability to charm the beasts with his singing). The dial on top of the clock is a planisphere. It indicates the hours and minutes on its outer edge and the position of the Sun and Moon. Beneath the hands, there is a sidereal disc, similar to a *rete* on an astrolabe, but without the star pointers. The dial plate beneath is engraved with two quite distinct sets of lines: the first for unequal hours and the second for Italian hours. The movement of this clock is exceptionally fine, having a most unusual form of blued and etched decoration on the steel back plate. It is quarter striking and has an alarm mechanism.

191 Attr. Jeremias Metzger

Astrolabic clock

Mixed media, *ca.* 1580

London, The British Museum, Dept.
of Medieval and Later Antiquities
[MLA CA1–2097]

This is one of a group of astrolabe
clocks dating from the second half of
the 16th century. Its movement is
probably the work of Jeremias
Metzger of Augsburg, though it may
have been constructed for another
maker to put in a case of his own
design. One of the main dials of the
clock is an astrolabe, with the fixed
latitude plate behind designed for
51°N. The dial has 24 hour numbers
and touch pieces for reading time
from the Sun hand by day and night.
The lower left dial is for the days of
the week. The dials on the right and
left sides are for the striking
indicators – quarter striking on the
right. The other side of the clock has
a manual dial for setting the epact, a
12-hour dial (now disconnected) and,
the main dial on this side, a larger
12-hour dial. This was probably
originally a 24-hour dial with a
sunrise and sunset feature, as found
on other movements by Metzger.
The case of the clock is original
except for the later dome decoration
and feet.

Also

• Caspar Buschmann, *Table clock*,
  1586, London, The National
  Maritime Museum [ZAA 0011]

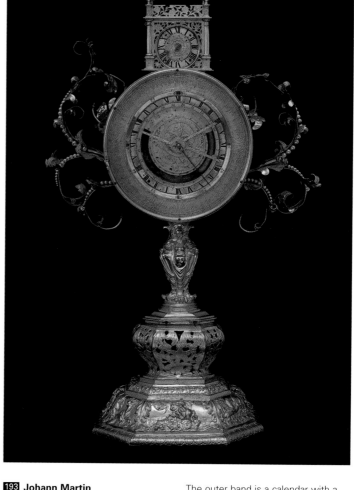

192 'A.S.'

Table clock
Mixed media, *ca.* 1600
Munich, Bayerisches
Nationalmuseum [67/41]

This quarter-striking clock has a fire-gilt brass and bronze case and has dials on all four sides. One side has an astrolabe dial scribed for the latitude of 50°N – (Frankfurt or Prague?), with the usual Sun, Moon and dragon hands for indicating eclipses. The dial below the astrolabe indicates the day of the week by associating ruling planet-gods with each day. The dial on the opposite side has a 24-hour dial with changing day and night lengths and with indication of hours and minutes.

Below this dial there is a smaller one for the actual length of day and night. On the other two sides are calendar dials (six months for each scale) including saints' days, the Dominical Letter, the Golden Number, the epact and dates of Easter from 1600 to 1687. Below these two dials, there are dials for regulation and alarm setting and indicators for the hour and quarter striking. Clocks of this type are the mechanical equivalent of the complex polyhedral sundial, in which numerous faces provide a wide range of information. Not surprisingly, this clock was probably made in Augsburg, the town in which so many of the elaborate gilt-brass dials were manufactured.

193 Johann Martin

Monstrance clock
Mixed media, 1669
Munich, Bayerisches
Nationalmuseum [R 3371]

This fine quarter striking clock was the masterpiece constructed by Johann Martin when he sought permission to work in Augsburg in 1669. The central astrolabe dial is scribed for latitude 48°N (the latitude of Augsburg). Around the outside of the astrolabe dial there is a circle indicating hours of daylight and darkness (now disconnected), and beyond this there are circles for the lunar calendar (from the astrolabe's Moon hand), 24 hours (from the astrolabe's Sun hand) and minutes.

The outer band is a calendar with a peripheral pointer. The dial at the top is the 12-hour dial. On the face of it, the combination of reliquary and clock might seem an odd one. It would seem to support the view that, in this instance, the concept of a 'clock' as a timekeeper was less important than its role as a luxury goods item. Clocks were expensive commodities – especially when they were gilt and bejewelled. As such, they were seen as appropriate vessels for other 'valuable' objects, such as fragments of a saint's bone or a piece of the True Cross.

194 **Isaac Habrecht III**
Clockwork celestial globe
Mixed media, 1646
London, The National Maritime
Museum [GLB O174]

Dating from 1646, this celestial globe
is modelled on earlier clockwork
globes of the great South German
makers. It is also very fine – in the
quality of the globe, the mounting
and in the mechanism – and is as
well made as any from the 17th
century. Nevertheless, it is a late
Renaissance piece and is relatively
old-fashioned for the date. The
accuracy with which the gearing
mechanism drives the sidereally
rotating globe, and the Sun and
Moon bands which revolve around it,
is no better than globes made a
hundred years earlier. In fact, the
globe itself is somewhat older than
1646, apparently having been fitted
by Habrecht with the movement and
mounting perhaps 40 years after its
original manufacture. The dial at the
top of the globe shows local mean
time and the phases of the Moon.
The quarter striking movement with
verge and balance-wheel control
(inside the globe) is wound through
the southern axis of the globe with
an ingenious two-position winding
square. There was originally an
orientation compass mounted at the
top of the meridian ring. The globe
can be positioned in relation to the
horizon ring, according to the latitude
of the viewer, so that, when
correctly set and orientated properly
with the compass, the position of
the celestial globe directly reflects
the night sky, thus aiding the viewer
in his identification of the brighter
stars.

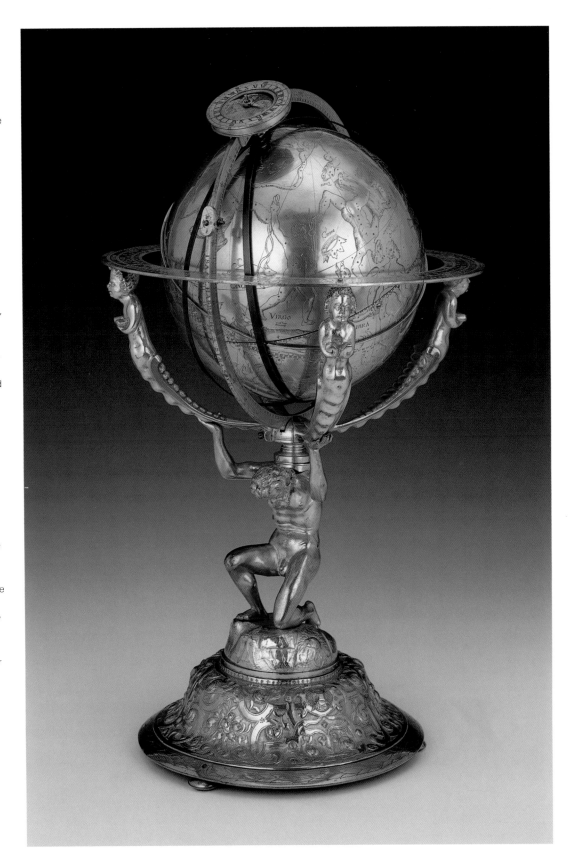

195 **William Hughes**

Musical chaise watch with automata
Gilt brass, enamel, glass, crystal and
gold, *ca.* 1790
National Trust (Anglesey Abbey)
[AA/H/54]

Watches and clocks of this kind
were made specifically with the
Middle and Far Eastern markets in
mind. The principal customers for
such articles were East India
Company officers, who needed them
as 'gifts' to persuade the Chinese
merchants and customs officials in
the Far East to facilitate trading and
to enable Company representatives
to buy the silks, porcelain and spices
they needed at home. Watches and
clocks which appealed most to the
Chinese were those with highly
decorative cases, with musical work,

and especially with plenty of action
on the dials, including fast moving
second hands and automata. This
chaise watch is an excellent example
of the genre and was apparently
originally the property of the emperor
of China himself. In the late 19th and
early 20th century, the watch
subsequently belonged to a
renowned horologist, Percy Webster.
The engraved gilt metal case is set
with coloured pastes and has a
Bilston enamel plaque inset in the
back. The white enamel dial has an
automated scene of figures crossing
a bridge over a waterfall, which
operates while music plays on six
bells. The lower dial shows hours
and minutes, with a central second
hand. The engraved gilt brass
movement has a verge escapement.

196 **Attr. Workshop of Jaquet-Droz**

*Hanging birdcage clock with
automata*
Brass, enamel, porcelain and
feathers, end of 18th century
LeLocle, Château des Monts, Musée
d'Horlogerie [no. 11]

There have been clock and
watchmaking workshops scattered
amongst the mountains near
Neuchâtel from at least the beginning
of the 17th century. Clocks and
watches from the areas around
Neuchâtel, such as LeLocle and La
Chaux-de-Fonds, are based on a
series of distinctive horological traits:
an interest in precision timekeeping
and miniaturization and a fascination
with clockwork automata. All these
features are brought together within
the framing aesthetic of an almost

romantic dedication to decoration.
The abundant use of precious metals,
gemstones, pearls, crystals and
enamels is characteristic of
Neuchâtelois timepieces. Many
clocks contain astonishingly
complicated automata. In this case,
the clock resembles a bird cage and
is meant to be hung from the ceiling
of the room. The base of the cage
contains a simple enamelled dial that
is intended to be read from below.
The bird can be programmed to sing
seven different tunes on a small
organ with ten pipes. The bottom part
of the birdcage is decorated with
small enamel plaques depicting
landscapes. These are interspersed
with miniature porcelain vases and
statuettes set into niches.

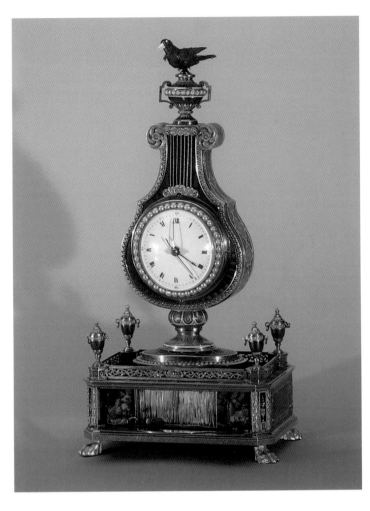

197 Pierre Jaquet-Droz and Jean Frédéric Leschot
Miniature table clock with automata
Gold, enamel, pearls, glass, brass and feathers, end of the 18th century
LeLocle, Château des Monts, Musée d'Horlogerie [no. 26]

Typical of Neuchâtelois manufacture, this small, lyre-shaped clock is constructed from gold, parts of which have been enamelled or inset with miniature pearls. The automata are activated every hour on the hour, when the bird moves up and down and appears to sing and, in the front of the rectangular base, two small panels open and the effect of a shimmering waterfall is created by the rotation of a series of twisted rods of glass. At the same time, a swan traverses the scene, gliding across the surface of the pond. When it arrives at the left of the scene, it dips its head into the 'water', and the automation stops until the next hour is struck.

Also
• French (Paris), *Singing-bird clock*, *ca.* 1880, Faversham, Harris (Belmont) Charity [H.154]

198 Parmigiani Fleurier
Hunter watch, 'Leda'
Pink gold, white gold, rhodium, crystal, sapphire and 29 jewels, 1996
Fleurier, Parmigiani Fleurier

The tradition of hand-crafted precision clock and watchmaking still survives in the Neuchâtel region of the Jura Mountains in Switzerland. As in previous ages, the manufacture of distinctive 'wholesale' pieces exists side by side with exquisitely decorated unique creations, designed for the most exclusive and discerning clientele in the world – the modern-day equivalents of previous generations of clients which included Chinese emperors, Indian princes, North American tycoons and the kings and queens of Europe. The 'Leda', for example, is completely representative of the quality and style one has come to expect of the Neuchâtelois tradition. It is an oversized 8-day watch. Its hunter case is 18-carat pink gold. The dial cover is hand-engraved with a scene depicting Jupiter visiting Leda in the form of a swan. The back cover has a scene depicting a classical *tempietto*, its silhouette reflected in the shimmering surface of a small lake. Above the *tempietto*, eight stars representing the constellation of Gemini hang in the night sky. The twins Castor and Pollux were the by-product of Leda's amorous encounter with the god. The dial itself is onyx with pink gold chapters. The number 12 is represented by the Greek letters ιβ (*iota* and *beta*). The movement continues the Greek motif in its decoration with acanthus-like palmettes and a key-patterned perspectival border. The watch comes with a presentation support: a white gold basketwork cushion, fringed with pink gold twisted cord.

199 **Peter Johannes Klein**
Astronomical-geographical table clock
Mixed media, 1738
Dresden, Staatliche Kunstsammlungen, Mathematisch-Physicalischer Salon [D.iv.d.3]

This fine and interesting clock has an exceptionally early and unusual feature – a clockwork terrestrial globe mounted in the back of the clock so that the north pole faces the viewer. The globe itself remains stationary and the Sun and equatorial hour circle revolve around it. In this way, the movement of the clock mimics the temporal effects of the daily rotation of the Earth on its axis. The outer dial has the 24 hours marked on it in Roman numerals, and the gilt-brass Sun pointer, attached at XII midday on the inner equatorial circle, points to local mean time. Under the Sun pointer there is an annual calendar circle which gets advanced one day in relation to the pointer with every 24-hour rotation of the central dial. The prime meridian of the globe appears to run through the Canary Isles, a common reference meridian at this date. Shrouding about half the globe is a black-painted quarter-sphere mounted in the equatorial hour circle and able to pivot about its central axis. This shows the areas of the globe in darkness at any particular time throughout the year, a very early example of the feature. The front dial of the clock shows the usual hours and minutes, with a mock pendulum in an aperture in the dial centre. The clock, which is in a case finely veneered with tortoiseshell, is quarter striking.

200 Pierre II du Hamel

Watch in the form of a hare
Mixed media, *ca.* 1660
Geneva, Musée d'Horlogerie
[AD 198]

Before the introduction of the balance spring to watchwork, all forms of watch were as much articles of jewellery as anything else, either worn around the neck or carried in a pouch, with a short ribbon or chain attached. None could keep time better than plus or minus about 20 minutes a day. Some were made as novelties, usually in the form of other unrelated objects, such as fruit or animals. This little hare is typical. The dial is read by hinging the body of the hare open to disclose the single-handed dial underneath. Such novelties were known as 'form watches' and continued to be made until the early 20th century.

201 John Merigeot and John Monk

Christopher Pinchbeck's astronomical clock
Mixed media, *ca.* 1768
Lent by Her Majesty the Queen
[2821]

This clock is probably one of the most complex astronomical clocks of its age. Made for King George III, and partly to his design, it was undoubtedly influenced by the slightly earlier four-dial astronomical clock made for the King by Eardley Norton (also still in the Royal Collection). This clock is, however, considerably larger. Although it is signed by Christopher Pinchbeck, the clock seems to have been technically designed by clockmakers John Merigeot and John Monk. Pinchbeck probably acted as the commissioning agent. The King's Architect Sir William Chambers was also involved in the design, and a preliminary drawing of this clock by him is preserved in Sir John Soane's Museum in London.

202 Jacob Auch

Astronomical pocket watch
Wooden box, gold and enamelled silver, 1790
Zurich, Uhrenmuseum [1679.71]

Jacob Auch, one of the foremost German watch and clockmakers of his day, specialized in creating highly accurate and complex astronomical clocks and watches. This double-sided gold watch combines mean solar time and calendar information on one side of the watch with a small planetarium giving astronomical information on the other. The mean time side has a 24-hour dial around the periphery with minutes at the top, seconds and days of the week at the left, months and calendar on the right and a regulation dial beneath. The planetary dial is heliocentric, and has Mercury, Venus and the Earth with its Moon circling the Sun. Mercury is given an eccentric orbit with epicyclic gearing and the Earth turns with the complete dial plate beneath. The ecliptic is engraved around the outside of the dial, enabling the position of the Sun, Mercury and Venus to be seen from the Earth's point of view.

John Constable, *Study of clouds
above a wide landscape* (see **234**)

THE Depiction OF Time

Human beings are makers – makers of things, of concepts and of images. Most of what man makes is intended to help him survive. The manufacture of tools is often seen as one of the first signs of intelligence. But another kind of tool necessary for survival is the abstraction of experience into concept. A creature who lives without this ability constantly has to invest precious energy into reacting to external stimuli. The being that can conceptualize can turn experience into knowledge and shepherd vital resources until they are truly needed.

Conceptualizing time is difficult. Nevertheless, philosophers, poets, theologians, scientists and artists have all spent countless hours trying to define – to capture – the essence of time. There is something deep in the Western psyche that constantly pushes us towards definition. The process begins with metaphor. Once the metaphor is established, the images quickly follow.

Images of a personified time took a very long time to become codified. In most non-Western cultures, for example, there is no single god or goddess of time. 'Time' is a quality, an essence, a commodity or a chore that is handed out to one or other of the gods to husband. Time may be measured by the Sun or the stars, but it is not created by them; nor is it deified.

Throughout the ages, artists have responded to time in ways that are both personal and culturally determined. If one were to write a potted history of the European representation of this aspect of time, it might start with a desire to represent the creation of time. This would then be followed by a desire to depict the sequential narrative and the eternal – two basic elements of both pagan and Christian belief in the West. As soon as one creates a history, the need to find a way to represent the past arises. The depiction of the passage of time, however, develops relatively late in the West. Perhaps not until the fifteenth century, with the rediscovery of how to utilize the effects of light and shade more precisely, can one begin to re-create those aspects of the natural world that evoke the temporal.

Time Personified

203 Roman

Mithraic relief

Marble, 2nd–3rd century AD
Modena, Galleria e Medaglieria
Estense [no. 2676]

Even though a number of scholars have written with great authority on the subject-matter of this relief, there is still no agreement as to who the central figure is, what the relief might have been understood to represent and for what purpose it was used. The combination of elements is so unusual that no-one, to date, has been able to provide a satisfactory explanation of the whole. The relief depicts a nude, young man, who appears to be bursting forth from an egg in a blaze of flames, with one half of the egg shell forming an inverted pedestal upon which he stands and the other sitting somewhat like a Phrygian cap on top of his head. The whole length of his body is encased in the coils of a large serpent, whose head rests on top of the upper half of the egg shell. The youth is winged and has three animal heads sprouting from his chest: a goat, a lion and a ram. He also appears to have cloven feet. In his left hand, he holds a staff and, in his right, a stylized bundle of flames to represent thunder. The whole figure is enclosed by a ring of the twelve zodiacal animals, with Aries placed at the top. In the four corners of the relief, there are depictions of the four winds, two of whom are bearded (possibly the autumnal and winter winds of Euros and Boreas) and two of whom are not (perhaps the spring and summer winds Zephyrus and Notus). The band of the zodiac and the boundary of the four seasons make it clear that this figure has something to do with time. Moreover, the feature of the three animal heads closely resembles the three-headed monster of Serapis (see **222** and **223**). Snakes, too, are often associated with time, their sinuous bodies said to reflect the undulating path of the sun along the ecliptic.

The image of 'Father Time' is so well known that one easily supposes that its form must date back to the mists of pre-history. In fact, the familiar figure of a bald and crippled old man, with wings on his shoulders and carrying a sickle or a scythe, a snake biting its own tail or an hourglass or clock, is a relatively modern concoction, cobbled together from bits and pieces over a period of about a thousand years.

The ancient Greeks and Romans did not have a personification of time *per se*. Instead, they depicted aspects of time in two ways. The first was with the figure of Kairos, a nude, winged youth who symbolized that brief, decisive moment during which one's fortune might change. A number of fragments depicting the god have survived, all of which are based on Lysippus's now lost bronze statue of Kairos – a winged nude youth, holding a pair of scales on a razor's edge.

The second antique image of time seems to have been imported from Persia and became part of the mystery cults associated with the god Mithras. This figure, alternately known as Aion or his Orphic counterpart, Phanes, symbolized the divine principle of the eternal in which time itself was the source of all the world's creative forces. Depictions of Aion vary. The Persian Aion is generally a male figure with a lion's head. His body is entwined with a large snake and he holds a large key in his right hand. The classical figure of Aion, however, is wholly human and shown standing within the 'circle of time', formed by the ring of the twelve signs of the zodiac. Depictions of Phanes (see **203**) tend to centre on the figure of a male youth, shown bursting forth in flames from a primeval egg. He is winged and his body is also caught in the coils of an oversized snake and he holds a thunderbolt in his right hand. According to the Orphic tradition, the primeval god, Chronos (Time), gave birth to Aither (the air), Erebus (the dark) and Chaos. Next, Chronos fashioned in Aither an egg, which split in two as Phanes, the first-born of the gods, sprang forth. The name Phanes derives from the Greek *phaino*, meaning 'I shine'. Phanes is variously identified as light itself, the Sun or as the bringer of light to the rest of creation. Despite its seductive complexity, however, the figure of Aion/Phanes failed to make much of an impact on the established hierarchy of the Greco-Roman pantheon.

The origins of our Father Time can be traced back to the Greek fondness for word association. Their word for time (*chronos*) was very close to the name of one of their oldest gods (Kronos). Kronos was the oldest of the Titans and the father of Zeus. He presided over the Golden Age of man and featured primarily as an agricultural god. He is usually depicted as an old man, with long hair and a flowing beard, holding a sickle. The Greek writer Plutarch provides the earliest surviving source making the claim that the similarity between the words '*chronos*' and 'Kronos' must be meaningful. As all the names of the gods are drawn from the primordial elements with which they have associations, he argues, then Kronos must be the god of time. The concept is often repeated and elaborated upon by later writers, but there is precious little evidence that classical artists ever depicted the old god specifically as a personification of time.

The pictorial formulae of depicting Kronos – or Saturn as he came to be called in Latin – as a god of time is largely due to the encyclopaedists, commentators and grammarians writing during the early centuries of the Middle Ages. For example, one of the commentators on Virgil's works, the fourth-century grammarian Servius, is the first to suggest that Saturn's sickle is a symbol of the way in which time cuts through all things. The mythographer Macrobius explains how his snake is actually a symbol of the year and the way in which it bites its own tail echoes the manner in which time devours itself. The well known tale of how Saturn ate his own children was also interpreted by the later medieval mythographers as an image of how time eats everything in its path or, to quote Ovid, "*tempus edax rerum*" – time devours all things.

204 **Artus I Quellinus**
Saturn
Terracotta, *ca.* 1650–64
Amsterdam, Historisch Museum
[BA 2308]

According to the myths, Saturn feared that one of his children might usurp his role as king of the gods. In order to avoid this fate, he demanded that Rhea, his consort, yield each of her newborn children to him to be eaten. It was only when the young Jupiter convinced his mother to substitute a stone for Saturn to eat, that the father of the gods vomited forth all the children he had previously eaten and was, indeed, overthrown. In the late Middle Ages, in 'moralized' versions of the verses of the Roman poet Ovid – the so called *Ovide moralisé* – readers were provided with explanations of the educational messages supposedly concealed within the poet's verses: here Saturn's cannibalism was presented as a symbol of how time eats away at all that it creates.

From a survey of all the world's cultures, it is surprising to note that only two have managed to deify or personify time in any sort of permanent way. The first is the Greco-Roman tradition which, during the Middle Ages, develops into the well known figure of 'Father Time'. The other is China. Shou Lao ('Old Longevity') or Shou Xing ('Star Longevity') are two names by which the Chinese 'god of longevity' is known. He seems to have originally been a stellar deity, associated with the bright star, Canopus (α Car) in the southern constellation of Argo. His original role was as a portent of peace. When Canopus was visible, there would be peace. When it was not, there would be war.

The iconography of the god developed slowly. In general, he is depicted as a short old man, with a huge, bulging forehead. One tradition has it that the shape of his forehead is supposed to resemble a penis – the 'font of all life' on Earth. Shou Lao's attributes include a wooden staff and a peach. The peach is actually the fruit of the *pantao* tree, a mythical tree that bears fruit only once every 3000 years. It then bears peaches only 3000 years after that. Shou Lao is also often accompanied by animals that are reputed to be very long-lived, such as the stork, and sometimes he is shown riding or standing beside his pet deer. As Shou Lao is the god who determines the date of every mortal death, he is sometimes shown carrying a scroll or a tablet. He writes down the moment of everyone's death when they are born; but, with cajoling, he has been known to juggle the numbers to ensure longer life for his supplicants.

205 Chinese
Shou Lao
Jade, 17th century
Cambridge, The Fitzwilliam Museum
[O.62.1946]

206 Chinese
Shou Xing with star mantle
Porcelain, 19th century
London, The Victoria and Albert
Museum [Circ. 256–1923]

The iconography for Shou Lao tends to fall into two categories. The more conventional images show him as an old man, with a bulging forehead, the 'peaches of immortality' and his staff.

Quite often, too, the material used to sculpt images of the god reflects his role as a god of longevity. For example, the Cambridge *Shou Lao* is made from jade, a stone known for its hardness and seen as a talisman towards longevity. Secondly, though less frequently, the god is depicted as Shou Xing and bears traces of his origin as a stellar deity. In the porcelain statuette from the Victoria and Albert Museum, for example, he is shown wearing a cloak covered with constellations. Shou Lao is one of the triad of lucky gods, which includes the god of happiness, Fu Xing, and the god of salaries, Lu

Xing. Although none of these gods have active temples or specific cults, they are extremely popular throughout China. Shou Lao, for example, is a regular invitee to the birthday parties of older people and infant children are often given auspicious red dresses, embroidered with the image of Shou Lao, on their first birthday.

207 **Chinese**

Shou (Longevity)
Gold leaf on silk backed with paper,
late 19th century
London, collection of G.S. Barrass

The character for *shou* has long held
an important place in the Chinese
imagination. Chinese tradition, which
enjoys finding figurative pictorial
formulas or 'explanations' behind the
form of characters, has it that the
original form of *shou* is composed of
two pictograms. The top part of the
character is a drawing of an old man;
the bottom part shows a hand
holding a wine vessel. The
combination of these two pictograms
evokes old age and celebration. The
combination may specifically allude to
the ritual wine vessels used as part
of the time-honoured tradition of
ancestor worship. Large scrolls
containing just the single character
are often hung to honour guests or as
part of birthday or anniversary
celebrations. This spectacular 'golden
shou' has an imperial provenance and
may well have been hung
somewhere in the Forbidden City in
Beijing during the final years of the
Qing Dynasty (1644–1911). Since the
19th century, it has become the
accepted convention that there are
exactly 100 different ways in which
the character *shou* can be written.
Perhaps not coincidentally, this
number also featured in one of the
most commonly used formulas to
invoke good fortune: *chang ming bai
sui* ('may you live 100 years'). Writing
the '100 *shou*' became a popular
calligraphic exercise. The characters
themselves were not hard to form
and the numerical coincidence of a
'double 10' seemed happily
auspicious. Even today, images of the
'100 *shou*' are regularly given as
birthday gifts, having been adapted to
the format of ball-point pens,
placemats, T-shirts, dish-towels and
so on.

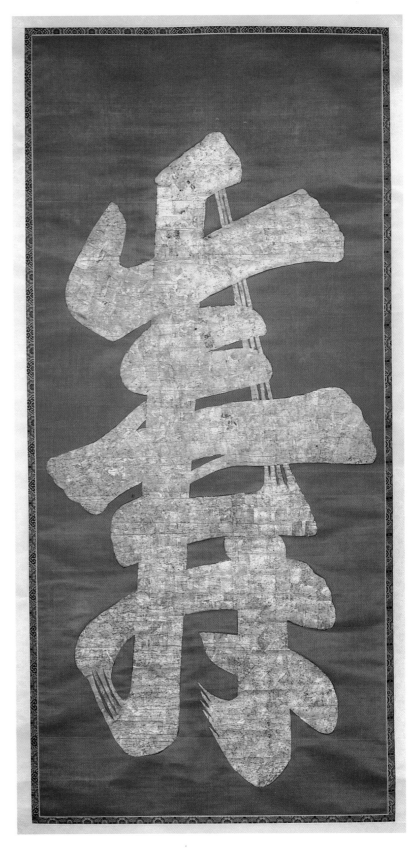

In the late Middle Ages, the image of Father Time comes into its own. The primary impetus seems to have been the desire to provide illustrations to Petrarch's *Trionfi* ('Triumphs'), which was composed in two stages between 1340 and his death in 1370. Petrarch's *Trionfi* generated a whole industry of illuminated versions of the text, as well as innumerable large-scale murals and tapestries. Its themes were depicted, too, on decorative plate, furniture, bridal giftware and clothing. Petrarch's Father Time was no benign agricultural deity. He was the destroyer of Fame – trampling over all the human vanities. In illustrations of the *Trionfi*, Time is invariably depicted as an old and bent man, often standing only with the aid of crutches. He is often winged to illustrate how *tempus fugit* or 'time flies'. He carries either a sickle or a huge scythe which, though it may have originally been intended as a farming implement for an ancient Greek agricultural deity, has now taken on a completely different significance. Some time during the Middle Ages, the scythe and sickle became the attribute of the 'grim reaper' – death. And, by extension, when Father Time is depicted holding one of these tools, he also takes on the role of a god whose task is to cut short the lives of men. Again, the medieval Father Time tends to be portrayed as a destroyer.

Another iconographic feature that first appears in the medieval representations of Father Time is the attribute of the hourglass or clock. As hourglasses and sundials had existed as timekeepers for centuries, one might speculate about why they seem to make a sudden appearance in the iconography of time at this point. One suggestion is that, in the course of the Middle Ages, the possession of personal timepieces became more common. A growing middle class gave rise to a growing need for 'consumables', one of which was timepieces for personal adornment. The other possibility is that, with a burgeoning merchant class, time itself became something of a commodity. Whereas a society whose wealth was based on agriculture or seafaring might be able to function with only the most rudimentary of seasonal clocks, a world that had to support 'businessmen' needed a more individualized means for keeping time. The third possibility is that, in a society in which urban centres play a key role, people often find themselves separated from the natural timekeepers of the Sun, Moon and the stars. Hourglasses and clocks are needed in an urban environment. A large public clock is often one of the first signs that a society is becoming urbanized. In Europe, public clocks began to appear during the second half of the fourteenth century – at almost the same time that Father Time gained his new attributes.

208 Nicolas Poussin

Study for *A Dance to the Music of Time*
Pen and wash drawing, *ca.* 1635–40
Edinburgh, National Gallery of Scotland [D. 5127]

The winged figure of Father Time sits at the right of the composition, playing music on a lyre to which four figures are dancing. As a group, the figures represent an allegory of human life and changing fortune. Individually, they are Poverty, Labour, Wealth and Luxury. Two small putti placed beneath Time's feet play with well known symbols of the *vanitas* – a soap-bubble pipe and an hourglass. On the left, there is a herm surmounted by the *bifrons* head of Janus, the tutelary god of the year, and in the sky the Sun god Apollo.

Also

• Michael Burghers after Nicolas Poussin, *A Dance to the Music of Time*, 1694, London, The British Museum, Dept. of Prints and Drawings [U.8–18]

209   Florentine

The Triumph of Time
Engraving, *ca.* 1460–70
London, The British Museum, Dept of
Prints and Drawings [1860.6–9.52]

Petrarch's *Trionfi* is probably the most
illustrated work of literature in Italy
during the 15th and 16th centuries. In
his text, Petrarch never actually
describes Father Time. Nevertheless,
the iconography of this scene
became fixed quite early – perhaps
some time during the 1420s or
1430s. This Florentine *Triumph of
Time* is wholly characteristic of the
genre. It represents Father Time as
an old man, bald and bearded,
standing with the aid of a pair of
crutches; he is also winged and borne
on a chariot drawn by two deer. The
text from Petrarch's poem beneath
may be translated: 'What more is this
our life than a single day,/ Cloudy and
cold and short and filled with grief,/
That hath no value, fair though it may
seem?/ ... And fleeing thus, it turns
the world around/ Nor ever rests or
stays nor turns again/ Till it has made
you nought but little dust.'

210   Giambattista Tiepolo

Father Time revealing Truth
Pen and brown ink, with two
shades of brown wash over pencil,
ca. 1755–60
Hamburg, Hamburger Kunsthalle,
[Collection Hegewisch]

For a short period during its history,
Tiepolo's composition was known as
'Time abducting Beauty'. The figure
of Father Time was easily identified
by his long-handled scythe, his wings
and the hourglass that is held by the
accompanying putto. The woman,
beautiful as she might be, is Truth
and her image has been drawn
directly from the iconographic
handbook of Vincenzo Cartari, where
Truth is described as semi-nude, her
modesty covered with a semi-
transparent veil, with a glorious Sun
shining beside her and her foot
resting on a terrestrial globe. The idea
of Truth being the daughter of Time
(*Veritas filia temporis*) comes
ultimately from *The Attic Nights* (XII,
XI, 7) of Aulus Gellius. Time's bringing
Truth out of the darkness is a
common formula in the Greek
comedies and also appears in the
writings of the Roman Stoic
philosopher Seneca (*De Ira*, II, II, 3).
The Church Fathers also adopt the
idea, attaching it to Matthew 10:26
("There is nothing concealed that will
not be disclosed, or hidden that will
not be known").

Omnia vincit Amor, vincit mox tempus Amorem.

Franciscus Perrier pinxit et sculp.   Cum Privilegio Regis

A Paris; Chez la veufue de deffunct Perrier, ruë des foffez St. Germain vis a vis l'Hoftel de Sourdys

211 **François Perrier**

Time clipping Cupid's wings
Engraving, *ca.* 1630–70
London, The British Museum, Dept. of
Prints and Drawings [1874. 8-8. 969]

Father Time is depicted as an elderly,
but well muscled, man. His head is
bald at the back and he is winged.
His scythe lies at his feet along with
an hourglass. Grasped between his
strong arms is a wriggling infant – not
to be eaten on this occasion, but only
to be chastened. The small child is
Cupid, and his fallen quiver full of
arrows lies at Time's feet. Time
himself is taking a pair of shears to

the young god's wings, clipping them
so that he will be restrained in his
flight. Perrier has appended the
message 'Love conquers all, but
Time conquers Love'.

Also

• James MacArdell after Anthony van
Dyck, *Time clipping Cupid's wings,
ca.* 1760, London, The Victoria and
Albert Museum [E.469–1959]

• French (after Jean-Baptiste Stouf),
Love disarming Time, 1789, Oxford,
The Ashmolean Museum
[NBP 314]

212 **François Perrier**

Frontispiece to *Eigentlyke
Afbeeldinge, van Hondert der
Aldervermaerdste Statuen of Antique-
Beelden Staande binnen Romen*
Printed book, Amsterdam 1638
London, The British Museum, Dept.
of Prints and Drawings
[1895.10-31.28]

Like many of the French painters of
his generation, François Perrier
travelled to Rome during his youth in
order to study the sculptural
monuments of Rome's glorious past.
He published a series of prints after
his drawings prefaced by a clever
allegorical scene which was obviously
his own invention. It shows the aged

god of Time using his scythe as a
crutch and gnawing on the broken
stump of what was, at the time, the
most famous fragment of antique
statuary – the Belvedere *Torso*. The
underlying message of Perrier's title
page is that even though an artist
might yearn to believe the
Hippocratic dictum that *ars longa,
vita brevis*, the truth was starkly
different. Time was a destructive
force that was still actively assailing
the treasures of the past. The
Ovidian theme of Time devouring the
accomplishments of the ages
(*tempus edax rerum*) was a common
theme for artists during the period,
but the literalness of Perrier's
rendition is unprecedented.

213 **Pompeo Batoni**

Time ordering Old Age to destroy Beauty
Oil on canvas, dated 1746
London, The National Gallery
[NG 6316]

Although the iconography of the painting is relatively clear, the painter himself made his intentions explicit in a letter. Writing to his patron, the Lucchese collector Bartolomeo Talenti, on 4 July 1474, Batoni insists that the composition will come completely out of "his own imagination" (*della mia fantasia*). Later, in August of the same year, he reiterates his promise, stating that "they come neither from fables, nor from histories, but instead from my own invention" (*di invenzione mia*), and he then describes the painting thus: "You can see Time, seated, [and shown] in the act of commanding Old Age ... who scratches the face of a beautiful girl of singular beauty". Father Time, who occupies the foreground of the picture, is easily identifiable by his relative age, a balding pate and a lovely set of feathery wings. He also holds a rough, wooden hourglass in his left hand. Old Age is shown as a leathery old crone, whose wrinkled face and scrawny neck contrast starkly with the peachy smooth skin of the young girl. The large circle made by the arms and hands of the three figures knits them together in a ghastly kind of dance, lending an air of inevitability to the whole proceeding.

Also

• Pompeo Batoni, *Time revealing Truth,* oil on canvas, *ca.* 1745–46, Providence RI, Rhode Island School of Design [59.065]

3 · THE DEPICTION OF TIME

214 Charles-Guillaume Manière
Globe clock with Father Time and Astronomy
Mixed media, 19th century
Lent by Her Majesty the Queen
[30005]

A number of French clocks dating from the 17th and 18th centuries have decorative themes on their cases elaborating the iconography of time. This clock depicts a winged Father Time who appears to be menacing a female figure – usually identified as Astronomy – with his scythe. She wards him off with a defensive gesture suggesting that this scene perhaps depicts a rather important point in history, in which the science of astronomy tells the grim harvester that she is now the patroness of the passage of time and that his indiscriminately destructive urges need to be curbed. The hours and minutes are contained within two bands which revolve around the surface of the globe once every 12 hours. The Roman numerals indicate the hours and the Arabic numerals indicate the minutes. The time is shown by the point that is bracketed between one of the struts supporting the horizon ring of the globe and the tip of Father Time's scythe.

Also

• North Italian, *Majolica dish with figure of 'Chronico'*, late 15th century, private collection

215 William Hogarth

Time smoking a picture
Engraving and mezzotint, March
1761
London, The British Museum, Dept. of
Prints and Drawings [S.2-156]

In this print, we see Hogarth using
the well known image of 'Father
Time the Destroyer' as the basis of a
satire attacking the studio practices
of some of his contemporaries. For
ages, artists had relied on the
application of a final layer of tinted
glaze to bring the range of colours
into closer accord. As time passed,
however, quite often the medium in
which these tints were suspended
would darken and discolour. In 18th-
century England, when collecting
Italian Old Masters was the vogue,
the leisured classes came to develop
a taste for the apparent 'moodiness'
of these age-darkened works – to
such an extent that living artists
began to 'age' their paintings

artificially in order to find favour
within the market. There are stories
of Sir Joshua Reynolds mixing
varnish-brown into his paint surface
to achieve the desired effect and a
marvellously blunt description by
William Gilpin, in which he recounts
how he had "used to hold [his
paintings] over smoke till they had
assumed such a tint as satisfied my
eye". But the practice of 'smoking
pictures', as it was known, enraged
Hogarth. In his print, Father Time is
'smoking' a painting – but the
message is made quite clear by the
fact that the end of his scythe has
also pierced the surface of the
canvas: ageing and smoking pictures
does not improve them, it destroys
them. The Greek quotation that runs
along the upper frame of the
painting, taken (somewhat mangled)
from the *Florilegium* of Johannes
Stobaeus, reads: 'For time is not
such a great craftsman, but he
weakens all he touches'.

216 Hans von Aachen

Allegory with Father Time
Oil on copper, 1680s
Stuttgart, Staatsgalerie [2130]

On one level, the meaning of this
painting is relatively clear. Father
Time, his long, naked body stretched
out across the top of the painting like
a medieval personification of the
plague, is the great leveller. In his left
hand, he holds an hourglass
(surmounted by a foliot from a clock)
of which the sands have clearly run
out. His wings are black. The
significance of the other figures in
the painting, however, is less self-
evident. Two figures on the left are
Venus and Cupid – but whether the
child rushes to his mother with
affection or for comfort is uncertain.
The middle figures could be either
gods or men. The figure on the right
seems to be Ceres, goddess of the
harvest. In the background, Minerva
is shown as a statue amongst the

crumbling ruins of a once great
temple. With her left arm, she
mimics Father Time's gesture.
Instead of an hourglass, though, she
holds a small statuette – perhaps a
Victory. In fact this painting is
probably an allegory about the
triumphant victory of the Viennese
over the Turks in 1683. In relation to
von Aachen's other works on exactly
this subject, painted while he was
court painter for Rudolf II in Prague,
this seems to make sense. Various
turbans litter the foreground.
Minerva is shown seated beside a
turban and some rather tattered
battle flags. There is also a discarded
scimitar in the foreground. The
central figures can now be read as
Mars triumphant over a vanquished
Turk. The overall message of the
painting, then, is not about human
vanity, but about how – with Time –
the righteous victor will prevail over
his adversary, and love and bounty
will return to the land.

**217 Giovanni Antonio Pordenone
and Niccolò Vicentino**

Saturn/Chronos
Chiaroscuro woodcut made from four
blocks (tan, two shades of green and
black), *ca.* 1531–32
London, The British Museum, Dept.
of Prints and Drawings
[1860. 4–14.113]

This image of Father Time is
attributable to the Venetian painter
Pordenone. It records part of his
design for the façade of the Palazzo
d'Anna on the Grand Canal in Venice.
The iconography of the figure of
Saturn fits with his role as a time
god. He leans backwards, his feet
encircled by a serpent biting its own
tail. In his outstretched right arm he
holds a foliot – the part of a clock that
makes the 'tick-tock' of a clock
mechanism. The substitution of a part
of a mechanical clock for the more
customary attribute of an hourglass
seems to appear first in the middle
years of the 15th century. One of the
earliest representations of this
iconographic change appears in the
late 15th-century *Triumph of Time*
attributed to Jacopo del Sellaio (in the
Museo Bandini in Fiesole), where an
aged Father Time is shown standing
on top of a foliot and verge
escapement that is attached to a
large, 24-hour clock dial. The dial is
supported on a slim rod which is
being gnawed by two dogs – one
white and one black – symbolizing
the constant alternation of day and
night. Pordenone's print of Father
Time seems to have enjoyed a wide
diffusion during the next two
centuries, as it was popular as a
source for the sculptural decorative
work on clock cases during the 17th
and 18th centuries (see **218**).

218 French

*Boulle marquetry mantel clock with
Father Time*
Mixed media, 18th century
Lent by Her Majesty the Queen
[30013]

The figure of Father Time on this
French clock must have been based
on the polychrome woodblock print
designed by Pordenone (**217**). The
reclining posture and slight tilt of
Father Time's head strongly recall the
print – as does the fact he is depicted
holding a foliot. The slightly awkward
posture of the winged cherub placed
at the top of the clock may also
reflect the gesture made by the putto
in Pordenone's print. The clock has a
later, two-train hour striking
movement, supplied by Vulliamy in
about 1834. The chapter numerals
are painted on individual enamel
plaques. The base is richly decorated
with ormolu mouldings and the ebony
plinth is richly inlaid.

219 Crispijn de Passe after Maarten de Vos

The Use and Abuse of Time
Set of four engravings, 1580s
London, The British Library, Dept. of Prints and Drawings [1868.6–12. 2054 to 2057]

220 Philip Galle after Pieter Bruegel

Temperance
Engraving, ca. 1560
London, The British Museum, Dept. of Prints and Drawings [1868.3–28.369]

In Bruegel's image, Temperance is presented as the central virtue governing all of the seven Liberal Arts: Arithmetic (in the lower left, with men counting), Music (in the upper left), Rhetoric (the players on the stage in the upper left), Astronomy (the man measuring the Moon with a pair of dividers), Geometry (a man measuring the Earth, surveyors and architects), Dialectic (the men disputing in the middle right) and Grammar (the schoolteacher with his pupils in the lower right). It is her regulating and moderating influence that allows the other arts to flourish. The attributes she bears concern regulation: the clock she wears on her head alludes to the natural rhythms of life which

should not be exceeded; the bridle in her mouth refers to restraint in speech; the glasses she holds in her hand allow her to see more clearly and her foot is resting on the blade of a windmill – the symbol of man's control over the wayward winds. The inscription can be translated: 'We must look to it that, in the devotion to sensual pleasures, we do not become wasteful and luxuriant; but also that we do not, because of miserly greed, live in filth and ignorance.' The virtue of Temperance also underlies the theme of Maarten de Vos's series on *The Use and Abuse of Time*, based on the idea that as man works as a penance for Original Sin, so work itself brings one closer to God. In the first print, Father Time stands beside an image of Diligence and Assiduity – a tidy woman engaged in the activities of spinning and reading. The second print is entitled 'The Image of Happiness and Wealth'. Since all is as it should be, the Earth is bountiful and peace reigns. In the third print the ill effects of Wealth have led to 'Negligence and Inactivity'; and, in the final print, 'The Image of Misery and Poverty', all is ruined. The formerly prosperous countryside has been reduced to ill-thatched hovels full of half-naked peasants.

Past, Present and Future

221 **Titian (Tiziano Vecellio)**, *Allegory of Prudence, ca.* 1560–70, oil on canvas, London, The National Gallery [NG 6376]

When Titian's great allegorical painting was first identified and published during the 1920s, it was thought to be a somewhat unconventional depiction of 'the Three Ages of Man'. Very soon afterwards, however, two of the great art historians of the period – Fritz Saxl and Erwin Panofsky – recognized that the subject-matter of the painting was not as straightforward as had originally been suggested. They noted that its iconography or significance rests on understanding several merged pieces of information, cleverly intertwined so that formally the picture makes a unified aesthetic whole.

The central section of the painting depicts three men, who seem to correspond to Old Age, Middle Age and Youth. Above these heads, however, there is a Latin inscription, arranged in three parts, so that each part corresponds to one of the heads. Above the elder man's head is *EX PRAETERITO*; above the mature man's head is *PRAESENS PRUDENTER AGIT* and above the youth's head is NI *FUTURU*[M] *ACTIONE*[M] *DETURPET*. The whole can be translated as: "From [experience of] the past, the present acts prudently, lest it spoil future action".

The Latin quotation itself is drawn indirectly from the work of one of the well known encyclopedists or mythographers of the late Middle Ages, Pierre Bersuire. "Prudence," he tells us, "consists of memory of the past, the ordering of the present and the contemplation of the future." From this, one can see that the combination of heads and inscriptions in Titian's painting represents the cardinal virtue of Prudence. Indeed, in other works of art, one finds a number of three-headed personifications of Prudence, for example, an *Allegory of Prudence* ascribed to the workshop of the fifteenth-century Florentine sculptor Desiderio da Settignano (see **226**). However, the faces in the relief seem all to be vaguely the same age. The fact that Titian chose to represent three distinct Ages to depict Prudence suggests that the old man is, to a certain extent, meant to be understood as a representation of the past; the mature man should be associated with the present and the youth with the future. Or, as another medieval encyclopedist, Fulgentius, described it: "Prudence is composed of three faculties – Memory, Intelligence and Foresight – of which the respective functions are to conserve the past, to know the present and to foresee the future. Prudence surveys the tripartite span of life."

There is something in the manner and style in which Titian has painted the three different heads that helps to convey this message. The old man is painted in a loose and brushy style, suggesting the vagaries and indistinct memories of past times. The mature man is presented frontally and is forcefully modelled with a strong play of light and shade that tends to make him, somehow, more palpable – more 'present' – than his two counterparts. The youth is placed so that he is facing the strong source of light that enters the painting from the left – as if facing the dawning of a new day.

Besides the idea of a three-faced Prudence, there was an equally strong tradition for representing Time itself with three faces. Since Aristotle, time has been repeatedly conceptualized as having three distinct parts – the past, the present and the future. In the writings of the Scholastics of the early Middle Ages these concepts were more closely analysed and, eventually, visualized. The three-faced image of Time seems to appear first during the late fourteenth century in France, and was soon widely disseminated.

The top half of Titian's painting, then, is a rather complex allegory involving the Three Ages of Man, a medieval representation of the virtue of Prudence and the three-faced figure of Time. What about the odd combination of animal heads along the bottom of the picture? To the left, beneath the old man, is a wolf; in the centre, there is a lion and to the right is a dog. The source of these heads can be traced back to Hellenistic Egypt and the hybrid god Serapis. Serapis had an accomplice, a terrible monster with three heads – a wolf, a lion and a dog, whose body was encased by the coils of a huge python. It is not known whether this beast carried any symbolic meaning for the Hellenized Egyptians, but the fourth-century Roman grammarian and philosopher Macrobius describes the monster thus: "The lion, violent and sudden, expresses the present; the wolf, which drags away his victims, is the image of the past, robbing us of memories; the dog, fawning on its master, suggests to us the future, which ceaselessly beguiles us with hope". This 'time-monster' then reappears in the handbooks of Vincenzo Cartari and Pierio Valeriano in the mid-sixteenth century (see **223**). Moreover, the image seems to have enjoyed a particular vogue in Venice.

Although his sources can be identified with some certainty, it is more difficult to say exactly what this image might have meant to Titian. There are two possibilities: either this picture was devised by some clever patron who supplied a detailed programme for the different parts of the picture or, perhaps, the picture was of an intensely personal nature and refers to some significant event in the artist's own life. Whereas neither possibility rings wholly true, Panofsky's suggestions in support of the latter argument are certainly intriguing and must, for the meantime, serve as the most plausible explanation. Panofsky believed that the painting was autobiographical. Not only did he point out similarities between the face of the old man and the *Self-portrait* by Titian in the Prado, but he suggested that the features of the mature man in the middle of the picture closely resembled a portrait thought be of Titian's second son. Moreover, the age of the younger man coincides with that of Titian's nephew, Marco Vecelli. According to Panofsky, therefore, the image of the three men represents the 'past, present and future' of Titian's own household during the 1560s and reflects a time when Titian was prudently setting his affairs in order and rationalizing his fortunes to pass on to his son and nephew.

222 Greek (Lycia)

The three-headed monster of Serapis
Bronze, 1st or 2nd century AD
London, The British Museum, Dept.
of Greek and Roman Antiquities [GR
1866.10–6.1, Bronze 949]

In the most famous shrine to the
Egyptian god Serapis, there was a
statue created by the sculptor
Bryaxis which showed the god in all
his majesty and accompanied by a
strange three-headed monster. The
monster had the hairy body of a dog
and the heads of a lion, a wolf and a
dog attached to its neck. Around the
length of his torso, a huge snake's
body was coiled. Although there is
no known connection, most Greek
and Latin writers associated the
three-headed watchdog of Serapis
with the more widely known
watchdog of the underworld, Pluto's
three-headed dog Cerberus. Plutarch,
for example, says the two are
essentially the same beast –
overlooking the fact that Serapis's
beast has the extra features of a
large snake coiled around its body
and three different animal heads (*De
Iside et Osiride*, 78). The later Roman
and medieval writers continued this
tradition, but added that the image
symbolized the three faces of Time.
For scholars of the Renaissance,
then, the monster of Serapis was an
allegorical embodiment of time.

223 Pierio Valeriano

The three-headed image of Serapis,
from *Hieroglyphica, sive de Sacris
Aegyptiorum, Aliarumque Gentium
Literis Commentarii*
Printed book, Basle 1556
London, The British Library, Dept. of
Printed Books

Throughout the Renaissance, there
was a fascination with the idea of
the Egyptian hieroglyphs. Writers
and philosophers of the period
believed that these pictograms
contained essential truths, too
powerful or too subtle to be
declaimed solely in words. Perhaps
owing to its Egyptian ancestry and
certainly benefitting from its late
medieval glosses, the three-headed
monster of Serapis regularly
reappears in the iconographic
handbooks of the late 15th and 16th
centuries. It first appears as an
allegorical device in the 1499 edition
of Francesco Colonna's
Hypnerotomachia Poliphili, and it is
mentioned twice in the *Hieroglyphica*
of Pierio Valeriano.

Also
• Francesco Colonna,
Hypnerotomachia Poliphili, Venice
1499, London, The British Library,
Dept. of Printed Books

224 Hellenistic (Asia Minor)

Hecate
Parian marble, 2nd–3rd century AD
London, The British Museum, Dept.
of Greek and Roman Antiquities
[GR 1877.5–13.1, Sculpture 2161]

Hecate, the triple goddess of the
underworld, was often depicted with
three faces. In this representation,
she is shown as a priestess with a
smouldering altar on each side. The
figure on her right side holds a nail
and a hammer, the centre figure
bears two torches (similar to images
of Selene, the Moon goddess with
whom Hecate is often identified),
and the figure on the left has a
dagger and a serpent in her hands.
Hecate was often associated with
witchcraft as she was able to see
into each of the three realms of the
known world – the celestial, the
mundane and the infernal. Although
there is no positive proof, it seems
probable that the image of the three-
headed Hecate served as a model
for later medieval representations of
Prudence.

225 Attr. Bartolomeo Bellano
Hecate–Prudence
Bronze with silvered eyes,
ca. 1480–90
Berlin, PKB, Bode Museum/
Skulpturensammlung
[A.N. 303/ 1942]

The identity of this figure is still
something of a mystery. The
statuette shows a slim woman
striding forwards. In her left hand
she holds a heart (which seems to
be a replacement – possibly for
another, now lost, attribute). In her
right hand, she holds a flaming torch.
Her head bears three faces of
women of different ages. To the
right the face of an old hag; to the
left, a young woman and, behind,
the face of a child. One's first
instinct is to see this figure as a
personification
of three-faced
Prudence similar
to Titian's *Allegory
of Prudence* – the
three faces not only
representing the three aspects of
Prudence, but also embodying the
three Ages of, in this case, Woman.
There are, however, no known
representations of either Prudence or
Time holding a flaming torch. This
fact has led some scholars to
suggest that this figure represents
the goddess Hecate, the daughter of
Perseus and Asteria. Hecate is often
associated with Selene, the goddess
of the Moon, and is usually referred
to as the 'triple goddess' in that she
represents the Earth, the Moon and
the underworld. Numerous classical
depictions of Hecate show her as
either a triple bodied goddess or a
female figure with three heads or
three faces. She is mainly associated
with sorcery and witchcraft and, as
such, seems to have special control
over crossroads – places notorious
for evil and mishap.

**226 Attr. Workshop of Desiderio
da Settignano**
Allegory of Prudence
Pietra serena, *ca.* 1460
London, The Victoria and Albert
Museum [3004–1856]

As the inscribed legend on the
bottom explains, the relief depicts a
three-faced image of the Cardinal
Virtue of Prudence. The three faces
all share the same neck and the
same tousled thatch of hair. The two
side figures are beardless, while the
central figure sports a small beard.
Erwin Panofsky was the first to
explain the significance of the figure,
noting its relationship to the
iconography of Titian's *Allegory of
Prudence* (see **221**). Oddly,
however, in this rendering there
seems no allusion to the virtue of

Prudence relying on knowing the
past, the present and the future. The
three faces seem to have similar
features and there is no obvious
difference in their ages. There is,
however, something rather
reminiscent of contemporary
representations of Christ in the half-
closed eyes and slightly forked beard
of the central figure that could lead
one to speculate on possible
associations between the *trifrons*
Prudence and the Christian concept
of the tripartite Holy Trinity. As there
is no trace of such a connection in
the sources or commentaries of the
period, however, it seems that any
interpretation based on this possible
association would be ill-founded.

Joos van Cleve, *St Jerome in his study*, oil on wood panel, *ca.* 1524–30, Cambridge MA, Fogg Art Museum, Harvard University Art Museums

The Image of Vanitas

Efflorescence and Evanescence

Ivan Gaskell

Vanity of Vanities, saith the preacher; all is vanity .... Fear God and keep His commandments: for this is the whole duty of man. For God shall bring every work into judgement, with every secret thing, whether it be good, whether it be evil.

Ecclesiastes 12:8 and 13–14

Human skulls, frail bubbles about to burst, snuffed candles – this is the stuff of *vanitas* imagery: a visual articulation of the consciousness of the end of time. The end of time is personal – each life must end. It is also apocalyptic. At the moment of the final entry of God's creation into a state of timelessness, some souls will be judged forever blessed in His sight, others will fall outside of it and be damned for eternity. Both the personal and the universal are linked, for the peril of each soul is that, only in this finite lifetime, can one prepare for the Last Judgment when that decision of infinite import – to be saved or cast out for eternity – is made.

The theme of judgment and redemption is central to the Christian Church. One vital step towards redemption is the rejection of everything that is of this world – the human body, riches, fame, power *etc*. The generic term for all these worldly goods is 'vanities' – examples of man's vain hope of finding everlasting consolation in the evanescence of this world. If one consults a seventeenth-century biblical concordance, for example, there will be nearly a hundred entries for 'Vanity &c'. The message that everything of this world was but vanity formed the basis of endless homilies in the Roman Catholic world. It also thundered down from ten thousand pulpits in the newly Protestant lands of Europe and North America. Moreover, the *vanitas* became one of the prime subjects for painted panels and canvases, hewn stone statues and cast bronzes. But where did the *vanitas* imagery come from? Did visual articulations of *vanitas* themes themselves go through a cycle of blooming, flourishing and decay?

The first *vanitas* images appear during the fifteenth century. Of course, images of death and decay had appeared throughout the classical period and the Middle Ages as part of tomb decorations, but the appearance of self-contained works of art referring to the transience of human life seems to be a Renaissance phenomenon. Theories abound as to why people in the fifteenth century seem to associate themselves increasingly and explicitly with an acknowledgement of their own mortality. People were neither more nor less mortal than they had been previously. Changing

The message that everything

demographics, pandemic disease patterns and shifts in devotional practices have all been called upon to try to explain the more or less sudden appearance of *memento mori* – reminders of death – as, for example, a standard component of fifteenth-century portraiture. But, as none of these hypotheses seems fully convincing, it might be worth considering that the cause of this phenomenon (as far as there may be one) might lie elsewhere.

An ancient proverb attributed to the Greek philosopher Thales – 'man is a bubble'

As many others have noticed, during the fifteenth century in both southern and northern Europe, artists seem to become increasingly interested in creating 'naturalistic' depictions of the world around them. From the Florentine obsession with single-point perspective to the Burgundian fascination with the minutiae of nature, the fifteenth-century artist seems committed to creating something 'live' on his canvas or manuscript page. Artists employed the 'reality effect' to bring life to death. For example, Christ had never looked so 'dead' as he does in Giovanni Bellini's crowning panel for the Pesaro Altarpiece (Pinacoteca Vaticana, 1471–74). The theme of *Ecce homo* – 'Behold the Man' – finds new resonance in Hans Holbein's *Dead Christ* (Kunstmuseum, Basle, 1521), where the dead Christ is shown, stiff with rigor mortis, lying on the stone slab of his own tomb, his hands tinged with gangrene and the wound in his side gaping dryly open. If one of the central acts of Christian piety was the meditation on Christ's death, certainly both these images provided a new pictorial stimulus.

If, on the other hand, we judge the efficacy of a pictorial invention by the extent of its repetition, Albrecht Dürer's peculiarly brooding *St Jerome* (Lisbon, Museo del Arte Antigua, 1521), was especially successful in promoting the iconography of the *memento mori*. In an engraving made the same year, Lucas van Leyden adapted the prominent skull and the melancholy saint's finger pointing ruminatively at it. The workshops of Marinus van Reymerswaele and Joos van Cleve also produced numerous elaborations of Dürer's invention. In one of Joos's finest autograph versions (Cambridge MA, Fogg Art Museum, 1524–30) the biblical tradition of the *vanitas*, personified by the translator of the Bible into Latin, meets its classical equivalent. An ancient proverb attributed to the sixth-century BC Greek philosopher Thales – *homo bulla* ('man is a bubble') – has been inscribed on the architrave of the wall niche.

One of the most strikingly inventive articulations of the *vanitas* theme in the first half of the sixteenth century is found in a painting by Jan Sanders van Hemessen (Lille, Palais des Beaux-Arts, *ca.*

complemented by the Latin inscriptions about the glass and on the phylactery over the figure's arm. The artifice of the startling floating skull and of the tiny beaker on the window sill beyond draws the viewer further into the picture – wondering at the detail of an implausible reality made visible. In a painting such as this one, art is no longer simply the means of representing a set of ideas about human frailty. Art, the mirror of life, is fully implicated in the consideration of mortality.

The theme of death and transience in sixteenth-century art could take epic as well as intimate forms. Nowhere is this to be seen more frighteningly displayed than in Pieter Bruegel the Elder's *Triumph of Death* (Madrid, Museo del Prado, *ca.* 1562). Here, the well known themes of the triumph of death and 'the dance of death' combine to form a truly terrifying image. An army of the dead arises and, to the sound of a tolling tocsin, booming kettle drums and a screeching hurdy-gurdy, overwhelms the living. There is no defence and no escape for soldiers or young lovers, emperor or peasant. In what is perhaps the most terrifying corner of the painting, two exiguous skeletons hack down the remains of blasted trees in a wasted landscape that stretches to a horizon where columns of distant smoke clot the blood-red sky. No living thing is left standing in this epic epitome of desolation.

By the end of the sixteenth century, all the ingredient themes of *vanitas* imagery were well established and easily recognizable by most viewers of the visual arts. In 1599, the great Dutch graphic artist Jacques de Gheyn II produced an iconographically compendious version – *The Allegory of Transience* (London, British Museum). A picture of the Last Judgment is flanked by a peasant and an emperor – both equally subject to death's dominion; a naked boy blows bubbles in the center of the composition while, in the foreground, two decayed cadavers are laid out flanked by a fuming urn of incense on one side and a vase of flowers – doomed to fade – on the other. Once this legion of objects had become established as part of people's visual vocabulary, each could be used individually.

of this world was but vanity thundered down from ten thousand pulpits

1535). Only half of what must have been a portrait diptych survives. It depicts a figure with butterfly wings (itself indicative of the passing of time) holding not a human skull, but a looking glass in which a fantastically floating skull is reflected. The winged figure points to this mirrored image which, implicitly, belongs to the portrait sitter in the now missing panel. The glass, held by fleeting time, reflects the truth of the sitter's mortality and the sentiment is

This is exactly what happened during the late sixteenth century, giving rise to an entire *genre* of still-life paintings in which these objects – skulls, broken glasses, snuffed candles, certain arrangements of flowers – were progressively abstracted to stand alone as painted reminders of transience.

The *vanitas* was not confined to still life in the seventeenth century.

It existed side by side with a far more direct pictorial confrontation with death. The same Jacques de Gheyn who allegorized the theme so comprehensively also sketched a pen-and-ink drawing of his intimate friend, the great Dutch poet, art theorist and artists' biographer, Carel van Mander, as he lay upon his deathbed in Amsterdam in September 1606. He drew van Mander not once, but twice, from slightly different angles, on the same sheet (Frankfurt, Städelsches Kunstinstitut). The variant images denote a doubling of effort, suggesting an anxiety on de Gheyn's part not to allow the moment to pass without making the fullest possible visual description. Why?

Obviously, a pictorial confrontation with a dead subject was analogous to portrait painting, and, even though the inclusion of *memento mori* in portraits suggests that the sitters were certainly aware of their own mortality, part of portraiture's appeal was its ability to fix human appearance permanently. In this respect, art seems to have the ability to cheat death. It is interesting to note that

most sophisticated viewer. The late sermons of the poet and divine, John Donne, were much preoccupied with death. He delivered one of the better known sermons, 'Death's Duell, or a Consolation to the Soule, against the Dying Life and the Living Death of the Body', in the very last stages of his existence here on Earth. Reportedly, Donne was so visibly deteriorating that many of the congregation felt that he, following the model of Christ himself, was dying just for them. Later that year, Donne posed for the stone effigy to be placed on his own funerary monument in St Paul's Cathedral. He posed standing in his shroud.

As the century progressed, many of the more inventive artists sought ways to reinvigorate themes of death and transience. The Leiden artist, Gerrit Dou, internationally famous in his own lifetime for his highly detailed and finished scenes of daily Dutch life, tried several variants on the *vanitas* theme. In particular, he promoted a secularized derivative of the penitent St Jerome theme, painting a number of scenes of non-specific, bearded male penitents in monkish

In spite of claims for art as a means of cheating death, even art itself is mutable

the claim is increasingly made throughout the course of the sixteenth and early seventeenth centuries that *vita brevis ars longa* ('life is short, but art is long'). We see it exemplified by the Utrecht painter Herman van Vollenhoven in his *Painter in his studio painting the portrait of a couple* (Amsterdam, Rijksmuseum, 1612). The painter engages our gaze with his own while pointing to a double portrait on an easel before him and the sitters beyond. The male sitter also regards the viewer while laying his hands upon a human skull, beside which is an hourglass. Here is a painter's demonstration of art overcoming death by promising to outlast mutable human life. Such claims for art formed the subject matter of a number of paintings during the period and formed the basis of increasingly elaborate conceits during the seventeenth century. David Bailly's *Vanitas with a young painter* (Leiden, Stedelijk Museum de Lakenhal, 1651) is one of the most elaborate. Here the painter has depicted his younger self pointing to an array of objects with *vanitas* associations, among which is a second self-portrait, showing him aged as he was when he painted the allegory.

As the seventeenth century progressed, the *vanitas* still lives became largely formulaic in terms of the ideas they embodied, however competently executed they may have been. Meanwhile, the inclusion of *vanitas* accessories, such as the skull, in portraits becomes equally formulaic – almost as if it were a conventional gesture towards piety rather than an inventive expression of a sense of doom that might arrest the contemporary viewer. Nevertheless, there are a few truly exceptional images from the period that manage to shock even the

habits accompanied by the requisite reminders of human evanescence, including the obligatory skull and hourglass (see, for example, *The praying monk* in Dresden, Gemäldegalerie Alte Meister, *ca.* 1635). There is no specific Roman Catholic piety attached to such works. They are simply a record of the conventional, fictionalized, generic 'penitent'. Dou also explored the *homo bulla* theme in a *Boy blowing bubbles with vanitas still life* (Tokyo, National Museum of Western Art); and when in, 1658, he was asked to provide a self-portrait for the most prestigious collection of its kind in Europe, that of Cardinal Leopoldo de' Medici in Florence, Dou depicted himself in a suitable *vanitas* pose – resting his right hand on a skull and indicating an hourglass with his left (Florence, Galleria degli Uffizi).

The treatment of the themes of *vanitas* and death by the learned and innovative painter, Jan Steen, has been overshadowed by his reputation as a comic artist. Yet, as Steen's art amply demonstrates, comedy can be just as effective in addressing the human condition as any other mode. *The bookkeeper of death* (Prague, Národní Galerie, 1662–63) is surely a very personal work. It was painted when Steen himself had just recovered from a serious illness and had recently lost his young son, Cornelis. He depicts himself standing in an interior behind a boy holding an hourglass. To one side, there is a male figure, seated at a desk writing in a ledger. In the shadow of the doorway, Death appears, holding a wailing boy by the hand. For a more public audience, however, Steen raised the curtain on *The world as a stage* (The Hague, Mauritshuis, 1665–67). Alluding to

Pieter Breugel the Elder, *The Triumph of Death*, oil on panel, *ca.* 1562, 117 × 162 cm, Madrid, Museo del Prado

the contemporary conceit that sees the follies of humanity enacted in the world as though upon a stage, Steen depicted the interior of a tavern in which numerous figures play out stock roles as lovers, gamblers, self-deluders, gluttons and sloths. Steen's approach, however, is always Erasmian – he never censors, he merely observes. This is simply the human condition. Nevertheless, above the scenes enacted, discreetly peering down from the loft, there is a small boy blowing bubbles with a skull at his elbow.

For all the inventiveness of exceptional artists such as Jan Steen, the *vanitas* was set to implode. Some artists realised that, in spite of claims for art as a means of cheating death, as *vita brevis ars longa* might suggest, even art itself was mutable: both were transient. This notion found its most sophisticated visual expression in *trompe l'oeil* paintings where the decay of art objects themselves was the subject. For example, the Antwerp artist, Cornelis Norbertus Gysbrechts, regularly depicted a *vanitas* still life with the usual ingredients of a human skull, an hourglass, a guttering candle and bubbles, but with the surface of the painting itself rendered as though it were partly cut away from its stretcher. In one version (Kingston-upon-Hull, Ferens Art Gallery; see above, p. 14), he added a knife – still inserted between canvas and wood – with the detached part of the canvas frayed and falling forward. To complete the indictment of the false

claims of art, the painter's brushes and palette have been included in the lower foreground. Yet this painted demonstration of art's decay has itself survived for over three hundred years – illustrating, perhaps, the ultimate irony of the *vita brevis ars longa* dictum.

By the eighteenth century, the claims of *vanitas* imagery on the European and American imaginations, and the claims of art to cheat death, were as good as exhausted. Skulls still found their conventional place on New England colonial tombstones and the 'Day of the Dead' was still celebrated in what would become Mexico, but people dreamed of immortality in a different register. Yet, for some, death could be conceived as a welcome release from, as Jonathan Swift put it, in his own epitaph, "savage indignation". His portrayal of those who cannot die – the Struldbruggs in *Gulliver's Travels* (1726) – remains one of the most terrifying representations of human decay without end in Western imagining. In spite of Swift's warning, however, immortality of the body is a modern ideal, best expressed by the efforts of modern medicine: conquer time, cheat death, and live forever. As we enter the third millennium of our era, we might instead do well to feel the skull beneath our own flesh with some of the same enthusiasm as our predecessors.

Christoph Amberger, *Matthäus Schwarz the Elder*, oil on panel, 1526, 73.5 × 61 cm, Madrid, Thyssen–Bornemisza Collection

Time and the Portrait

Lorne Campbell

Hans Eworth, *Mrs Wakeman* (detail), oil on panel, 1566, 91.6 × 71.2 cm, private collection

It need not take long to make a portrait. A portrait photograph can be taken in a fraction of a second; drawn portraits can be maunfactured in a few moments. In 1509 Cranach snatched a piece of charcoal from a stove and drew on the wall of an inn a portrait of the Emperor Maximilian I, which everyone recognized and at which they marvelled. Hilliard could make a portrait of Elizabeth I of England "in white and black in four lines only ... by the Idea he hath, without any pattern".

Some portrait-painters, however, work very slowly or inefficiently. Sir Thomas Lawrence delayed so long over a portrait of the Earl of Mexborough's wife and baby son that the Earl lost patience: "'Well', said Sir Thomas, 'I've been a long time, I'll allow; but I've got well forward with Lady Mexborough: it's the baby wants finishing. Now if Lady Mexborough would kindly bring the baby and give me another sitting, I really will finish.' – 'Well, Sir Thomas,' said Lord Mexborough, 'my wife will be happy to give you another sitting whenever you like, but the baby's in the guards!'" In fact the portrait was exhibited in 1821, when the baby was eleven. Ingres began his portrait of Mme Moitessier (London, National Gallery) in 1844 but did not finish it until 1856. When Picasso painted Gertrude Stein in 1905–06, he had eighty or more sittings but finally completed the portrait in the absence of the sitter.

Antonis Mor, in contrast, was said to have painted his portrait of his friend Hubert Goltzius in an hour or less. The picture, now in the Musée d'art ancien in Brussels, is done with consummate skill. Here and in similar portraits by Mor, the heads could well have been painted in less than an hour. Mor may have owed something of his success as a portraitist to the speed at which he could work. In England, Van Dyck worked "on several portraits in the same day, at extraordinary speed". He "never worked for more than one hour at a time on any one portrait .... When the clock struck the hour, he got up and bowed to his sitter, to indicate that that was enough for that day, and arranged another appointment: after which his valet came to clean his brushes and give him another palette while he received another sitter who had the next appointment." Godfrey Kneller boasted that "he has had fourteen persons sit to him in a day".

Such feats of speed and skill were remembered because they were exceptional. Most sittings proceed in an orderly fashion. The sitter comes to the artist at the same hour, so that light conditions will be similar, and at intervals of two to three days, so that paint applied during the previous sitting will have dried. Three sittings for a painted portrait seem to have been normal in England in the

Kneller boasted that he had had

Portraits were often dated to the day – and occasionally to the hour and minute

seventeenth and eighteenth centuries; but no portraitist can work according to rules, for he is constrained by circumstances and by the behaviour of each sitter.

It is usually the sitter who chooses when exactly a portrait should be painted. In 1626, however, the Grand Duke Ferdinand II of Tuscany was suffering from smallpox. It was presumably his advisors rather than Ferdinand himself who decided that Sustermans should paint his portrait on the seventh and ninth days of his illness. Some sitters chose to be painted on their birthdays. Matthäus Schwarz, for example, a prominent citizen of Augsburg, had himself painted on 20 February 1526, when he was "exactly twenty-nine years old". From the pendant at his neck hang a watch and an hourglass. Though similar references to the passage of time are often included in portraits, Schwarz's interest in chronology may have been unusual. The author of the first treatise on double-column bookkeeping to be published in Germany, and also of a pictorial chronicle of his life and clothes, Matthäus Schwarz was interested in astrology as well as chronology. When he was painted by Christoph Amberger, at 4.15 pm on 22 March 1542, his horoscope was included and his exact age was recorded: 45 years 30 days 21¾ hours. The significance which this time held for him is now mysterious, but Amberger did not complete his pendant portrait of Matthäus's wife Barbara Mangolt until 21 August 1542, five months after her husband's picture. That was her thirty-fifth birthday. The choice of date for Matthäus's portrait was without doubt equally deliberate.

Because many sixteenth-century sitters chose to be depicted at the age of thirty or thirty-three, it has been proposed that they wanted to be shown at a 'perfect' age. It was calculated that Christ was thirty or thirty-three when he was crucified and resurrected; and that the rest of humanity would be resurrected at the same perfect age. On 3 April 1551 Hermann von Weinsberg, a lawyer and wine-merchant of Cologne, recorded that he had reached the age of thirty-three (his birthday was on 3 January), the same age at which Christ had been crucified; and that, at the General Resurrection, when all men would come before Christ's Judgment Seat, everyone would appear to be thirty-three – even if they had died in infancy or in old age. Later in the same month and at the end of the following month, he commissioned from Barthel Bruyn the Elder portraits of his widowed mother (then fifty-three), his wife (then thirty-nine) and himself.

Portraits were often dated to the day – and occasionally, as in the case of Amberger's portrait of Matthäus Schwarz, to the hour and

fourteen persons sit to him in a day

minute. Those times presumably indicated when the artists finished their portraits, rather than when they started them. Such inscriptions attest to the fact that the sitters looked exactly so at those specified times. Whether such statements can be taken as literally true, however, remains an unanswerable question.

In the interesting inscription of Eworth's portrait of 'Mrs Wakeman', the unknown sitter speaks: *I am now come to those ripe years at last ... I once was young and now am as you see.* The portrait is dated 1566, when the lady was thirty-five. "Now" must mean the time at which the portrait was completed; but a portrait is out of date as soon as it is finished. Eworth's picture is dated in both senses of the word. In 1553 Mary of Hungary sent from the Low Countries to Mary Tudor a portrait by Titian of Philip II of Spain. She apologized for it and admitted that it was three years old but stressed that it had been considered very like at the time. "One can make conjectures on the progress that he will have made by adding three years to his age ...."

The principal function of a portrait is of course to commemorate the person represented. This is simply stated on a pair of portraits of a man and woman of the Gozzadini family painted at Bologna in about 1485: 'in order that our features may survive'. Dürer saw in portraiture one of the main purposes of art: "It preserves the likenesses of man after their deaths". His patron the Emperor Maximilian I wrote: "He who during his lifetime provides no remembrance for himself after his death ... is forgotten with his passing bell, and therefore the money that I spend on my remembrance is not lost ...." His sentiments are echoed in the inscription on the frame of the Judd Memorial of 1560 (**300**), where the sitters speak (spelling modernized):

> When we are dead and in our graves
> And all our bones are rotten
> By this we shall remembered be
> When we should be forgotten.

Though portraits are usually reckoned to be immutable memorials of transient beings, they are occasionally seen as impermanent, fragile and unreliable. The young Elizabeth I sent a portrait of herself to her brother Edward VI with a touching letter: "... the face, I grant, I might well blush to offer, but the mind I shall never be ashamed to present. For though from the grace of the picture the colours may fade by time, may give by weather, may be spotted by chance, yet the other not Time with her swift wings shall overtake, nor the misty clouds with their lowerings may darken, nor Chance with her slippery foot may overthrow .... I shall most humbly beseech Your

Majesty that, when you look on my picture, you will vouchsafe to think that, as you have but the outward shadow of the body before you, so my inward mind wisheth that the body itself were oftener in your presence ...."

More than three centuries later, in 1891, Oscar Wilde imagined the jealousy which Dorian Gray felt for his portrait: "Why should it keep what I must lose? ... Oh, if it were only the other way! If the picture would change, and I could be always what I am now!" In his novel, Wilde was able to grant Gray's wish.

"If the picture would change, and I could be always what I am now!"

People who commission portraits often worry that they will be accused of ostentation. Many have tried to justify their actions by finding moral purposes for portraits. Portraits could be used as aids towards self-knowledge and self-improvement and in preparing for the inevitability of death. Eworth's 'Mrs Wakeman' reflects upon her lost youth and beauty:

> My childhood past that beautified my flesh
> And gone my youth that gave me colour fresh
> I am now come to those ripe years at last
> That tells me how my wanton days be past
> And therefore, friend, so turns the time me
> I once was young and now am as you see.

"So turns the time me": I am changed by Time and contemplate the changes wrought by Time. Thomas Whythorne, an English musician who died in 1596, had himself painted at least four times between about 1549 and 1569. He wanted "to see how I was changed" and realised that "as my face was altered, so were the delights of my mind changed". On his third portrait he caused to be written:

> As Time doth alter every wight
> So every age hath his delight.

He believed that the sitters should contemplate their portraits to consider "how they ought to alter their conditions" and prepare for death. He believed that people should leave their portraits to their friends and to their children to encourage them to learn from their moral examples.

In 1529 Lucas Furtenagel represented the painter Hans Burgkmair and his wife Anna engaged in just this kind of contemplation. Anna holds a mirror, its handle inscribed, in German, 'Hope of the world'. On its frame are written, in Latin, 'O Death' and, in German, 'Know thyself'. The mirror reflects not the heads of the Burgkmairs but two skulls. In the top right corner a German inscription warns: 'Such was our shape in life; in the mirror remains nothing but this'. What exactly that means is not clear. The inscription is perhaps a piece of pseudo-profundity which certainly encourages the spectator to take

an interest in the tragic lady and the mysterious reflections.

A less sophisticated but more accessible picture is the Judd Memorial, also known as the *The Judd Marriage*, which is dated 1560. It depicts William Judd, then aged forty-seven, and his wife Joan, then aged twenty-eight. William was a skinner of London and a merchant of the Staple of Calais; his uncle Sir Andrew Judd, also a skinner and Stapler, had been a prominent Muscovy merchant and Lord Mayor of London and had died in 1558. William's wife Joan Cromwell alias Williams was a great-niece of Thomas Cromwell, Earl of Essex (executed in 1540) and first cousin to the father of Oliver Cromwell. William and Joan are identified by their coats of arms and by their initials. Very little is known about them; it has not yet been discovered why they commissioned the picture or where it was placed. In 1560 they appear to have been living in the London parish of Allhallows, Lombard Street, where William rented from the Skinners' Company a tenement in Gracechurch Street.

The couple brood upon mortality. They rest their hands on a skull, supported on a tomb-like chest. It is decorated with two shields, one showing the quartered arms of France and England, the other showing the arms of the Staple. Over their heads is written *Behold our end* and on the chest are the lines:

> The Word of God hath knit us twain
> And Death shall us divide again.

In front of the chest lies the naked corpse of a man: not, perhaps, the body of any particular person but rather another reminder of mortality. Below is the inscription:

> Live to die
> And die to live
> Eternally.

These lines are often found on tombstones. Above the skull is a candle, with the inscription *Thus consumeth our time*; the candle stands between two woolsacks, recalling Judd's activities as a Stapler, and two metal vases of flowers. "For all flesh is as grass, and all the glory of man as the flower of grass. The grass withereth, and the flower thereof fadeth away. But the word of the Lord endureth for ever."

Both William and Joan were alive when the picture was painted. They were conscious not only of the need to provide some commemoration for themselves but also of the benefits to be derived from contemplating death and the transience of human life.

Lukas Furtenagel, *Hans Burgkmair and his wife Anna*, oil on panel, 1529, 60 × 52 cm, Vienna, Kunsthistorisches Museum

PONDELOK · UTOREK · SREDA · STVRTOK · PIATOK · SOBOTA . BULGARIAN DAYS OF THE WEEK : NEDELYA ·

Jean-François Millet, *A peasant grafting a tree*, oil on canvas, 1855
Munich, Neue Pinakothek

Seasons and Moments

Time and Nineteenth-Century Art
John House

Robert Braithwaite Martineau, *The last day in the old home*,
oil on canvas, 1862, London, Tate Gallery

By the mid-nineteenth century, time had become a central issue in wholly new ways in the lives of populations across the Western world. In everyday life, the speed of railway transport transformed the experience of the immediate environment. Instead of the gradually shifting views seen from horse-drawn transport, the railway passenger was offered a kaleidoscopic sequence of new and unexpected perspectives; and the dramatically shortened travel times seemed to make the world a far smaller and more accessible place. In broader terms, too, the increasing rapidity of change – technological, industrial, social, economic, political – meant that for the first time radical alterations in world view were to be anticipated within a single lifetime. The inhabitants of the modern city could no longer expect to pass on to their children what they themselves had inherited from their parents. And, in cosmic terms, the discovery of 'deep time' revolutionized the understanding of man's place in the universe, at the same time as challenging any literal interpretation of the Judeo-Christian narrative of creation.

These developments posed problems for the visual arts. Academic art theory, in the eighteenth and nineteenth centuries, was clear about one thing: painting and sculpture were media that should deal with space – but not time. In contrast with the written word, they were concerned with single moments, not with the narration of stories through time. The German critic Lessing's celebrated essay *Laocoön* (1766) was only one of many statements of this theoretical position. In France and England, the very similar conclusions of Diderot and others were equally influential. Lessing's insistence on the "purity of the moment" in painting became a metaphor for the art of painting itself. On this account, everything that was extraneous to the physical substance of the painting was illegitimate, and contaminated the purity of the pictorial experience.

In academic theory, 'the moment' presented in paintings was an ideal moment – one that distilled the essence of the subject depicted. The subject of a painting should be immediately understood, from the combination of figures, gestures and expressions, and all the elements in the composition were meant to contribute to this central point. Yet, in a sense, a temporal dimension was admitted. The chosen moment should enshrine the whole narrative. The artist, Lessing argued, "must therefore choose the one that is most pregnant, and from which what precedes and what follows can be most easily gathered" (Lessing 1914, pp. 91–92).

Within this theory, some further temporal markers were permitted. Details might be included that indicated earlier or later stages in the story, provided that they did not disturb the fundamental temporal and thematic unity of the scene. Lessing also emphasized that time was needed for the viewer to contemplate the work of art: "The

Commentators reiterated the novel stresses and pressures of urban life

longer we gaze, the more must our imagination add; and the more our imagination adds, the more we must believe we see" (Lessing 1914, p. 19).

Lessing's principles were, of course, prescriptive and not descriptive. In his own historical context, he was seeking to locate painting and sculpture within the broader spectrum of the arts, and to impose a theoretical framework that consolidated their status. For Lessing, the pictorial arts were, by definition, more restricted than poetry, because of their limitation to the single moment.

A closer examination of the history of art in the eighteenth and nineteenth centuries, however, reveals a far more varied and flexible relationship between the arts, and far more diverse attitudes towards the roles that time might play in the pictorial arts. This essay examines a number of moments in the nineteenth century when artists engaged most directly with these issues, and asks in what ways these can be related to changing attitudes towards time itself.

Rural life and agricultural labour were central themes in painting in the mid to late nineteenth century all across the Western world. In these images, time plays a key role. The patterns of everyday country life are played out within clearly defined temporal frameworks – day and night, the seasons of the year and the stages of life. The patterns of labour are determined by the times of day and the seasons, passed on from generation to generation. In Jean-François Millet's *Peasant grafting a tree*, a peasant father grafts a tree while the mother looks on, holding the baby who will benefit from the father's providence. Here, further temporal dimensions are added – by the old cottage behind the figures and by the picture's evocation of a line from Virgil's Eclogues, "Graft your pear-tree, Daphnis, and your descendants will gather the fruit", a line that Millet seems to have had in mind when he conceived the picture. On occasion, this same notion of repeated patterns of life could be harnessed to a negative view of rural life. In Gustave Courbet's *Stonebreakers* (1849–50; destroyed; formerly Dresden), as Courbet himself insisted, the young man is destined to take over the task of the bent old man who breaks the stones – a pessimistic vision of manual labour as an inescapable fate. The optimistic vision of human life working in harmony with nature, however, was overwhelmingly the more common in the nineteenth century.

The visions of time in these pictures present an immediate paradox. They consistently evoke cyclical views of time, based on endless repetition and the idea of a steady-state world. They virtually never represent the new agricultural machinery and practices that were

beginning to transform the countryside. By contrast, the pictures were designed to be viewed in an urban context – in art exhibitions or dealers' galleries in the cities where the impact of constant change was most inescapable. Why were images of 'time's cycle' so popular in contexts dominated by 'time's arrow'?

This popularity must be viewed as a form of imaginative escapism. Commentators during the period reiterated the novel stresses and pressures of urban life, emphasizing the insecurities and constant changes that destabilized the everyday experience of the city dweller. Urban capitalism offered the possibility of self-advancement; but, alongside it, there was the ever present fear of ruin. By contrast, the standard vision of the countryside was one of solace and reassurance. And, whereas the historian must, of course, expose this vision as a myth, it was, nevertheless, an immensely persuasive myth within nineteenth-century urban culture.

In their own way, paintings of peasant life preserved the unities advocated by the academic theorists. Each scene presented a distilled moment that sought to epitomize a particular aspect of rural life and work. A very different approach was adopted, however, primarily in England, by artists seeking to give visual form to the ever changing nature of urban life. Here, the example of William Hogarth was paramount. His series of prints and paintings, such as *The Rake's Progress* (1733), enabled stock narratives of social and moral ambition and fall to be presented in visual form, in a sequence of images that each summed up a key point in the inevitable path to ruin. Augustus Egg and William Powell Frith followed this example. The most ambitious attempts to reconcile moral narrative with the limits of painting, though, were pictures that sought to embody such complex narratives in a single image. William Holman Hunt's *Awakening conscience* (1853–54; London, Tate Gallery) is the most celebrated of these, but another picture, Robert Braithwaite Martineau's *Last day in the old home* of 1862, shows still more clearly the network of interconnecting narratives that could be synthesized into a picture that ostensibly presented a single moment in time.

In Martineau's picture we witness the final ruin of an old aristocratic family. The mother pleads with the father not to lead their son astray, as she scans the newspaper for apartments to rent. The old family servant hands back the keys of the house to the sorrowing grandmother. There are many markers of pasts and futures. The coat of arms and the ancestral portraits, one labelled *Holbein*, tell of a long family history, and the racing notebook in the father's hand reveals the immediate cause of their ruin. The cold, dead hearth speaks of the ending of their shared family life. An auction catalogue on the floor and the tickets on all of the objects in the room, except the racing picture, indicates that the future holds the sale of all the family heirlooms. And the father trains up his son in the debauchery through which he has ruined himself – its vanity symbolized by the bubbles in the champagne with which father and son toast an ancestral portrait on the wall. Beyond the immediate network of narratives, there is a broader story of change that preoccupied the Victorians – the decay of the hereditary aristocracy and its replacement by the unstable world of market capitalism.

For pictures such as Hogarth's or Martineau's, narrative signs were indispensible. These devices did not go unchallenged, however. Edouard Manet, in particular, deliberately worked against such legibility. As contemporary viewers immediately realised, the compositions, gestures and details of pictures such as *The balcony* (Paris, Musée d'Orsay) cannot be interpreted in terms of a clear, coherent story. We are given no idea of the relationships between the figures or their reasons for being on the balcony. Manet's subversion of the conventions of legible detail must have been deliberate. His friend Antonin Proust recalled his mocking interpretation of just such a detail in a picture by Alfred Stevens: "Alfred Stevens had painted a picture of a woman drawing aside a curtain. At the bottom of this curtain there was a feather duster which played the part of the useless adjective in a fine phrase of prose or the padding in a well turned verse. 'It's quite clear,' said Manet, 'this woman's waiting for the valet'" (Proust 1897, pp. 202–03).

Manet's pictorial language was designed to evoke a sense of immediacy, with human gestures presented as if caught in passing. It was through this language that he sought to convey a sense of modernity. The effects he achieved, however, were worked out with great care and were often the result of extensive reworking. A comparable sense of immediacy was pursued by different means in some of the landscapes of the Impressionist group in the early 1870s. Throughout the previous century, landscape painters had made informal oil sketches out of doors of transitory natural effects. But, in paintings such as Monet's *Impression, sunrise* (Musée Martmottan, Paris), exhibited at the first Impressionist group exhibition in 1874, the capturing of such effects became the primary purpose of a picture presented to the public as a finished work of art. Even so, a rapidly brushed canvas as *Impression, sunrise* must have taken longer to paint than the duration of the very transitory effect it depicts: of the Sun rising through mist. But the painting's technique, seemingly so spontaneous and unlaboured, evokes the idea of the moment caught in passing.

In a sense, paintings such as this sought to erase the sense of passing time from the picture. Most Impressionist landscapes of the 1870s, however, were more elaborate in their execution and more complex in their subjects, and were engaged with notions of time in a number of ways. Monet's *Railway bridge at Argenteuil*, for example, raises these issues especially clearly. The foliage on the trees shows that it is summer and the play of sunlight and the crisp shadows indicate a particular time of day (though the careful finish of the picture shows that much of it must have been executed after that precise effect had passed). The presence of the railway bridge reveals the 'progress' of technology into the countryside – of the 'arrow' into the 'cycle' – while the crossing of the two trains on the bridge invokes the precise timekeeping needed for railway timetables (a particularly vexed issue at this date, before the standardization of time). In contrast to this, the sailing boats and the two figures on the bank suggest the more flexible, subjective notion of leisure time, unconstrained by such strict schedules.

"It takes a great deal of work to succeed in

In their later works, the Impressionists were less concerned with the processes of modernization, and Monet, in particular, became increasingly preoccupied with the most fleeting effects of light. Yet, paradoxically, the results were very unlike *Impression, sunrise*. He explained his aims while he was working on his series of *Grain Stacks* in 1890 : "I'm working away at a series of different effects (of grain stacks), but at this time of year the sun sets so quickly that I can't keep up with it .... I'm becoming so slow in my work that it makes me despair, but the further I go, the better I see that it takes a great deal of work to succeed in rendering what I want to render: 'instantaneity', above all the enveloping atmosphere (*enveloppe*), the same light diffused over everything, and I'm more than ever disgusted with things that come easily, at the first attempt" (Wildenstein 1979, III, p. 258).

His ostensible subject was 'instantaneity', but his increasingly slow working methods and his dissatisfaction with things that came 'at the first attempt' meant that his finished paintings were the result of protracted reworking and elaboration in his studio.

Increasingly, too, Monet came to see that this involved a different conception of the picture, and of the viewer's relationship to it. He discussed this in 1892 with the American painter Theodore Robinson: "He said that he regretted he could not work in the same spirit as once, speaking of the sea sketch Sargent liked so much. At that time anything that pleased him, no matter how transitory, he painted, regardless of the inability to go further than one painting. Now it is only a long continued effort that satisfies him, and it must be an important motif, that is sufficiently inspiring – 'Obviously, one loses on one side if one gains on another, one can't have everything. If what I do no longer has the charm of youth, then I hope it has some more serious qualities, and that one might live for longer with one of these canvases'".

These comments show that Monet saw a clear link between his working methods and the viewer's experience: only paintings enriched over a period of time would offer the material for prolonged contemplation.

As quoted by Robinson, this comment related to the viewing of individual canvases. The temporal experience of Monet's paintings, however, was further transformed when they were exhibited in series, as he had intended from 1891 onwards. As he explained to an interviewer, they "acquire their full value only by the comparison and succession of the whole series" (Bivanck 1892, p. 177). We

rendering what I want to render: 'instantaneity'"

know frustratingly little about the way in which his series were installed, but the fragmentary evidence we have suggests that they were not presented in temporal sequences that followed either the changing seasons or the times of day. Instead, Monet presumably arranged them in what he saw as aesthetically coherent groupings, for viewers to explore as they pleased.

Both his series exhibitions and his wish that the viewer "might live for longer" with his paintings reveal an attitude to time that categorically separated the experience of the picture from the 'instantaneity' that was its ostensible starting point; each viewer was to explore the painting in his or her personal time. There are clear analogies here with Symbolist notions of the nature of artistic experience as a process of gradual apprehension, and with Henri Bergson's concept of *durée*, used to characterize the individual's experiences of the world. Stéphane Mallarmé was a close friend of Monet's, but there is no evidence that Monet was aware of Bergson's ideas, first propounded in 1889.

The theme of subjective time in painting reached its most complex formulation in the work of Paul Gauguin in Tahiti in the 1890s. *The dream* (1897; London, Courtauld Institute) can stand as a sort of manifesto for this. Gauguin described the painting in a letter to a friend: "Te Rerioa (the Dream), that is the title. Everything is dream in this canvas; is it the child? is it the mother? is it the horseman on the path? or even is it the dream of the painter!!! All that is incidental to painting, some will say. Who knows. Maybe it isn't" (*Lettres* 1930, p. 66).

The painting presents us with a range of options – of figures whose dreams we might imagine. Beyond this, the figures and animals on the walls and the carved cradle offer us further spaces for contemplation, both through their actions and through their evocation of an old, 'primitive' culture. The dreams of the figures, we feel, embody these deep traces of inheritance and memory. Yet, as Gauguin was keen to emphasize, all these options came within the control of the painter, whose creative vision we are invited to explore.

Gauguin's intended audience was in Europe; and *The dream* draws on the longstanding tradition of dreams of the 'Other' – of remote times and spaces that offered an imaginative escape from the European dominance of 'time's arrow'. As we have seen, painting in the nineteenth century might engage with the present, whether through narratives of the moral dilemmas of modern society (as in Martineau's *Last day in the old home*), or through the immediate visual experiences of the modern world (as in the earlier work of the Impressionists). Yet by the end of the century, both Monet and Gauguin in their different ways were re-emphasizing the need for painting to transcend the frameworks of linear history and clock time, in order to engage with the viewer's imagination. In one sense, this marks a return to ideas closer to the academic idealism of the beginning of the century, since it insisted that the viewer's experience should grow from the contemplation of the ideal moment distilled within the painting. But the nature of the experiences invoked at the end of the century was wholly different. The work of art no longer appealed to a shared canon of values that united the aesthetic with the ethical. Instead, it treated the experience of the painting as something timeless. Its appeal was now individual and private; every viewer was now invited to explore it in their own subjective time.

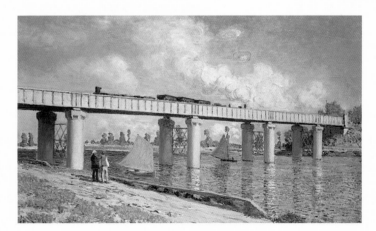

Claude Monet, *The railway bridge at Argenteuil*, oil on canvas, 1873 London, The Helly Nahmad Gallery

3 · THE DEPICTION OF TIME

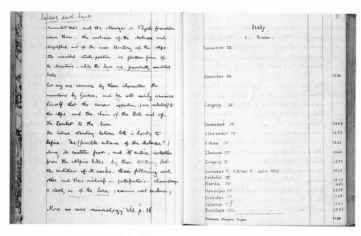

John Ruskin's Rock Book, 1861–62, Ruskin Foundation,
Ruskin Library (University of Lancaster)

"These Fragments I have Shored against my Ruins"

Marcia Pointon

Joseph Wright, Miniature portrait of *The Duke of Wellington*, mounted in a
frame containing locks of hair, 1808, private collection

William Dyce's painting *Pegwell Bay: A Recollection of October 5th 1858* (painted in 1859 and exhibited at the Royal Academy in 1860; 274), seems to epitomize nineteenth-century preoccupations with time. To begin with, the title given by Dyce to a painting of a scene which he originally sketched in 1857 both signals a precise date in the calendar and, at the same time, raises the question of memory. Secondly, 'astronomical time' appears in the comet represented in the sky. 5 October was the day when the English public was advised to look for what was expected to excel all visible comets since that of 1811. Thirdly, both 'diurnal time' and 'geological time' are represented by the activity of collecting shells and stones. Making a personal collection focuses the eye on the detail. This sort of activity reassuringly suggests what can, at least at some level, be grasped and known. Astronomy, on the other hand, represented the great unknown.

The compelling clarity of *Pegwell Bay* has led to speculation about Dyce's use of photography. When compared with W.P. Frith's *Ramsgate Sands: Life at the seaside* (1854), for example, in which a microcosm of society is assembled on a stage-like space, *Pegwell Bay* approximates more closely to the poignancy of a single moment. Dyce disposes his figures to give the impression of the desultory and the accidental, an impression which is, of course, elaborately orchestrated. The relative smallness of the foreground figures and the pools of reflective water at the front of the picture produce the sense of a distant focal point – forming a gulf between viewer and object. Optical devices were popular at this period and the stereoscope, in particular, creates a similar effect of distancing.

On another level, *Pegwell Bay* is an intensely personal memory of a specific time and a specific place. The full title reads rather like a page from a diary and the use of the artist's family as models reinforces this autobiographical strain. Personal time is mapped on to a demonstration of geological and astronomical time which is, in turn, articulated within a spatial organization dependent upon single-point perspective. The women in the foreground of *Pegwell Bay* in their crinolines and shawls are abrasively contemporary, but this contrasts sharply with the apparent timelessness of the beach on which they find themselves. The time of fashion, then as now, is an

Every step taken towards adulthood was a step

absurdly brief moment of futility and worldly delusion, while the appearance of the cliffs changes so slowly the change is unrecognizable to the human eye. Further, women's time is often understood to be a cyclical time of reproduction, rather than of production in the economic sense. Even more than the women, though, it is the figure of the child that stands in a special

One of the aims of public museums was to organize time and space

relationship to its environment. This figure is picked out in head-and-shoulders against the reflective wet sand.

Children in western culture are powerful presences in an adult world. In particular, the promotion of the child as a being endowed with special prescience – a subject in his or her own right, with privileged and intuitive access to natural forces (or to divinity) – is a creation of the era following on the Enlightenment. Wordsworth's *Prelude* (1799–1805) posits childhood experience as formative of the adult subject. And, for William Blake, every step taken towards adulthood was a step away from childhood's innate knowledge of immortality. Dyce's son is not represented alone at Pegwell Bay, but the artist has, nonetheless, contrived by pictorial means to establish his separateness from the surrounding adults. Poised for the future, absent-mindedly holding his spade, this small child, with feet firmly apart, stares over and beyond the viewer.

The motivating passion of virtually every field of nineteenth-century scholarship was the endeavour to discover a linear sequence through which to make sense of the past. We think, in particular, of European historians of this era: Ranke, Froude, Macaulay, Marx. But nineteenth-century geologists were, similarly, determined to establish the relative dates of geological phenomena. "For if no scale of time be known, the problem of the history of the successive conditions of the globe becomes most deplorable"(John Phillips, 1837). The only way to deal with this, it was thought, was to adopt as a scale for measuring the Earth's relative antiquity the series of stratified rocks which could readily be seen, for example, on beaches such as Pegwell Bay. Fossils and spectacular geological specimens, gem-stones and precious minerals, were part of the *Wunderkammern* that princes and antiquarians established throughout Europe in the seventeenth and eighteenth centuries. And, as public museums developed in the nineteenth century, one of their aims was to organize time and space (the distant in history and the foreign in geography) in explanatory displays that not only identified all specimens but also placed them within a time frame that was known and linear.

away from childhood's knowledge of immortality

John Ruskin is now primarily remembered as an art critic and social theorist, but his earliest passion – one which remained with him to his death – was for mineralogy. Ruskin's drawings of gneiss rock and other geological formations justly elicit admiration for their exactitude, but he did not simply draw them for drawing's sake. Ruskin collected minerals on a large scale. He purchased from

professional dealers as well as finding specimens himself and spent hours sorting and labelling rocks, many of which he subsequently gave to museums and educational institutions. The collecting and study of minerals was one of the ways in which Ruskin managed his acute anxiety about time – he regularly calculated how many working days of his life he could reasonably anticipate remained to him and he never failed to write his diary. Minerals were, as Phillips claimed, a key to understanding the Earth's surface and, thereby, a tool that could be used to calculate the passage of global time. But for Ruskin – and, no doubt, for many others – minerals were also objects of aesthetic contemplation, analogues for moral virtues like firmness and clarity and reminders of the connections between the inanimate and the animate worlds. On the one hand, they belonged to the material world. On the other, they could be collected, handled and scrutinized. Ruskin accorded to minerals a capacity to evoke human and personal histories. For example, he named the mineral specimens that he donated to the Natural History Museum in 1887 after individuals who had been important to him in his life: Bishop Colenso of Natal (after whom the Colenso diamond, stolen in 1965, was named), Sir Herbert Edwardes (a soldier and civil servant in the Punjab whom Ruskin admired and after whom he named a ruby) and Joseph Couttet, the alpine guide who was Ruskin's companion and nurse during several of his youthful visits to the Alps and to Italy, after whom a piece of rose quartz was named. Functioning, therefore, at one level as autobiography within the public exhibition space, these minerals provide an apt demonstration of the way in which geological specimens collected ostensibly for the advancement of knowledge were also constituted as cultural artefacts signifying human as well as geological time scales.

Some stones that Ruskin collected were very rare. But, unsurprisingly, since he taught the virtues of authentic traditional workmanship, Ruskin also loved ordinary pebbles. Such stones invoked for him histories that elided the human with the geological. Lecturing at the London Institution in 1876, Ruskin held out both his hands to his audience. On his right palm lay "a little round thing" and on his left "a little flat one". The first was a pebble and the second a sovereign. The latter launched him into the themes of empire, value and the economy. The former, the black pebble, led him to recall a meeting with the great geologist James Forbes – a man seemingly made of mountain flint in his inaccessibility and taciturnity. To these two objects, Ruskin added a further black pebble, one that "used to decorate the chimney-piece of the children's play-room" in his aunt's house in Perth when he was seven "just half a century ago". With this third pebble, Ruskin exposed the bond between two sorts of time: the human span of time evoked by the image of his self at age fifty-seven and his self at age seven, and the mineralogical span of time that forms the life of the pebble. The span of time invoked by the memory of his own childhood was, of course, a mere flicker by comparison with the mineralogical time of his pebbles. But childhood – his own and that of others – was more than a rough and emotive yardstick in Ruskin's discourse; it was a framework for the emotional management of time.

Mineralogy, for Ruskin, was also a form of historical enquiry – and it is certainly no accident that Ruskin used the same book that he had used to chronicle the kings of Europe for his translations of passages from a German geological text and to work out a colour system for classifying minerals. He used the blank sheets of this book to inscribe mineralogical lists and set them alongside the names of thirteenth-century popes and kings with their lineages. An attempt to impose order on the apparent chaos of medieval kingdoms and fiefdoms is juxtaposed with an attempt to impose order on the physical world (see illustration on p. 198).

When someone is born or dies, those around take stock; beginnings and endings occasion a looking backwards as well as a looking forwards. Material artefacts – jewellery and other mementos – accommodate the need to mark such moments which were, customarily, inscribed in family Bibles. Thomas Mann, whose novels are suffused with temporal imagery, opens Part II of *Buddenbrooks* with the energetic head of the family firm recording the birth of his latest-born child in a gilt-edged notebook with an embossed cover. Throughout the novel, this family chronicle reappears as births, deaths, marriages and divorces are recorded. It is the material object through which the passage of time as it impacts upon the generations of the family is inscribed. As a counterpart to the family to which it belongs, it is a revered object. As the end of the dynasty, and the close of the novel, approaches, the sensitive, musical and sickly last-remaining scion of the house, Hanno, finds this notebook left open on the writing desk. His eyes rove over the names in the family tree, of which his own is the last. Then, taking his ruler, "he made with the gold pen a beautiful, clean double line diagonally across the entire page ...". In response to his enraged father's demand that he explain his conduct, he replies merely, "I thought – I thought – there was nothing else coming". Shortly after this prophetic act of symbolic mutilation, the death of his father occurs, followed soon after by his own death from typhoid.

Like family chronicles, jewelled artefacts have always played an important part in marking rites of passage. In short, jewellery made of gold (mined and processed) and gem-stones (minerals excavated) connects its wearer to the geological time-span to which so much attention was devoted in nineteenth-century England. The human

frame for which jewellery is devised is predestined to death and decay, while the ornaments made of enduring minerals will outlive it by far. Therefore, the juxtaposition of human and mineral points up the comparative brevity of a human lifespan. The forms of jewellery reinforced this. Carried on a chain about the person, watch keys, for example, often bore funerary motifs – such as weeping willows and skulls – in the Early Modern period. These served as a reminder to those who possessed time-pieces that the passage of time led inexorably to the Day of Judgment. Jewellery could also support small-scale imagery (portraiture or emblematic motifs) and human hair (either in its natural form or teased into a figurative device). The use of human hair as a medium in portraiture was by no means uncommon in the seventeenth and eighteenth centuries and, by the nineteenth century, hairwork had become popular and commercialized (see **309**).

Hair in jewellery connotes either love or mourning. In each case, it is associated with conditions that are understood to transcend time. Instruction books for home hairworkers clearly indicate that this was a self-consciously commemorative discipline. As one contemporary text put it, "Hair is at once the most delicate and

"Teeth, bones, and hair, give the most

lasting of our materials and survives us like love". Characteristic examples of nineteenth-century hairwork include rings and lockets combining locks of hair, hair plaited as a backdrop to a scene with an urn and weeping willows, and initials and a motto. In other instances, the whole article is made of worked hair. Possessed of a lock of hair of a loved one, it was said, the subject "may almost look up to Heaven and compare notes, may almost say: 'I have a piece of thee here, not unworthy of thy being now'." Touch is an important vector of memory and so objects designed to be handled and worn close to the body have particular resonances in relation to the human experience of time.

As Sir Thomas Browne tells us, "Teeth, bones, and hair, give the most lasting defiance to corruption". Our hair changes colour and falls out with the passage of time, and cutting and saving locks of hair was (and is) a common way of marking the passage from infancy to old age. In the case of the Duke of Wellington, an unusually elaborate example has survived. Locks of the Duke's hair were preserved and at some later date incorporated into an ornamental frame surrounding his miniature portrait painted by Joseph Wright in 1808 (illustrated on p. 198). An inscription on the reverse indicates that the portrait was painted just before Wellington left for the war in Spain. This artefact constitutes an essay in time written with ingredients that each, distinctively and independently, stages a particular time in one man's life. The first lock of hair (in the left-hand compartment) is dark brown and, according to the inscription incised beneath it – *1808 A.W.* – commemorates the year in which Wellington was created Lieutenant-General following his victory over the Danes. At the foot of the frame is a lock of light-brown hair under which are engraved the words *18th June 1815 Copenhagen*, the date of Wellington's victory at Waterloo; this lock must come from the Duke's celebrated charger, Copenhagen. The third glass lozenge reveals a lock of completely white hair with the inscription *1852 W*. Nationally revered, Wellington died aged eighty-three on 14 September 1852. As one biographer has remarked, it is hard to know where Victorian sentimentality ends and the feeling for an almost godlike being begins. It was even rumoured that Lord Clanwilliam asked for a relic not only of hair but also of the hand that wrote the Waterloo dispatch.

To gather these fragments of a life and to enshrine them around an image of the man in his prime is an act of piety. It is also an act of veneration which invites subsequent generations mentally and imaginatively to match these different forms of evidence. The hair, in its actuality, evokes the scale of the human body itself. By contrast

lasting defiance to corruption"

with this almost coarse manifestation of the organic, the delicately limned portrait offers a window into nostalgic identification with a heroic and personalized past. The three ages of hair, and the symbolic uniting of the man with his horse, authenticate that past.

The introduction of portrait images into bracelets and necklaces became popular during the reign of Queen Victoria who, in 1840, had a portrait of Prince Albert made up into a bracelet with a gold band and who, in 1850, presented to Lady Sarah Spencer on her retirement as governess to the royal children a bracelet made up of miniature portraits of her five charges. A less costly way of linking together material souvenirs of the children of a family was to make up jewellery comprising connected miniature glazed cases, each containing a lock of hair and inscribed with the appropriate child's name (see **309**). In one such instance, the nine children born over eighteen years to the Tabuteaus, a Huguenot wine merchant and his wife who settled in Dublin, were commemorated with identical lozenges suspended from a gold chain which terminated at either end with a pin surmounted by a flying bird in gold. Each lozenge contains a lock of hair with the child's birth date and name engraved on the reverse. The eldest, Ethel, was born 18 November 1871 and the youngest, Rupert, on 20 March 1889. Unlike the Duke of Wellington's portrait, this piece was designed to be worn on the body – probably as an ornamental chain to fasten a shawl, though it may have started life

as a bracelet – and it would be easy to mistake the hair, gleaming under glass, for some semi-precious stone. The interchangeability is, perhaps, not accidental. It was (and is) commonplace to compare children to jewels, with the implication that they are beyond price. The result of a careful collecting and saving of locks of hair – for the jewellery must have been made up after the birth of the last child to achieve the necessary symmetry – this piece functions in many ways, some ornamental and practical and some symbolic. Forming part of a woman's apparel, it is also (through its formation) a mute family chronicle. It is mute in so far as each segment stands for and materially represents the existence of a child whose identity can only be discovered once the chain is removed and the inscriptions on the reverse sides of the lozenges are deciphered. It shares this capacity to provoke participation with relics and museum objects that invite a *rapprochement* from the viewer. It is mute also because it tells only a fragment of the story. Unlike reliquaries, which are associated with death, this series of mini-shrines occludes all but the celebration of birth. Yet these locks of hair, caught in a time warp and enshrined in long-lasting gold, paradoxically invoke the very inevitability of death which they are structured to defy.

Human beings perceive of themselves existing in something called time and this medium is measurable and communicable, but only by reference to those things that are external to the mind. Externalities comprise both the subject's own body – and the processes of ageing that we can discern – and the artefacts with which societies endeavour to make sense of the world in which they live. The museum, in which nineteenth-century philanthropists and governments invested so much, not only represents an institutionalized device to 'mark time' (and thus to stay still) but is also a striving to make sense of the individual by annexing it to the social. The physical make-up, the apparent functions, and the resonances of a series of objects – in private 'museums' of family history or in the public domain – have been explored in this brief account. Each is a fragment provoking viewers today to question, to touch and to feel. Each object represents in some way the human effort expended upon capturing and securing our natural surroundings – the organic and mineral world we inhabit and which outlives each one of us. None of the artefacts discussed in this essay, some celebrated and others little known, was designed to be useful. Instead, each was produced in the expectation that it would operate in a specially designated mental space, that it would in some sense commemorate an individual, an epoch or an event. That these artefacts remain at some level meaningful within the context of an exhibition and an historical account made so many years after their creation offers, perhaps, the most compelling evidence of a widespread conviction that – against all odds – memory lives.

3 · THE DEPICTION OF TIME

Marcel Duchamp
To be looked at (from the other side of the glass) with one eye, close to, for almost an hour, 1918
Oil paint, silver leaf, lead wire and magnifying lens on glass (cracked), mounted between two panes of glass in standing metal frame, on painted wooden base, overall height 255.8 cm
New York, The Museum of Modern Art, Katherine S. Dreier Bequest

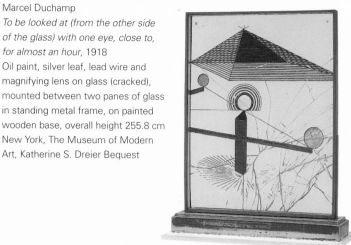

Art and Time in the Twentieth Century

Dawn Ades

Giorgio de Chirico, *Gare Montparnasse (The melancholy of departure)*, oil on canvas, 1914, 140 x 184.5 cm
New York, The Museum of Modern Art, Gift of James Thrall Soby

At the turn of the twentieth century, there were a number of radical changes in the way the universe could be explained that affected the relationship between experience and the visual arts. To begin with, time itself was shattered into different components. Einstein's general theory of relativity, Freud's work on memory and on dreams, Henri Bergson's philosophical attempts to relate consciousness to the material world in the notions of *durée* and intuition, or Marx's use of time to analyse labour and production in capitalism – each world view proposed a different scale for measuring time; scales which seemed to contract and expand according to criteria that owe as much to imagination as to scientific analysis. Added to this new diversity, there were also genuine concerns (based on a slight misunderstanding of the second law of thermodynamics) that the 'clockwork' of the physical universe itself was running down. This, coupled with the deep-seated anxieties generated by the general collapse of a rationally ordered world view during the aftermath of the First World War, led to a cultural pessimism about the decadence of civilization – an idea that gripped the public imagination then in the same way that chaos theory does today.

Many have claimed that photography and the cinema dealt a mortal blow to the old modes of expression. For example, André Breton, writing in 1921 in connection to an exhibition of Max Ernst's photomontages and collages, stated: "Today, thanks to the cinema, we know how to make a locomotive arrive in a picture. As the use of slow motion and fast motion cameras becomes more general, as we grow accustomed to seeing oaks spring up and antelopes floating through the air, we begin to foresee with extreme emotion what this time-space of which people are talking may be. Soon the expression 'as far as the eye can reach' will seem to us devoid of meaning, that is, we shall perceive the passage from birth to death without so much as blinking …" (Breton 1948, p. 177).

The possibility of a Fourth Dimension filtered into

Whereas new modes and media, hitherto strange to art, have emerged in the twentieth century, many of which are directly inspired by ideas about 'time' and 'tempo', the old modes of painting and sculpture have been remarkably resilient. The ways in which time has been implicated in art and the ways in which it has been represented, however, is a more complex matter.

Painting, unlike music or poetry, does not have a natural temporal extension. It cannot automatically command a time of viewing. Its apparent immediacy is an illusion; there are no unambiguous means to retain the spectator's attention, short of the introduction of movement. In 1918, Marcel Duchamp created a work that connects this concern to his long-standing interest in "*horlogisme*" and, in particular, his interest in the relationship between the science of perspective and the kind of time measurements marked by a sundial's gnomon. The shape of the pointed rod in the centre of his little glass picture, *To be looked at (from the other side of the glass) with one eye, close to, for almost an hour*, is derived from a sundial, and the circles and radiating lines recall dialling diagrams, such as those in Emmanuel Maignan's *Perspectiva Horaria* of 1648. Above the tip of the gnomon, there is a circle, perhaps embodying the 'gaze' of single-point perspective. Overall, the painting is about the passage of time itself. When Duchamp completed the work, he suspended it from a balcony, where the shadows cast by its elements might function like a sundial.

The possibility of a Fourth Dimension also filtered into the thinking of the avantgarde artistic circles of the period. Duchamp was fascinated by these theories, avidly reading Poincaré and Jouffret. In his notes, he speculates what this Fourth Dimension might be and how it might be explained in terms of our perception of the three known dimensions. Duchamp also speculates about "Virtuality as 4th dimension": as an echo is virtual sound, so a reflection is a virtual image. In his notes on deviations of time in a space-time continuum, he produced the idea of a "Clock seen in profile so that time disappears, but which accepts the idea of time other than linear time" (Duchamp 1973, p. 101).

Duchamp's notes are a wonderful mingling of scientific and fantastic speculation, set out in brief but suggestive propositions. They also bear on entirely new ways of making works of art. His 1914 work, *Three standard stoppages*, is closely related to these speculations. It is a unique object that creates its own genre – a mixture of elegant scientific experiment, game with chance and constructed object. Three metre-lengths of string held horizontally were dropped from a

the thinking of the avantgarde

metre high, and were then glued down. Templates made from the curved shapes they created produced "new units of measurement". It is surely coincidence, but Duchamp's method does seem to anticipate Stephen Hawking's description of one aspect of general relativity: "bodies always follow straight lines in four-dimensional space-time, but they nevertheless appear to us to move along curved paths in our three-dimensional space. (This is rather like watching an aeroplane flying over hilly ground. Although it follows a straight line in three-dimensional space, its shadow follows a curved path on the two-dimensional ground)" (Hawking 1988, p. 34).

The Fourth Dimension was also widely discussed in relation to the

Cubist paintings of Picasso and Braque, as an explanation of their ambiguous relationship to the real world of the senses. Both the shifting perspectives in these paintings (with the suspicion that the painter's eye has wandered round the figure) and the ways that planes both intersect and define volumes creating highly ambiguous spatial relationships find echoes in Duchamp's *Notes*. For example, one might cite: "On the vanishing line in perspective there are several vanishing points …. By analogy, there will be several vanishing lines all belonging to the same vanishing plane" (Duchamp 1973, p. 91). On the other hand, these still difficult paintings could also be understood as a reaction against the 'snapshot' character of the photograph. Photography's instantaneous image is apparently immediately accessible to the viewer; its transparency seemed to short-circuit the contemplative mode of appreciation or the effort of decipherment demanded by a painting. Cubist paintings tease and contradict our attempts to read the image: they force us to spend time, investing in a promised resolution which is forever frustrated. We try to read these image as clues towards a represented reality, only to come up again and again against un-anchored or abstract forms or marks of the brush. There is no resolution.

Henri Bergson, the most influential philosopher in Paris just before the First World War, sought a metaphysical solution to the problem of psychological time. Without dismissing the value of science and analytical reason, he argued that they were "incapable of grasping the facts of creativity and of time" (Kolakowski 1985, p. 6). Bergson's notions of intuition and *durée* touched a chord for many artists. '*Durée*' was Bergson's term for what he called 'real time' (as opposed to the 'time of physics'), in which each moment of consciousness is different and is not reversible: "Each moment carries within it the entire flow of the past" (Kolakowski 1985, p. 3). Unlike the matter of the past which perishes, memory does not, so *durée* is the continuity of consciousness in the memory.

Henri Matisse's *Notes of a Painter*, for instance, must be informed by Bergson's notions of *durée*, when he says that "Underlying this succession of moments which constitute the superficial existence of beings and things, and which is continually modifying and transforming them, one can search for a truer, more essential character, which the artist will seize so that he may give to reality a more lasting interpretation" (Matisse 1978, p. 37). Matisse desired to "reach that state of condensation of sensations which makes a painting. I might be satisfied with a work done at one sitting, but I would soon tire of it; therefore I prefer to rework it so that later I may recognise it as representative of my state of mind" (Matisse 1978, p. 36). Matisse's painting, through condensation of both colour and line, aimed at a "broader meaning, one more fully human", which would unite intuition and memory. "Movement

4 · THE DEPICTION OF TIME

seized while it is going on is meaningful to us only if we do not isolate the present sensation either from that which preceded it or from that which follows it" (Matisse 1978, p. 37).

For Matisse, there was no question of literal movement in a work, but, in conjunction with the machine aesthetic that emerged in the early twentieth century – most notably with the Futurists – this was to become a real challenge for the artist. The Futurists held that a new consciousness was born of speed, modern cities, long-distance communication and novel techniques like X-rays. Simultaneity should replace the single, static image, while the beauty of the machine had rendered the classical canon obsolete. The important 1912 *Manifesto of Futurist Sculpture* by Umberto Boccioni celebrates mechanical movement: "We cannot forget that the ... engaging and disengaging of two cogwheels, the fury of a fly-wheel or the whirling of a propeller, are all plastic and pictorial elements, which any Futurist work of sculpture should take advantage of. The opening and closing of a valve creates a rhythm which is just as beautiful as the blinking of an animal eyelid but infinitely more modern" (Boccioni 1973, p. 64). Despite this rousing appeal, Futurism still sought to represent movement by static pictorial or sculptural means, rather than incorporating it literally.

In fact, the first 'moving sculpture' was a private gesture within a very different context, possibly inspired as an ironic comment on the logical failure of Futurist sculpture. Duchamp constructed his *Bicycle wheel* of 1913 for the studio as a kind of pleasant distraction. It was, he said, like "looking at the flames dancing in a fire-place" (Schwartz 1997, p. 588). Aware of Futurist rhetoric, and also sensible of the aesthetic attraction of the machine, Duchamp here aborts their utopian ambitions. Not only is there neither visible labour nor obvious representational content in this 'collage-object', but the wheel turns aimlessly, without function and in a void. It is suggestive of the disproportion between human time and the time of industrial production analysed by Marx: "Machinery constitutes a sort of industrial perpetual motion, which would go on reproducing without pause did it not encounter in its human assistants certain obstacles – their bodily weakness and their wills" (Marx 1942, p. 428).

Naum Gabo's *Kinetic sculpture* of 1920, the first to utilize mechanical power, produced in the context of the Constructivist experiments in Russia, is at the other axis of the modernist relation to the machine. The 1920 *Realist Manifesto* by Naum Gabo and his brother Antoine Pevsner conceived three-dimensional construction according to new criteria, proposing the "activation of space" through new forms and materials, replacing solid mass by planes and conveying energy through kinetic rhythms. Among the artists

influenced by them was László Moholy-Nagy. Moholy pioneered an apparatus for "painting with light" in space, his *Light-prop* or *Light-space modulator*, of 1930, which became the subject of his abstract film: *Light-Play: Black and White and Grey*.

Subsequently, movement has been incorporated into sculpture in diverse ways – from the random, 'natural' sway of Calder's mobiles to the kinetic constructions of the 1960s. Movement may be mechanically controlled, its repetitive character producing tension, climax, and even humour. Artists have also explored the active involvement of the spectator in various ways. In the 1960s, the Brazilian artist Helio Oiticica fabricated capes to be worn, and his compatriot Lygia Clark made rubber 'scarves' to be draped in therapeutic experiments round the body – each operating in natural, unstructured time.

Movement within a work on film or video creates its own time, with repetition, pulsation, slow motion and so on. The American Joseph Cornell took a B-movie in the 1930s and recut it (adding a pink filter), running sequences backwards and repeating others, making a totally disorienting dream-like film. Installations now often include video or film; and, conversely, video artists can also use settings and

The experience of working in time is visibly

constructions to locate their narrative (as in Bill Viola's work). The South African artist William Kentridge has used a triptych formation to place his extraordinary sequences of drawings and film fragments either side of a central "communicating vessel".

In modern art, the presence of clocks, watches and hourglasses stands for many things beyond the traditional reminder of mortality and the fleeting nature of human life. One artist who does use the theme of *memento mori* is Joseph Cornell, who made several variations, all known as *Untitled (Sand-fountain)*, of a type of box-construction containing sand and a broken glass goblet, and sometimes other objects like shells or bark. Some of these actually function, like a cross between an hourglass and a child's game: the sand pours from a compartment in the upper part of the box into the cup of the glass, from whose uneven edges it then spills over to form shifting dunes on the floor of the box. The sand is restored to the top of the box by tilting it upside down. Beside the metaphor of the hourglass, Cornell sets up another, paradoxical, association, through the title and the sand 'cascade' itself, to water. The idea of a water fountain, the 'fountain of youth', the fountain of life, haunts the soft dry grains and broken container, mournful remains of the 'sands of time'. Old-fashioned maps of the heavens are often pasted on to the rear wall of Cornell's boxes. He loved these images as examples of the constructive power of the human imagination, rather than as bizarre instances of out-of-date scientific theory. In another *Untitled* box of 1948–50, Cornell carefully arranged white watch-faces on the intersections of a geometrical grid behind a white cockatoo. Stripped of their numbers and hands, the watches are rendered useless, but also resemble the Moon.

Single clock-faces in paintings by de Giorgio De Chirico and Max Ernst carry highly ambiguous meanings in relation to physical and psychological time. In De Chirico's *Gare Montparnasse (The melancholy of departure)*, 1914, a clock prominently mounted in a brick tower dominates the façade of the station and street – a familiar and reassuring object from the modern world of train timetables, punctuality and long-distance communication. The contradictory perspectives in this painting, however (none of the angles of the walls, towers and arcades meet at the same point), produce a highly unstable space and a nightmarish experience of a new relationship between space and time.

Perhaps the most famous representation of 'time' in twentieth-century painting is Salvador Dalí's *Persistence of memory*. Dalí painted this in Paris in 1931, not long after he had joined the

encoded in many works of art

Surrealist movement. In the midst of painting a canvas of his favourite landscape (the rocks and bay at Port Lligat in Catalonia with a leafless olive tree) Dalí was left alone one evening after a dinner at which there had been a particularly ripe Camembert cheese. Instantaneously he 'saw' the solution to his landscape: "I saw two soft watches, one of them hanging lamentably on the branch of the olive tree" (Dalí 1968, p. 317). The impetus behind this seemingly irrational image lies in Dalí's avid interest in physics, which was just as great as his interest in Freud and psycho-analysis. As he writes in his 1935 essay 'The Conquest of the Irrational': "Today the new geometry of thought is physics, and if space, as Euclid understood it, was nothing more to the Greeks than a very distant abstraction, inaccessible still to the timid three-dimensional continuum that Descartes was to announce later, in our time space has become, as you know, that terribly material, terribly personal and significant thing which weighs us all down like authentic *comedons* [blackheads] .... Salvador Dalí, in 1935, is no longer content to make auto-amorphism for you out of the agonising and colossal question which is that of Einsteinian space-time, he is no longer content to make libidinous arithmetic out of it for you, no longer content, I repeat, to make flesh of it for you, he is making you cheese of it, for be persuaded that Salvador Dalí's famous soft watches are nothing else than the tender, extravagant and solitary

paranoiac-critical camembert of time and space" (Dalí 1968, p. 422).

The cheese/clock is a metaphor for the curvature of Einstein's space-time. Later, after the Second World War, Dalí added another twist to his soft watch by representing it, in *Disintegration of the persistence of memory*, exploding into fragments (New York, Museum of Modern Art). This was prompted again by scientific research into the atom, and, above all, by the explosion of the atomic bomb.

Automatism produced a kind of free drawing practice which is the visible marking of the time of its making. André Masson's drawings, for example, are an acknowledged equivalent to the Surrealist practice of automatic writing – "the dictation of thought, in the absence of any control exerted by reason, and outside any moral or aesthetic considerations" (Breton 1969, p. 26). Masson, aware of Paul Klee's graphic practice of 'taking a line for walk', conceived of these drawings as the visible unfolding of an image, in which the process is as important as the finished work. He allowed an ink-laden pen to run over the surface of the paper, and then chose whether or not to make visible the suggestions he 'saw' within the lines. To this extent then, they differ from children's drawing, for he would start without a conscious subject, and would experience the images as 'revealed'.

Automatism took many forms within Surrealism, but it is in these drawings that the experience of their making in time is visibly encoded. A similar effect can be recognized in the work of Jackson Pollock, who acknowledged the influence of the Surrealists as well as that of the Native American sand-painters. The web of lines in Pollock's later paintings, overlapping and turning in from the edges to enclose their space, are also the record of that activity. There is, despite the apparent scrawl, an almost hallucinatory clarity in these canvases. Although one cannot literally locate a beginning and an ending, there is still a strong awareness of the lines crossing and lapping over one another, which gives a sense of temporal activity.

Joseph Cornell, *Untitled (Cockatoo clock)*, mixed media, *ca.* 1948–50, Chicago, collection of Mr and Mrs E.A. Bergman

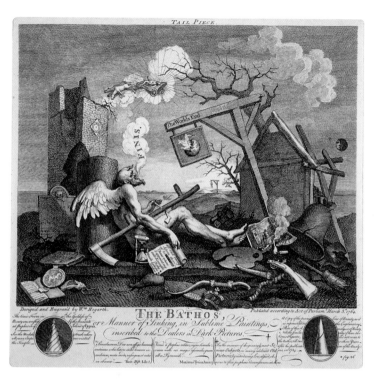

227 William Hogarth

*The Bathos, or the Manner of
Sinking in Sublime Paintings,
inscribed to the dealers in Dark
Pictures*
Etching and engraving, March 1764
London, The British Museum, Dept.
of Prints and Drawings
[1868.8–22.1629]

When Hogarth first advertised his
print in April 1764, he described it as
a "Tail-piece", indicating that he saw
it as his final engraving – one that
might serve as the colophon to a
bound edition of his collected works.
It was intended not only as a final
comment on what Hogarth saw as
the deplorable state of life and the
arts in England in the 1760s, but also
as a final salvo against the theorists
and aesthetes with whom he so
violently disagreed. At its simplest,
The Bathos depicts the end of time.
Father Time is shown collapsed, his
pipe, scythe and hourglass all
broken. With his last breath, he
exhales *FINIS*. Scattered around him
is the debris of life: a playbook open
to its last page with the stage
direction *Exeunt omnes* highlighted;

a statute of bankruptcy for Mother
Nature; a discarded shoemaker's
model; a broken bow, broken crown
and a worn-out wig brush; at the
right, there is a torn and burning
copy of *The Times* and, in the middle
distance, a ruined belltower with a
clock face with no hands and a
collapsing tavern, of which the trade-
sign signals *The World's End*. Even
the Sun god Apollo has collapsed,
the wheels of his chariot shattered
and his horses dead. As the
accompanying text makes clear,
however, this image is not about
exhaustion, dying or giving up. In
1757, Edmund Burke had published
*A Philosophical Enquiry into ... the
Sublime and Beautiful*, in which he
codified contemporary feelings about
the superiority of those *vanitas*
paintings that, with dark and moody
tones, created a feeling of the
sublime in the viewer. Hogarth
thought this was all nonsense; and
the purpose of *The Bathos* is to
show how the ruin of man's
achievement is not, in any manner, a
catalyst towards the sublime.

One tends to associate *vanitas* painting with sixteenth- and
seventeenth-century Holland, but the tradition of recalling
the imminence of death in the midst of life has a much
longer tradition. The term *vanitas* comes from the Latin translation
of Ecclesiastes (12:8), warning that "*vanitas vanitatum, omnis
vanitas*" ('vanity of vanities, all is vanity') – everything that man
does or achieves in this life is in vain. *Vanitas* imagery has an
equally long tradition. Skulls and skeletons can be found on ancient
Greek funerary stelae and Roman sarcophagi often bear the image
of a putto leaning on an extinguished torch to symbolize the 'light
gone out' with the deceased's passing. The most compelling images
of death, however, date from the late fourteenth century. Seemingly
influenced by a new strain of penitential guilt which arose during
the early years of the century and was certainly exacerbated by the
grisly experiences surrounding the Black Death, which ravaged
Europe between 1347 and 1351, the image of the decayed and
putrefying corpse becomes standard, highlighting the transience and
ultimate worthlessness of every aspect of the material world.

The ultimate message of the *vanitas* image, though, concerns
time. Transience is a manifestation of time. The grains of sand
counted by the hourglass or the relentless ticks of the watch
measure moments of earthly time. And the true aim of every
Christian is to recognize these temporal signals for what they are –
insignificant relics of an inconsequential state of being – and to turn
towards the eternal and timeless glory of God's kingdom.

228 Pieter Claesz

Vanitas – still life
Oil on oak panel, 1656
Vienna, Kunsthistorisches Museum
in Wien, Gemäldegalerie [9035–1656]

229 Abraham van der Schoor

Vanitas with skulls
Oil on canvas, 17th century
Amsterdam, The Rijksmuseum
[SK-A-1342]

The style, construction and subject-
matter of the *vanitas* painting
changes throughout its long history,
but during the 17th century certain
components seem to reappear with
regularity. The most common feature
is the human skull as an obvious
symbol of human mortality. Another
is the burning or extinguished
candle. Quite often, symbols
specifically referring to transience are
included, such as flowers or music.
Timekeepers are used as allusions to
time's passing. The sandglass has
established associations with the
iconography of time, but the
significance of the watch seems
slightly more complex. Like the
sandglass, the watch symbolized
tempus fugit and the irreversible
nature of time's passage. But, during
the 17th century, watches were still

sufficiently rare to be considered
precious. The inclusion of a watch
within a *vanitas* painting was not
only an allusion to time, but also to
the vanity of all possessions.

230 Edwaert Colyer

Trompe l'oeil with letters
Oil on canvas, 1703
Leiden, Stedelijk Museum de
Lakenhal [S.1195]

Amongst the most prolific *vanitas*
painters working during the second
half of the 17th century, Colyer
specialized in dense compositions in
which visual and textual references
to the vanities of this world often
carried a specific topical allusion.
Frequently, he creates a scene in
which a series of sumptuous objects
– globes, gold plate, musical
instruments – are set beside a book
which is opened to a description of
the apocalyptic vision of St John or
beside an illusionistically rendered
broadsheet concerning the vanity of
the world. The idea behind these
compositions seems to be that one
must push beyond appearances and
actually 'read' what is being
presented in order to appreciate fully
the intent of the picture. Another

device Colyer used repeatedly was
the 'pin-board' to which a number of
pieces of paper – letters, playbills,
printed scraps of news – were
tacked. On the one hand, these
paintings present pretty images of
the bric-à-brac of contemporary life.
On the other, the very nature of the
subject-matter underscores how
much of what occupies our time is
truly ephemeral – the newest
geographical survey, the most recent
of His Majesty's speeches, the
playbill from last night's
entertainment and the comb with
which we combed our hair this
morning. One can either see these
paintings as light-hearted studies or,
remembering the times in which
they were painted, understand that
Colyer's real message here concerns
the extent to which all the artefacts
and mementos of our daily lives are
really no more than flotsam and
jetsam. In a world where everything
is *au moment*, nothing is permanent,
nothing is important.

231 **Alexander Cozens**
*Before the Storm, No. 20
(Circumstance)*
Oil on laid paper, 1770s
London, Tate Gallery [TO 1949]

One of the defining features of the scientific revolution of the 18th century was a renewed belief in the value of empiricism. In the Newtonian age, nature herself was seen as subject to a series of immutable mathematical and physical principles. The keys to unlocking the mysteries of nature were observation and experiment. By tireless examination, one could begin to discover what was 'true'. Numerous artists of the period were effected by the spirit of the times and looked for rules by which a cannon of accurate depiction might be established. Physiognomy, botany, geology, meteorology and even a nascent form of psychology all played a role in the new science of painting. Cozens was one of the foremost champions of this new way of depicting nature. In his collection of twenty-seven *Circumstances*, Cozens codified the effects of nature into three types: accident (wind, fog, storm *etc*), seasons, and characters reflecting the time of day.

232 **Joseph Mallord Willam Turner**
The New Moon: or 'I've lost my boat, you shan't have your hoop'
Oil on mahogany, 1840 (exhibited)
London, Tate Gallery
[N 00526/TO 7319]

Constable and Turner are often portrayed as the two different faces of early 19th-century painting. Constable is seen as the earnest pupil of nature; Turner as the poetic genius. Turner may have claimed that he "drove the colours about till he had expressed the idea in his mind", but the innumerable sketches and preparatory drawings that survive prove that, for Turner, painting was anything but an immediate response to stimuli. It was an extremely slow and thoughtful process, with each work emerging from a hard-won battle, crafted specifically to provide the effect of immediacy and spontaneity. In his *New Moon*, Turner captures several aspects of what goes towards making up 'the immediate'. Astronomical time is featured in the appearance of the New Moon; the time of day by the effect of the light; the running of the tides by the glistening wet sand; and the evanescence of childhood by the depiction of running boys and dogs and the ghastliness of sibling rivalry.

During the second half of the eighteenth century, the popularity of the *vanitas* painting waned. Images reminding the viewer of imminent death had little appeal for the bold and confident men and women of the enlightened, post-scientific revolutionary era. In northern Europe in particular, new trends in thinking shifted man's fixation on *post mortem* salvation and redirected his attentions towards the glories of the natural world. One might say that, during this period, the strictures of religious orthodoxy were loosened and man became more free to examine the world around him; but this would be missing the point. Instead, people began to look for the manifestation of God's goodness in the created world itself. One clear symptom of this change is a renewed interest in the 'marvel of the moment'.

One way to see British landscape painting is as an evolving story of painters struggling with the challenge of how to depict 'now'. Effective landscape painting must combine the recognizable features of the land with a sense of freshness and immediacy. It must appeal to the viewer's nostalgia for a specific place, while also conveying a sense of constantly changeable weather. The natural world is never static: the leaves must flutter, the clouds scud and the brook babble. In the first 'stage' of British landscape painting, the artist was concerned with rules and models. Representatives of the second 'stage' abandoned the rule book and looked towards experience. By painting and repainting skyscapes, for example, one would be able to reproduce its effects. The third 'stage' certainly passed through the first and second stage, but felt that an understanding of the verities of nature could not be gained either through the head or through the eye: the essence of landscape was in the heart. The aim was not to re-create 'the look' of nature, but to evoke its 'feel' through paint.

233 **John Constable**

Cloud study with horizon of trees
Oil on paper mounted on board,
dated 27 September 1821
London, Royal Academy of Arts

234 **John Constable**

Study of clouds above a wide landscape
Watercolour and pencil,
dated 15 September 1830
London, The Victoria and Albert
Museum, Dept. of Prints, Drawings
and Paintings [240–1888]

From about 1820, Constable started making a series of studies of clouds. Between 1821 and 1822 alone, he made well over 200 cloud studies of the skies over Hampstead Heath in north London. From his own writings and comments, it seems that this particular fascination with clouds developed from a combination of artistic concerns. Prompted by the comments of an anonymous critic of his painting of *Stratford Mill*, Constable recognized the weakness of the clouds painted there and set about to remedy it. For, as he noted in a letter to his friend John Fisher, "That Landscape painter who does not make his skies a very material part of his composition – neglects to avail himself of one of his greatest aids .... It will be difficult to name a class of Landscape, in which the sky is not the *key note*, the *standard of Scale*, and the chief *Organ of sentiment* .... The sky is the *source of light* in nature – and governs everything." Rather than just perfecting a single idealized sky, as most earlier artists would have done, Constable began by studying the changing effects of the sky with an almost scientific intent. He also read contemporary treatises on meteorology, including Foster's *Researches about Atmospheric Phaenomena* (1836). His impetus was, as he himself proclaimed: "We see nothing until we truly understand it". For Constable, the essence of the thing was in knowing how it changed. In this regard, his cloud

studies are a real meeting between the eye and the brain – the eye captures the momentary impression of a particular effect or configuration and the brain fills in the 'how' and the 'why' which helps to fix an image that succeeds in being as fresh and immediate for future generations as it was for the artist himself.

Also

• John Constable, *Study of Hampstead Heath looking towards Harrow, 27 September 1821*, London, Royal Academy of Arts

• John Constable, *Study of sky effect: Noon looking north-east, 26 September 1833*, London, The Victoria and Albert Museum, Dept. of Prints, Drawings and Paintings [202–1888]

235 Salvador Dalí

Soft watch exploding
Ink and pencil, 1954
St Petersburg FLA, Salvador Dalí
Museum [Dali EL 134.1980]

Contemporary science and, in particular, physics had always been one of Dalí's intellectual passions. His well known pictures of 'soft watches', for example, were direct evocations of Einstein's theories of relativity. The soft watch represented the "Camembert of time and space" and the destruction of our assumptions about the fixed nature of the cosmos. In some of his later works, dating from after 1945, well known images taken from the paintings of Raphael and Piero della Francesca are redrawn as if exploded or atomized. These derive directly from the impressions left by the explosion of the atomic bomb over Hiroshima. After Hiroshima, as one sees in Dalí's *Soft watch exploding*, even the relative calm of the

Einsteinian universe had been shattered. The numbers of the watch's dial jump from its face as the watch begins to disintegrate. With the Bomb, the structure of matter itself had been decomposed. Accordingly, the representation of matter had to change or, as Dalí himself described it: "To the continuous waves of Raphael, I have added discontinuous corpuscles to represent the world of today". For Dalí, the scientific discoveries of the 20th century meant that the artist/philosopher – who must transcend the real in order to represent the true – now faced the challenge of re-creating the substance of the universe itself. In his 1951 'Mystical Manifesto', for example, he says, "Since the Theory of Relativity substituted the substratum of the universe to the ether, dethroning and thus bringing back time to its relative role already accorded it by Heraclitus when he said that 'time is a child', and also [by] Dalí when he painted

his famous 'soft watches' – since the whole universe seems filled with this unknown and delirious substance, since the explosive equivalent mass-energy [*sic*], all those who think, outside the Marxist inertia, know that it is up to the metaphysicians to work out precisely the question of substance." The most ambitious of these physics-based paintings is his *Anti-protonic Assumption* of 1956, where the Assumption of the Virgin is 'explained' in terms of a combination of quantum mechanics and theories concerning the collision of matter and anti-matter.

236 Jasper Johns

Face with watch
Intaglio print (artist's proof), 1996
West Islip NY, Universal Limited Art
Editions, Inc.

Many of the prints and paintings that Johns created during the 1980s and 1990s are directly related to his series of four large paintings entitled *The Four Seasons* (see 075), in which a number of autobiographical elements are drawn together to form an allegorical commentary on the passage of the seasons and its relationship to the Four Ages of Man. In this print, surreal elements recalling Dalí, Miró and Duchamp have crept in. The stylized features of a human face have been scattered to the four edges of the image, while the white face of a watch remains intact, hanging from an illusionistic nail – compact, impassive and, presumably, still ticking its way into the future. In the centre of the piece, there is a 'work of art' displayed, based on one of Johns's own artistic trademarks, cross-hatching. These elements, though personal, also refer to the tradition of the *vanitas* painting with its allusions to the five senses, to the passage of time and to the problematic status of artistic achievement within the larger scheme of life and death – if *vita brevis*, does it really follow that *ars longa*?

237 Joseph Kosuth
Clock (one & five), English/Latin
version (exhibition version)
Mixed media, 1965/1997
London, Tate Gallery [TO 7319]

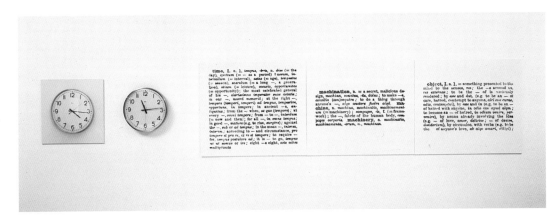

Most artists have had some sense
that those who looked at their works
were participating in the illusion that
was being created. There are
numerous stories from antiquity
about the painter who rendered
objects with such meticulous care
that later viewers were tricked into
believing the objects were real.
Several of the great perspective
paintings of the Italian Renaissance –
Masaccio's *Trinity*, Mantegna's
Camera degli Sposi or Giulio
Romano's *Fall of the Giants* in the
Palazzo del Te in Mantua – rely on
the viewer standing in a specific spot
in order to appreciate the effect the
artist had intended to create. From
the mid-1960s, Conceptual Art has
taken this process one step further.
Both the intellectual and physical
process of making a work of art – as
well as the process of viewing it –
were regarded as important. Since
the ideas imbedded in the concept of
the work itself cannot fade, each
piece remains alive and timely.
Kosuth's *Clock*, for example, is one
of a series of works in which he
approaches the 'reality' of an object
from a number of different angles.
The clock as an object exists. The
photograph of the clock, however,
has another kind of reality: it exists
both in the present and in a specific
moment of the past. The English
dictionary definition of 'time' exists
in our minds and is re-created as we
read the text. The additional
definitions of 'machination' and
'object' lead us to question the
relationship between the clock and
the passage of time. Kosuth saw
these pieces as a form of 'linguistic
anthropology'. Their unique nature
derives from the fact that they are
re-created anew each time another
viewer – with his or her personal
'cultural baggage' – steps up to
confront them.

238 Robert Rauschenberg
Soviet American Array III
Coloured intaglio, 1988–90
West Islip NY, Universal Limited Art
Editions, Inc.

Rauschenberg was one of a group of
American artists who, in the 1950s,
were profoundly influenced by the
work of Marcel Duchamp. 'Art' was
what the artist made. In his earliest
works, the Combines, Rauschenberg
constructed his pieces from
discarded bits of rubbish. The
significance of these works lay in the
way that the detritus of a rapidly
changing world could be brought
together, by the artist, to form a new
and meaningful whole. In the 1960s,
he began to experiment with multi-
media collages composed of
photographic transfers and painting
techniques drawn from Abstract
Expressionism. To quote Robert
Hughes: "The bawling pressure of
images... creates an inventory of
modern life, lyrical outpourings of a
mind jammed to satiation with the
rapid, the quotidian, the real". They
exist as an uncanny premonition of
the channel-surfing of the 1990s
culture in which everything is now,
immediate and infinitely
interchangeable. In a number of his
more recent pieces, Rauschenberg
has also included the image of a
clock, which makes them eerily
reminiscent of the 17th-century
vanitas painting. Life may be 'now',
but the clock is ticking.

Chinese, *Ancestor scroll of the Li family* (see **269**)

THE
T•Experience OF
Time

A lthough it is true that man has invested a great deal of effort in trying to understand the nature of time and has spent many a long day and night attempting to perfect means of quantifying its passage, there are a number of ways in which his life is governed by other types of clock that are so elemental that they nearly escape his notice.

The human body is run by a series of 'clocks' which regulate the rhythms of its organism. The first sound a foetus hears is the constant beat of its mother's heart. The normal adult heart beats about 75 times per minute. The menstrual cycle of the human female has an average period length of 29.5 days – the exact interval of a synodic month. Human sleep-wake patterns follow a cycle 24.8 hours long. This period is known as a circadian rhythm from the Latin words *circa*, 'about', and *dies*, 'day'. More than 1000 different circadian rhythms have been detected in the human body. For example, cell division is most active in the late evening. The production of adrenaline is at its lowest around 4:00 AM and reaches its peak at the time of waking. Our kidneys function most efficiently at midday.

The other sort of clock by which we are all governed is ageing. According to our individual DNA and an astonishingly wide variety of social factors, people age in different ways and at different speeds. Nevertheless, from the moment we are born we are all set on a one-way track. Always trying to superimpose measurables on the unknown, man has divided his own ideal life span into a number of discrete units by instituting specific rituals, each of which is used to characterize a certain 'age' or 'stage' in life. Age-related rituals are a cultural constant; but the particular ceremonies surrounding the naming of a child, baptism, circumcision, the initiation into society, marriage and the formalization of both maturity and 'old age' vary according to religion, nationality and socio-economic factors. Demographics also affect one's position on the ageing scale. As the general population ages, the period of childhood is being extended, emotional maturity delayed and the period called 'middle age' has moved from twenty-five to forty!

Lorenzo Costa, *A concert*, oil on panel, *ca.* 1490,
London, The National Gallery

Music and Time

by Iain Fenlon

Johann Maelzel, Beethoven's metronome (see **240**)

Whether silently read or performed, music exists only in time. As such it possesses the potential to be something that can be accurately measured, a language which, like all others, moves through time and is articulated within it. Although this is a basic characteristic of all music, from the court music of the Tang dynasty to Mozart's string quartets and beyond, the urge to quantify music as precisely as possible through written forms is not common to all musical cultures. Both African drumming and Indian ragas, for example, remained as oral traditions, passed on from generation to generation, until European ethnographers attempted to record them.

For the most part, the compulsion to specify music's temporal and other components (such as its dynamic or affective qualities, or the speed at which it should be heard), is the driving force behind the development of Western systems of music notation, languages made up of graphic signs and symbols occasionally combined with verbal instructions which, in the course of the last millennium, have become increasingly precise and potentially more prescriptive. A page from almost any late nineteenth- or twentieth-century score presents an extraordinarily complicated notational system in which the rhythm, pitch and duration of every note is exactly quantified in relation to those that precede and follow it. This, in turn, is combined with other kinds of instructions relating to dynamics, accentuation, speed and the technical way that a particular effect is to be achieved in performance. The result is a visual analogue of the music itself and, to a high level of specificity, fulfils the primary task of all musical notations: to fix music in time and to articulate and describe its temporal existence.

The earliest substantial body of music in the Western tradition, plainchant, is preserved in thousands of manuscripts which register regional and other variants according to local liturgical usage. As the official music of the liturgy of the Western Christian Church, plainchant was performed all over Europe in cathedrals, monasteries and churches. Chant is notated in neumes, graphic signs which indicate melodic movement or repetitions of the same pitch, but do not determine rhythm. The oldest sources are written *in campo*

Musical structures using canons, number,

aperto ('in an open field'), with the neumes placed higher or lower according to their pitch, but without the help of stave-lines to guide the eye (see **240**). Over a period of time, stave-lines were gradually added until the four-line stave became common. Although neumatic notation was 'reformed' and standardized in the nineteenth century, the four-line stave has remained in use ever since its introduction. Five-line staves can also be found as early as the ninth century, but it was not until the eleventh century that the version familiar to us from modern notation evolved. Guido of Arezzo, a Benedictine choirmaster whose name is linked to the development, coupled the five-line stave with a mnemonic device designed to speed up the process of learning or notating music. Based on the syllables of the hymn, 'Ut queant laxis' (sung on the feast day of St John the Baptist), this system allowed anyone familiar with it to memorize the scale as arranged on the stave: *Ut queant laxis* **Re** *sonare fibris,/* *Mira gestorum* **Famuli** *tuorum,/* **Solve** *polluti* **Labii** *reatum,/* *Sancte Iohannes.*

In common with much else in his writings, the purpose of Guido's system was essentially practical. The 'Ut queant laxis' melody could be used by a singer either to notate an unwritten melody on hearing it (by matching it to the correct sequence of tones and semitones) or to learn an unwritten melody on hearing it (by matching the written notes to the melody). The so called Guidonian hand, which Guido himself may have invented, visually associates syllables with pitches and was probably adopted by him as a tool for training singers. Taken together these two innovations revolutionized learning. The average time that it now took to learn the complete repertory of plainchants for the liturgical year was reduced from ten years to eight, and unknown melodies could be sung at sight. This practical emphasis in Guido's writings stands in stark contrast to traditional music theory, which tended to follow the writings of Boethius and, as Guido somewhat tersely put it, were "not useful to singers, only to philosophers". Guido's criticism, however, was entirely characteristic of the eleventh century.

During the early Middle Ages, all forms of measurement shared two characteristics. They were practical methods that had grown out of specific tasks, and they were still relatively approximate. The unmeasured music of plainchant and its notated polyphonic derivatives (such as *organum*), and Guido's consolidation of mnemonic systems, are clearly products of the same mentality. But then, around 1300, there was a dramatic shift in European attitudes towards the quantification of both time and space. In this process, the somewhat general and impressionistic methods of earlier

back. At the beginning of the fourteenth century, both London and Venice each had about 90,000 inhabitants, a steep increase over previous centuries. In such large urban centres, a new professional class of merchants and craftsmen bought and sold, counted and exchanged, in what the French historian Jacques le Goff called "an atmosphere of calculation".

At the beginning of the fourteenth century, Paris was the most important intellectual and commercial centre in Europe. By then, the university there had become so much an established part of the urban way of life that bright and ambitious men, many of whom were to become prominent as administrators and high-ranking churchmen, attended in droves. One of them was Philippe de Vitry, who was described by Petrarch as "ever the keenest and most ardent seeker of truth, so great a philosopher of our age" and was surely one of the leading intellectuals of the time. Educated at the Sorbonne and later secretary and adviser to three French monarchs, Vitry was also a poet and composer who wrote an authoritative treatise on the practice of music, the *Ars Nova*. Although some of the concepts presented in it had been formulated by others, Vitry's discussions of rhythm and notation greatly refined the notational (and hence the compositional) possibilities of mensural music because it was based on the recognition of five note values. Although few of Vitry's own motets have survived, they do establish a new structural scheme in which the tenor is sung in long notes while the remaining voices wheel around it at a much faster rate in a decorated melodic style that incorporates passages of rhythmic repetition (known as 'isorhythms'). Vitry's innovations were taken further by Guillaume de Machaut, the greatest master of the French Ars Nova, whose output was larger than that of any other fourteenth-century composer. Machaut's music is also the most stylistically and formally varied of the period. He made considerable use of isorhythmic technique, above all in his celebrated *Messe de Nostre Dame*, and also experimented with other structural devices. The poetic text of 'Ma fin est mon commencement', for example, is essentially an explanation of the overall structural matrix of the piece and of its resolution in performance: the tenor is generated by writing the upper voice in reverse, while the contratenor, written to last only

anagrams, isorhythms etc were only possible thanks to the precision of the new notation

centuries gave way to an interest in more precise forms of measurement. This development coincided with a rapid change within European society itself, caused by a notable expansion in the population and the urbanization which accompanied it. Between about 1000 and the outbreak of the Black Death in the 1340s, the population of Europe doubled in size, new towns were built and older ones expanded, and the frontiers of the world were pushed

half as long as the upper voice, is completed by repeating itself, but also in reverse. Musical structures using canons, number, anagrams, isorhythms and other notational conceits continued to be written with increased degrees of sophistication and ingenuity throughout the fourteenth and fifteenth centuries, culminating in the music of Dufay, Obrecht, Ockeghem and the young Josquin Desprez. Such techniques were only possible because of the precision of the notation system that had developed since 1300. They were also, however, a reminder of the prevailing contemporary view of music as a branch of mathematics, number in sound.

For much of the Middle Ages, the study of music was seen as St Augustine had seen it in the fifth century. Together with arithmetic, geometry and astronomy, the purpose of music was to develop an understanding of absolute number, the divinely ordered principle by which the universe was controlled. In this view, it was the theory as opposed to the practice of music that was prized precisely because it provided an intellectual understanding of the numerical relationships which are the rational bases of harmonies and rhythms. Its value lay in its demonstration of internal consistency, which in turn led to reflection on abstract number as the root of all rationality. It was Boethius who first coined the term *quadrivium* to describe the four main branches of learning: arithmetic, geometry, astronomy and music.

In this scheme of things, the study of music was emphatically not about the enjoyment of sound and even less about the ability to perform. Augustine explicitly denies the title of musician to those who could only play instruments, and his own references to the aesthetic appreciation of sound, or executive skill in music, make it clear that he regarded these as superficial aspects of a discipline which was basically intellectual. In these circumstances, it is not surprising that a body of musical theory was built up in which the essence of music was expressed in purely mathematical terms. For example, melodic intervals could be described as ratios, which were calculated by holding down the string of a monochord with one's finger; the musical 'ratio' between two different notes created when the string was plucked on either side of the finger corresponded to the ratio of the relative lengths of the string itself. In practice this was a refinement of Pythagoras's observations on the effects of striking an anvil with different sizes of hammer, thus producing different pitches that could be arranged in sequence. This led Pythagoras to the important discovery of the mathematical ratios which corresponded to the main intervals of the scale: 2:1 (octave), 3:2 (fifth), 4:3 (fourth) and 9:8 (tone). This grounding in mathematics implied that music was connected to other natural phenomena, and also to philosophical concepts such as the harmony of the spheres, which elaborated the notion that each of the planets produced a musical note which differed in pitch according to the speed at which they were revolving within their ring or sphere. Belief in a universe ordered by the same numerical proportions that produce musical harmonies also lies at the heart of the Platonic concept of the World-Soul, an exemplar for the physical world achieved through a kind of celestial monochord and so based on Pythagorean proportions. Such notions of cosmic harmony were

common in Europe until the sixteenth century and even beyond, culminating in a major statement of the doctrine in Kepler's *Harmonices Mundi* of 1619.

The *tactus*, a way of measuring music through hand motion, was first described in detail by Adam of Fulda in his treatise *De Musica* (completed in 1490), though the practice itself is undoubtedly older (see ▪241▪). In addition to silent hand movements, the marking of time in performance was sometimes done by beating the floor with a long stick or stamping on the ground. Such disruptive methods also had a long history. A few years after Adam's treatise, the *tactus* was also described by the composer and theorist Franchino Gaffurio, then working in Milan. According to his *Practica Musica* of 1496, one *tactus* was equal to the pulse of a man breathing normally, which suggests an invariable speed of metronome marking *ca.* 60–70. Later in the century, the first examples of tempo markings appear as written instructions according to which speed of the *tactus* would have been increased or decreased. In Luis de Milan's *El Maestro* (1536), such changes are indicated by the word *apriessa* ('quick') and *espacio* ('slow'), and inevitably distort the idea of a standard *tactus* derived from a natural phenomenon. Some later indications of tempo variation, such as *andante* ('at walking pace'), retain the connection between human functions and rhythms and the marking of music in time. It is interesting and possibly significant that the earliest written indications of speed come at about the same time as the first instructions about dynamics (in Giovanni Gabrieli's *Sonata pian e forte*). Both suggest a tension between the rigidities of the existing system of notation and a musical language which had been moving for some time in the direction of heightened expressivity, above all in Italy. In contrast to the intricate counterpoint of words and music, in which textual comprehensibility was often seriously compromised, the music of Monteverdi and Gesualdo is essentially rhetorical in its concern with meaning and sense as the motor of style. Some idea of this comes through from Vincenzo Giustiniani's description of the performances of the renowned ensembles of virtuoso singers at the courts of Mantua and Ferrara in the 1580s and 1590s: "They moderated or increased their voices, loud or soft, heavy or light, according to the demands of the piece they were singing; now slow, breaking off sometimes with a gentle sigh, now singing long passages *legato* or detached, now groups, now leaps, now with long trills, now with short, or again with sweet running passages sung softly, to which one sometimes heard an echo answer unexpectedly. They accompanied the music and the sentiment with appropriate facial expressions, glances and gestures, without awkward movements of the mouth or hands or body which might not express the feeling of the song. They made the words clear in such a way that one could hear even the last syllable of every word, which was never interrupted or suppressed by passages and other embellishments. They used many other particular devices which will be known to persons more experienced than I ....”

By the end of the sixteenth century, the performance difficulties of large-scale choral pieces and the various new kinds of operatic and theatrical works necessitated the introduction of some form of central coordination. Conducting, in its modern sense, is simply an extension of the practice of marking the *tactus* by hand movements, which can be seen in any number of earlier images of music-making (see page 214). It may also have been influenced by cheironomy, another way of shaping music in performance by using the hands to suggest the rise and fall of the melody. This ancient idea, which reached its most developed form in relation to unwritten musical traditions, was particularly relevant to the performance requirements of much late sixteenth-century Italian vocal music. As Giustiniani’s letter shows, the subtlety of the late madrigal style required flexible and constantly shifting speeds that varied according to the sentiment of the words. Further practical difficulties of coordination were caused with the arrival of opera in the Venetian public theatres beginning in the 1630s, bringing the need to maintain a sense of continuity and direction amongst the various performing forces that were spatially separated. At first, opera performances were directed from the keyboard (as were those of many other kinds of instrumental and vocal music from the sixteenth century onwards), but further refinements became necessary as both orchestras and choruses grew in size and continuo instruments, such as the harpsichord, disappeared from the pit. Batons were first used by conductors at the end of the eighteenth century and by the middle of the nineteenth century they had become commonplace. The first significant composer to use a baton was Weber; he used “a very small baton” with which he was able to exercise “the most perfect control over the band” at the premiere of *Der Freischütz* in 1821.

The metronome can be used both to establish a tempo and then to maintain it throughout the performance of a work. Although the apparatus was invented (in about 1812) by Dietrich Winkel, it was Johann Maelzel, a shrewd and experienced entrepreneur, who went on to exploit its commercial possibilities. Having appropriated Winkel’s idea, Maelzel patented it, and promptly despatched two hundred examples to composers throughout Europe. It was on this prototype mechanism of 1815–16, which is calibrated from 50 to 160, that Beethoven made his tempo markings (see page 214). These markings have caused considerable confusion amongst later scholars, not only because the composer sometimes changed his mind, but also because his publishers often misunderstood or deliberately altered his indications. Maelzel’s more refined metronome of 1821 (with calibrations that now ran from 40 to 208) became standard. It was further adapted in the twentieth century to help the synchronization

of music with film, and with the superimposition of different layers of a musical work in recordings. For these tasks, the maintenance of a constant metronomic pulse is essential. The success of Maelzel’s metronome was enormous. The apparatus became so familiar that, in the early twentieth century, it became a musical instrument in its own right; it appears in the opening of Maurice Ravel’s *L’Heure espagnole* of 1907–09 and is even more extensively deployed in Ligeti’s *Poème symphonique* of 1962, a work written for one hundred metronomes. The fruit of Ligeti’s love-hate relationship with the ‘happenings’ of the 1960s, this traditionally titled piece is clearly intended as an ironic commentary on contemporary electronic music and its implications for the relationship between composer, performer and audience.

Silence, often thought of as the antithesis of music, is in fact part of it. The use of silence for expressive effect is as old as music itself, but it is only in the twentieth century that it has been presented as the template upon which an indeterminate or aleatoric (or ‘chance’) work might be incised or even become the work itself. John Cage’s composition *4′33″*, in which an unspecified number of performers are required to remain silent throughout, is the classic example of the second category, his *4′33″ (no. 2) (0′0″)* of the first. As performed by Cage himself in the mid-1960s, this piece consisted of the preparation of sliced vegetables, which were then liquidized in an electric blender and drunk; the sounds of these actions were amplified to the audience through loudspeakers. And even though the title of the piece refers to a specific duration of time, Cage did not care how long the piece itself lasted. These are extreme examples of a trend to jettison the precision of the mensural system in favour of graphic or ‘action’ notations which rely on either pictures or expanded verbal directions. These alternative notations, usually intended to evoke a musical response from the performers through a process of analogy rather than direct instruction, have little use for exact indications of pitch and duration, though such moments may occasionally intrude. One consequence is that, to a greater extent than is true of more classical repertories, no two performances can be the same. Another is that such music can only achieve its ideological objectives by liberating itself from all past notational systems which had sought to fix sound in time. In this sense, these works are as unmeasurable as plainchant, the oldest and most long-lived of all Western traditions, which had sounded uninterrupted throughout the hours, days, months and years of the previous millennium.

239 Emanuel de Witte
Interior with a woman at the virginal
Oil on canvas, *ca.* 1665–70 (1667?)
Rotterdam, Museum Boijmans Van
Beuningen [Instituut Collectie
Nederland 2313]

During the 1650s, Dutch genre
painters became interested in
creating illusionistic representations
of interiors with views, and maybe
this painting is intended as nothing
more than a domestic interior.
However, it may also include
subsidiary themes connected to
contemporary ideas about music.
At least from the late Roman period
onwards, both physicians and music

theorists believed that there was a
musical component in the beating of
the human pulse. Galen, for
example, distinguished 27 different
varieties of pulse, explaining that
"just as music consists of high and
low notes arranged in proportion, so
does pulse consist of strokes of
greater or lesser speed and intensity;
and both music and pulse involve a
rhythmic pattern of intervals".
Stemming from this, music was
often used by doctors as a healing
agent, especially for illnesses that
were believed to derive from an
imbalance of the humours, such as
melancholy. Prescriptions were quite
precise with regard to modes and

tempos and also concerning the
instrument used. For example,
stringed instruments were believed
to be effective in dealing with the
feminine disease of *furor uterinus*;
and lutes, violins, guitars, the *lira di
braccio,* spinets, harpsichords and
virginals are all cited as particularly
efficacious in treating love sickness
in women. As each sickness had its
own characteristic pulse, playing
music of a specific rhythm, pitch and
tempo would invariably lead the
pulse back to normalcy. To this end,
doctors would often prescribe that
specific pieces of music were to be
played on a suitably stringed
instrument at regular intervals in

order to relieve the anxieties of
sensually or sexually excitable
women. The number of depictions of
young lovers shown at a virginal or
harpsichord in Dutch painting of the
late 16th and 17th centuries prove
that the association between music
and the passions was commonly
accepted; but the possibility that
these images might depict young
men and women attempting to
temper those passions seems to
have been overlooked.

Also
• David Bedford, *Musical score for
'The Garden of Love'*, 1970, private
collection

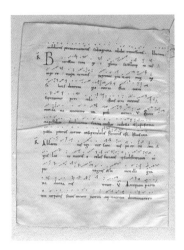

240 French

Antiphonal leaf
Ink on vellum, late 9th or 10th
century
The Schøyen Collection [Ms 96]

This unbound leaf bears text derived
from the Office of St Remigius,
which was celebrated on 1 October.
It is a fine example of staffless
notation in which indications of
tempo are lacking. The notation is
formed by neumes, or graphic signs,
that represent a movement in pitch.
These particular neumes closely recall
the type developed in the Aquitaine
and used throughout southern France
for much of the early Middle Ages.
Written neumes first appear in the
9th century, usually attached to
pieces of music that were less well
known. Neumes were, essentially, an
aide-mémoire. One needed the
guidance of an experienced
performer, who had already mastered
the tonality, tempo and inflections, in
order to understand exactly how the
piece was meant to sound. As most
music during this period in Europe
was part of a well prescribed body of
chants directly connected to the
celebration of the Mass, the
likelihood that any trained performer
would need more than the most
cursory visual reminders was slim.
For 20th-century scholars, who have
not benefited from the compulsory
eight to ten years' learning process
endured by most medieval singers,

the exact notes signalled by these
marks and the duration for which
they were meant to be sung remains
a mystery.

Also

- French?, *Early polyphonic musical
  notation*, 10th–11th century,
  London, The British Library, Dept.
  of Western Manuscripts [Additional
  Ms 36881]

- Franchino Gaffurio, *Liber Primus ...
  Theoricum Opus Musice Discipline*,
  4 October 1480, The British Library,
  Dept. of Printed Books

- John Tucke, *Notebook with notes
  on proportions*, 16th century, The
  British Library, Dept. of Western
  Manuscripts [Additional Ms 10336]

- Gregor Reisch, *Aepitome Omnis
  Phylosophiae. Alias Margarita
  Phylosophica*, 23 February 1504,
  The British Library, Dept. of Printed
  Books

- Lodovico Fogliano, *Musica
  Theorica*, Venice 1529, The British
  Library, Dept. of Printed Books

- Johann Nepomuk Maelzel,
  Beethoven's metronome, ca. 1815,
  Vienna, Gesellschaft der
  Musikfreunde in Wien

- Josiah Schmid, *Portable
  metronome with duplex
  escapement, ca.* 1840, Dresden,
  Staatliche Kunstsammlungen,
  Mathematisch-Physicalischer Salon
  [D.iv.a.315]

241 Johannes Ockeghem

Motet for 36 voices
Manuscript, 15th century
Paris, Bibliothèque Nationale de
France [Ms Fr. 1537]

In a number of images of music-
making, it is possible to see how
musicians throughout the ages
managed to maintain the right pace
for a piece of music through hand
gestures or the use of a stick or
some other instrument to tap out the
desired tempo against which the
notes were to be measured. In
Gregor Reisch's encyclopaedic tome,
the *Margarita Phylosophica* (or
'philosophical pearl'), for example,
the opening to Book V is illustrated
with a handsome image of a
personified Music (*Typus Musicae*),
surrounded by all the practitioners of
her art. Amongst the musicians, near
the centre of the composition, there
is a dapper young man shown resting
a long stick on his shoulder. He is
probably a representation of '*tactus*',
or the measured beat. During the
early Renaissance, one means for
maintaining the correct tempo was
by repeatedly beating a long stick on
the ground. In Ockeghem's
manuscript, there is a depiction of
the master conveying the tempo to
his pupils through the use of his
hands – an altogether quieter
method. Another image in which the
presentation of *tactus* is clear is
Lorenzo Costa's painting of *A
concert* in the National Gallery in
London (see p. 214). In an intimate
concert of this sort, when the
musicians performed as an ensemble
and were expected to face the
audience, the group relied on *tactus*
(literally, 'touch', from the Latin) to
communicate the beat. Both of the
singers are shown resting their hands
on a ledge. Their fingers are
positioned to show that they are
tapping the beat of the music.
Furthermore, the female singer rests
her hand on the shoulder of the man
playing the lute. She is conveying the
tempo to him by tapping on his
shoulder.

Zhou Pei Qun, *Chinese physician taking woman's pulse*, watercolour, 19th century, London, The Wellcome Institute Library

Does Time Heal?

Time in the History of Medicine

Ken Arnold

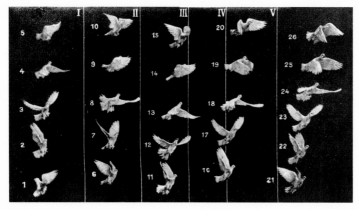

E.J. Marey, *The flight of birds*, photograph, 1890

From its myriad administrative styles to its biological core, medicine is today run by clocks. Not only do medical practitioners, like all professionals, work to frantically choreographed schedules, but scientists are now convinced that the human body itself functions according to the rhythm of a regular beat. An icon for this temporal basis of modern medicine is by tradition hung at the end of hospital beds – the chart that monitors a patient's temperature and other signs of well-being. Not only does keeping this record mark a strict regularity in the working lives of medical staff, but the very existence of the chart's horizontal dimension (the time-line) is a testament to the place of time in the modern biological body.

This was not always so. The inextricable intertwining of medicine and time was barely hinted at before the nineteenth century. Prefigured by strands of astrological thought, and elaborated early in medical history by specialists who investigated the nature and pattern of the human pulse, it was the increasing employment of instruments that produced two-dimensional charts by nineteenth-century physiologists that really brought time, in the form of a ticking clock, right into the heart of medical biology. The more people thought of human bodies in terms of well tuned chronometers, the more likely it seemed that strategies could be evolved to extend their longevity. The success of modern medicine in extending life has brought with it the twentieth century's fastest growing area of health-care: geriatric medicine. And at the end of a century increasingly obsessed with time in all its forms, some now dream of turning human bodies into time-pieces which just might never unwind.

The influence of astrology on the history of medicine around the world has a long and complicated history. Most of these traditions dictated that macro- and microcosms were both orchestrated by the rhythms of the heavens. Adhered to strictly, this view of the world suggested that no human action should be undertaken without prior consideration of the best time to do it: there were better and worse times to consult a medical practitioner, undergo an operation, take medication, and even pick the herbs used in a specific cure. During

Some now dream of turning human bodies into

the Middle Ages and Renaissance, elaborate practices of astrological diagnosis were developed, a number of which have recently resurfaced in today's complementary medical practices. The quintessential figure in the history of astrological medicine is Paracelsus, whose complex, not to say arcane, world view was animated according to the cyclical and temporal influences of astrological time. For Paracelsus, diseases disrupted the normal

Feeling the pulses was Chinese medicine's most important method of diagnosis

functions of life by setting up court in the human body, just as malevolent influences sometimes did in the body politic, in both cases according to a preordained astrological timetable.

Adherents to astrological medicine have always been convinced of the importance of time and periodicity for their discipline. For those more sceptical about such factors, one aspect of human physiology has persistently suggested that human life might somehow be animated by a regular beat: the pulse. In the West, the formal study of the pulse has ancient origins. Records indicate that Praxagoras of Cos was an early explorer of the topic, his interest being taken up with greater vigour by his pupil Herophilus of Chalcedon, who identified the pulse as originating in the heart, and who reportedly worked on a method of counting and even timing its beat. By the time of Galen, much importance had come to be attached to the observation of a patient's pulse.

In traditional Chinese medicine, the pulse was taken to provide a crucial indication of the circulation of the force *qi*, and was, therefore, accorded much significance. Feeling the pulses (a complex art involving the classification of the strength, duration, resonance *etc* of several types of pulse) provided Chinese medicine with perhaps its most important method of diagnosis. Descriptions of this Chinese tradition of feeling a patient's pulse in the writings of Jesuit travellers such as Johann Adam Schall von Bell and Ferdinand Verbiest in the seventeenth century helped bolster a revival of interest in the pulse amongst European medical scientists.

It was the revolutionary enquiries into the internal movement of blood by William Harvey that turned the study of the pulse into an important aspect of mid-seventeenth century science. Harvey's ground-breaking work involved measuring the amount of blood that flowed during a specific period, and thereby deducing that the volume of flow could only be produced by the circulation of blood. His studies were further elaborated by Stephen Hales, who attempted a quantitative assessment of blood pressure. Influenced by

time-pieces which just might never unwind

the idea that the body might usefully be conceived of as a finely adjusted mechanical device – a philosophy most richly articulated in the writings of René Descartes – the activities of Harvey and Hales signalled the innovation of bringing the accuracy of measurement into the field of medicine. The pioneer 'medical mathematician' Santorius was another who explicitly set out to apply instrumental observation to the pulse rate. A further attempt to create a medicine

based on number, weight and measure was offered by Sir John Floyer, specifically in *The Physician's Pulse Watch* (1707), which recommended measuring the pulse against a special watch with a second hand.

Pre-modern medical diagnosis commonly involved various standard routines. First, practitioners would listen to a patient's own description of his or her condition. Next, they would analyse the patient's general appearance. And only then would they undertake a clinical investigation, which might comprise three further measures: looking at samples of urine and/or faeces, examining the tongue and taking the pulse. Skilled practitioners no doubt did thus gain useful information about a patient's condition, but these exercises also served to satisfy a patient's expectations of serious professional concern. By and large, however, the increasingly scientific approach to understanding the pulse just outlined made little if any impact on the fashion in which such procedures were carried out. Fundamental changes had to wait for more than a century, when far-reaching technical and conceptual innovations profoundly affected the ways in which both medical investigators conceptualized the body and doctors related to their patients.

The nineteenth century saw the emergence of physiology as a new independent discipline, focused on gaining knowledge of the body through a systematic manipulation of its physical processes. One of the core claims for its scientific respectability rested on efforts gradually to replace the subjectivity of an observer's senses with the objectivity of technological mechanisms that, as it were, perceived phenomena directly from the human body itself. To this end a range of new instruments was introduced which allowed such phenomena as body temperature, blood pressure, heart activity and pulse rate to be automatically recorded in a form seemingly more truthful and trustworthy than traditional medical descriptions and nomenclature: namely, graphs and numbers. The charts churned out by these machines produced in turn clusters of data which were identified as characteristic of specific illnesses. The role of a doctor's personal understanding, intuition and memory, so fundamental to traditional medical practice, was thus gradually replaced by the more focused technical skills of reading and interpreting figures.

The sphygmograph used to record the pulse beat was just one weapon in this new arsenal of precise diagnostic tools. It was in 1835 that Julius Hérisson devised a sphygmomanometer, which displayed the pulse beat in a column of mercury. Karl Ludwig added a pen-and-drum arrangement to enable automatic recording. The French physician Etienne-Jules Marey was another to work on an improved instrument for measuring and recording the pulse. He presented his invention to the Académie des Sciences in 1860, an instrument which gained great notoriety from a dramatic public demonstration in front of Napoleon III.

At the heart of not only the sphygmograph, but also all the other devices that enabled continuous automatic recording of vital actions, was a more or less common mechanism that effectively translated bodily functions into continuous graphic traces recorded on revolving cylinders. Belief in the meaningful operation of these cylinders rested on a conceptual assumption of considerable significance, namely that a regularly beating time-piece somehow ticked away at the very core of the cluster of phenomena that were increasingly taken to define human life. It was an inexorable logic that further implied that if life's measurable signs could be traced along a time co-ordinate, then biological life itself might be a temporal phenomena. While adopted as only a tacit assumption for many medical scientists, this belief was fervently held as a fundamental scientific principle by Marey, who not only worked on refining the sphygmograph to measure pulse rates, but also developed an extraordinary range of other graph-based instruments for tracing the motions of the heart, the gait of humans and animals, as well as the flapping of bird wings. For Marey, the use of time-based mechanisms to produce graphic presentations of physiological knowledge was not just a convenient convention, but an expression of a core philosophical tenet: the human body was a motor in motion possessing its own internal physiological time. And, in an era when analogies to describe the body were increasingly drawn from the products of the Industrial Revolution, it seems likely that the gradual adoption of standardized clock-settings and the eventual establishment in 1884 of Greenwich Mean Time as a world standard might well have helped Marey and other late nineteenth-century physiologists in their arguments for establishing a standard physiological time within biology.

The idea of an internal body-clock has since become an extremely widely held assumption both within medicine and in modern culture more broadly. The philosophical conceit that humans have an innate measure of their own time can already be found in the writings of Marcel Proust and Oscar Wilde, both of whom invented characters whose internal clocks played tricks on external measures of time. Today, of course, we all confront the disjuncture between our own time and that of the world around us when we make long-distance journeys and have to deal with jet-lag. The scientific study of the physiological rhythms which control such phenomena as sleep/wake cycles, body temperature variations and hormone fluctuations is termed chronobiology. Chronobiologists have fairly recently determined that our body-clocks, which generate an approximately

24-hour rhythm, are physically located in a small part of the brain just behind the eyes. External time, they believe, is detected by the reception of signals through the eye, primarily activated at dawn and dusk.

The concept of a body-clock coupled with the perception that the medical professions can, with increasing confidence, make positive interventions in the workings of the body, or rather its malfunctions, has led to a modern reworking of an old idea: that of cheating death through old age. The idea of eternal youth has a history as long as Western thought itself, and particularly from medieval Western writing onwards there appears and reappears the conviction that a Golden Era when ageing was permanently delayed could somehow be reinvented. Similar inclinations to extend the body's clock can also be seen played out in the practices of preserving bodies through embalming, mummification and, most recently, in the cryogenic storage of corpses.

From at least medieval times, old age was held to be a distinct stage of life, with its own characteristic health and behavioural patterns. Nonetheless, the question of what determines ageing – primarily focused on the issue of whether it was a natural condition or a process which could be slowed, stopped or even reversed – has exercised many medical investigators. Until the early modern period, however, the question was treated more as an issue for philosophers than medical specialists. Early scientific interest in the area came in the form of statistical rather than anatomical or physiological enquiries. John Graunt amongst other mid-seventeenth century statisticians began to collect data on mortality and in organizing it according to times of the year and age began to speculate about what such figures might reveal. This work brought to the previously random or rather pre-ordained matter of human longevity a sense that it might be systematically understood. It also carried with it a strong medical imperative to distinguish more clearly between preventable and inevitable deaths, and to focus health practices on at least delaying the former.

In the wake of this computational tradition, medical attention came also to be focused on the physiology or pathology of ageing. Even by the early twentieth century, however, many of the issues surrounding old age were still poorly understood. Could the problems of ageing be prevented or simply alleviated through medical therapy? Was it a normal or pathological feature of human life? One man who chose to devote much of his life to these issues was the Austrian-American Ignatz Nascher (1863–1945), who coined the term 'geriatrics'. Research in the newer parts of the life sciences such as bacteriology came to be applied to the study of conditions found particularly in older patients, while neuropathological researchers such as Alois

Alzheimer (1834–1915) sought to distinguish senile brain diseases that particularly afflicted the old.

Despite the extraordinary medical advances of the last century or so, biological life expectancy has in the West increased remarkably little: eighty-year-olds today have about the same life expectancy as those of even three or four centuries ago. Actual life spans, on the other hand, have risen dramatically from an average of approximately 25 in the seventeenth century to about 75 today. (The greater longevity of females seems to have been evident in the West since at least the early Middle Ages.) The trend, therefore, seems to be towards reaching what has been termed the rectangular life expectancy curve, in which almost everyone reaches the age of between eighty and ninety. For some, this eventuality promises utopian visions of long healthy lives followed by swift and predictable deaths, with a consequent reduction both in the fear of death and the longing for eternity. Others see the prospect in less rosy terms.

commonest amongst infants and small children (and, of course, still is in many poorer parts of the world), the grim reaper is now likeliest to strike people after they are sixty. Most bizarrely of all, medical technology can now artificially perpetuate almost every sign of life without there being more than the shallowest sign that meaningful existence is continuing. Eternal life, it seems, has finally been achieved, but in the least desirable way imaginable.

The story of time within the history of medicine has led from early pioneering investigations of pulsation, through the evolution and subsequent ubiquity of a time-based understanding of vital life signs, to the eventual location of a body-clock within our brains – so that time has gradually moved to the core of modern biological life. Unsurprisingly, this evolving role for time within medical biology has also thoroughly influenced changing conceptions of death. The increasing expectation that life through to old age is a normal rather than minority experience has tended to suggest that death represents

Is old age a normal or a pathological feature of human life?

An unarguable consequence of this trend towards greater life-expectancy has been the growth of both medical research and clinical care devoted to the elderly, and at a more abstract level the gradual medicalization of old age. As medicine has become more successful at treating ailments associated with early and middle years, so its focus has shifted to old age. Thus, in the area of pharmaceuticals for example, one innovation after another, particularly with the increasing use of antibiotics in the last fifty years, has meant that more and more old people have recovered from acute illnesses. And, except in the area of birth-control, the majority of the most significant breakthroughs in medical technology since the 1960s have had their greatest impact on people above fifty years old.

As long ago as 1793, Benjamin Rush declared that, rather than old age, it is some disease or other that "generally cuts the last thread of life". Modern medicine has since effectively done away with the idea of dying from old age. What has replaced it is a myriad of minutely differentiated medical problems, which, cumulatively, demand an ever expanding proportion of health budgets (both private and public). Ironically, though probably the least respected medical specialism, geriatric medicine now takes up the greatest amount of a general practitioner's professional time. Paradoxically, modern Western medicine has helped to produce both more fit and active septa- and octogenarians than ever before, and (though less palatably and less visibly, since most of them are hidden in institutions) more chronically ill and bed-ridden elderly patients than ever before. It has also meant that whereas death was once

not just the absence of something internal to life, but also more particularly the absence of the proper services that modern medicine can provide to sustain life. At the start of a new millennium, we seem to have concluded that life is run according to a biological clock, and that medicine is charged with the duty of keeping it ticking.

I. Salinger, *A surgeon holding a naked female patient whilst trying to push away Death*, soft-ground etching, *ca.* 1940, London, The Wellcome Institute Library

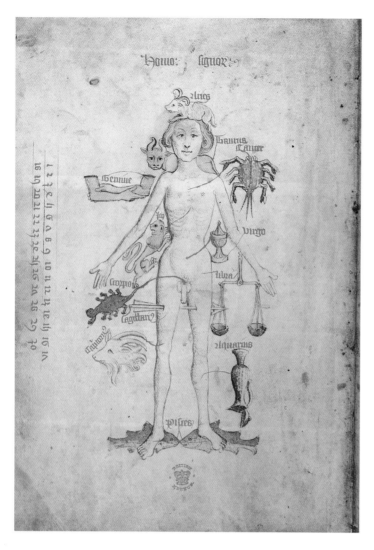

242 English (York)

'Zodiac man', from the *Guildbook of the Barber-Surgeons*
Manuscript, 15th century
London, The British Library, Dept. of Western Manuscripts [Ms Egerton 2572, fol. 50b]

The different parts of the body were believed to have 'cosmic sympathy' with the signs of the zodiac. For example, the head was ruled by Aries, the Ram; the neck by Taurus; the shoulders, hands and arms are ruled by the Gemini; the chest, liver and upper digestive system by Cancer; the heart by Leo; Virgo rules the intestines; Scorpio rules the genitals; Sagittarius the thighs; Capricorn the knees; Aquarius the shins and Pisces the feet.

Also

• Probably French, *Prognosticator*, *ca.* 1538, London, The Science Museum [1982.272]

• Byzantine, *Astrological Miscellany*, 15th century, Paris, Bibliothèque Nationale [Ms gr. 2419, fol. 1 r]

For those of us used to the trappings of a modern surgery or hospital ward, the concept of what role time might play in the practice of medicine is limited to machinery, such as the stethoscope or the electroencephalogram. But before the nineteenth century, the real timekeepers of medical practice were the stars.

Since antiquity, it was believed that the structure of the human body was a smaller reflection of the structure of the 'body' of the universe. The idea was based on an understanding that the laws which helped to define the order of the heavens (the macrocosm or 'great order') were identical to the laws that governed all life that existed on a smaller scale (the microcosm or 'small order'). Whereas the belief that the universe is governed by principles that run, unchanged, from the 'macro' to the 'micro' is still current in most scientific thinking of the late twentieth century (hence one of the problems in resolving differences between quantum mechanics and Einstein's theories about the nature of matter), the ways in which these different scales of being might be connected have changed.

Until relatively recently, the link between the macrocosm and the microcosm was astrology. The stars were seen as agents through which the essential qualities of all life on Earth were channelled. The movements of the heavens had direct bearing on the health of the individual. Hippocrates himself was credited with saying, "He who does not understand astrology is not a doctor, but a fool".

As surgeons were about to diagnose a patient or to carry out an operation – such as bleeding or cupping – they would first consult the stars to determine their orientation. To this extent, astrological medicine was extremely time-oriented. Every year, detailed astrological charts were published which noted the relative positions of the Sun and Moon against the heavens, outlining when it was safe to begin certain medical procedures on different parts of the body (see **059**). Each part of the body was associated with a particular zodiacal sign and a particular planet (see **242**). The science of knowing which part of the body was governed by which astral governor was known as melothesia. A successful operation required that the Moon and the Sun be properly oriented, or aspected, with regard to the zodiac, the other planets and the location of the illness. If they were badly aspected, then the bodily humours might become unbalanced – the baleful influence of Saturn would cause an increase of black bile and the onset of melancholy or an excess of yellow bile caused by some fiery constellation could lead to unquenchable fever. In particular, knowing the phase and aspect of the Moon was vital. The watery nature of the Moon increased the patient's flow of blood. Too much and the patient might bleed to death; too little and putrefaction would set in. Surgeons and physicians would often carry a small *aide-mémoire* with them on their rounds to remind them which sign ruled which part of the body, often in the form of the so called 'zodiac man' or 'melothesiac man' (see **242**).

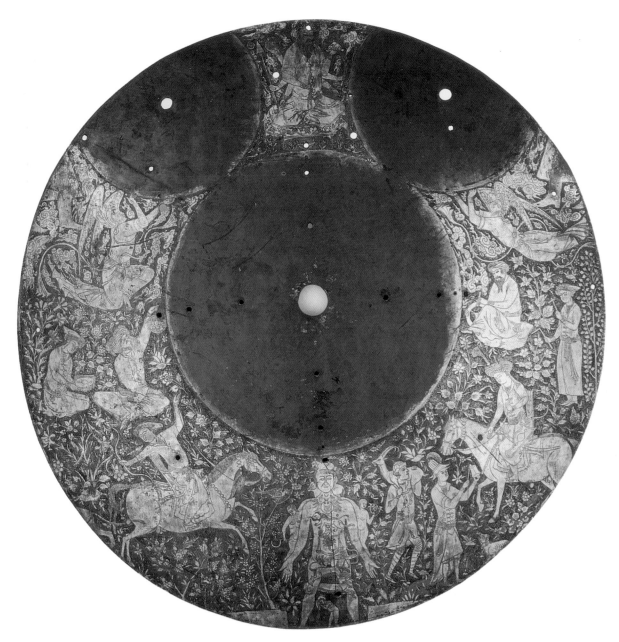

243 **Persian**

Astronomical disc
Engraved gilt copper, 1641–66
London, The Victoria and Albert
Museum [1577–1904]

The exact purpose of this disc is not
clear, but it must have formed the
base plate of an instrument, probably
used for time-telling and certainly
used for astrological purposes. The
surface is highly decorated with
figures. In the lower left and lower

right of the disc, there are two
mounted gentlemen shown enjoying
the hunt. Above them, along the
sides, there are depictions of
astronomers. On the left, one reads
a book while the second holds an
astrolabe. On the right, there is a
standing man, holding an architect's
square and a small mallet. Beside
him, there is a figure measuring off a
distance on some unidentified
instrument with a pair of dividers.
Above the astronomers, there are

two angels posed as if they were
once holding part of the instrument –
presumably rotating discs or
volvelles – that must once have been
attached to the plate. At the top,
there is a six-armed god holding a
rat, a spear, a sieve, a crown, a flail
and a hammer. These attributes
suggest that he may represent one
of the 108 versions of the
Bodhisattva Avalokiteshvara or may
be an unidentified Indo-Persian
hybrid god. At the bottom of the

disc, there is a 'zodiac man', his
body covered with the signs of the
zodiac. Of these signs, the Gemini
are shown holding a sickle and a
harp – a feature that indicates a
classical prototype. The tail of
Sagittarius, however, ends in a
dragon's head – a typically Arabic
feature. The presence of the 'zodiac
man' on this instrument underlines
the importance of astrological
medicine in the East as well as the
West.

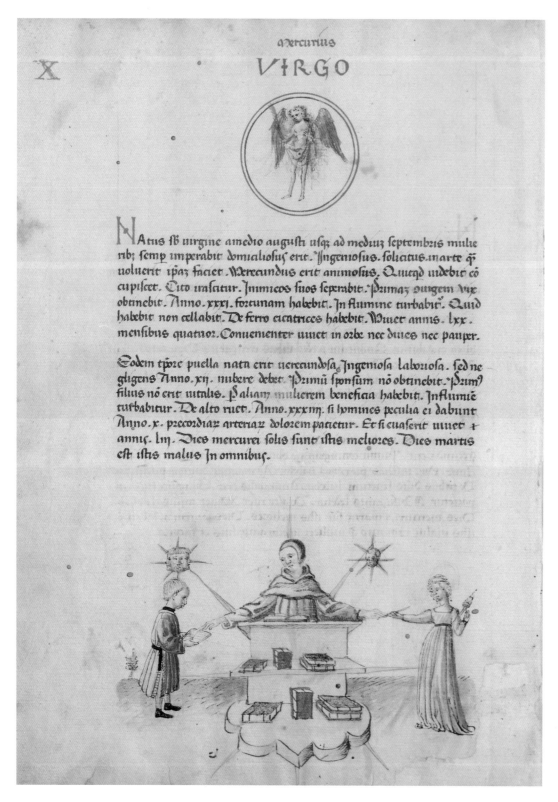

244 North Italian

Astrological Miscellany
Manuscript, 1st half of the 15th century
Modena, Biblioteca Estense [Lat. 697 (α.W. 8.20)]

As rulers over everything on Earth, the planets were also believed to rule specific ages during a man's life. In the Ptolemaic universe, the quicker planets were thought to be closer to the Earth and the slower ones further from it. Moreover, the closer planets – such as Mercury and Venus – were conceived as being somehow 'younger' than those, such as Saturn, which appeared to transit the sky at a sluggish rate. Planets were assigned those ages to which their movements best correspond. The Moon, which had the shortest orbit, circling the Earth every 29 1/2 days, was given rulership over very young children. Mercury, which completed its orbit every 88 days, was the guardian of school-aged children – young boys learning to read and young girls learning to spin. And so it went, through Venus ruling young lovers, the Sun ruling young adults in their prime, Mars ruling the warriors and housewives, Jupiter ruling the mature adult and Saturn as the guardian of the old and decrepit. As there were believed to be seven planets in the cosmos – five planets plus the two luminaries, the Sun and the Moon – this formed the basis of the well known formula of the Seven Ages of Man, each age being watched over by one of the planet-gods: "They all have their exits and their entrances,/ And one man in his time plays many parts,/ His acts being seven ages ..." (*As You Like It*, II, VII, 140–42).

Also

• North Italian (Padua), Prosdocimo de' Beldomandi, *Opere*, 15th century (1435), Oxford, The Bodleian Library [Ms Can.misc. 554]

245 Mssrs Kajima & Sons

Physician taking a lady's pulse during a home visit
Photograph, 19th century
London, The Wellcome Institute
[BRN 21476]

Measuring the pulse was a fundamental part of traditional Chinese medicine, and spread throughout South East Asia, Japan and parts of India. The purpose was to determine the patient's *qi*, or 'life force', by noting the pulse's strength, duration and resonance. To quote one 18th-century traveller to China: "When they are called to a Sick person, they first set a pillow under his Arm; then placing their four Fingers along the Artery, sometimes gently and sometimes hard, they take considerable time to determine the Beating, and distinguish the minutest Differences; and as according as the Motion is more or less quick, strong or weak, uniform or irregular, they discover the Cause of the Disease. So that without asking the Patient any Questions they tell him where his Pain lies, whether it is in the Head, Stomach or Belly ....". The writer, a Jesuit priest named du Halde, then provides a translation from a Chinese text, entitled *The Secret of the*

Pulse, written by "Wang-shû-ho, who lived under the Dynasty of the Tsin, that is, some hundred Years before Christ". The treatise describes not only the different kinds of pulses and the places on the body where these may be read, but also details how each season of the year has its proper pulse. For example, during the first and second Moon, the pulse has a long and tremulous motion, like a musical instrument. During the fourth and fifth Moon, the pulse races with fire. The treatise lists the different qualities of the pulse, such as full, deep, sharp, flying low, confined and empty, and the various diseases they denote. It also notes that pulses vary between the sexes and that the pulse in the left hand of male patients should be taken, while the right hand of women is used to measure their pulse.

246 Quirin Gerritsz van Brekelenkam

A doctor taking a girl's pulse
Oil on canvas, *ca.* 1650
London, The Wellcome Institute
[BRN 40476]

Even though the real significance of the pulse was not understood until William Harvey discovered the circulation of the blood, the practice of testing the pulse in the West can be traced back to the Egyptians. It formed a fundamental part of Galenic medicine. The heart and the arteries were seen as a huge bellows, drawing in air and blood by dilating the vessels and expelling fumes through contraction. As the purpose of the pulse was to expel vapour from the heart, the pulse itself could be used as an indicator of the emotional condition of the patient. Both love and anger, for example,

were known to dilate the heart and open up the bodily passages. Taking the pulse is depicted in numerous paintings made in Holland and Flanders during the 17th century. Some of these are straightforward: the doctor, having examined the urine of the patient, tries to test his diagnoses by comparing his findings with the state of the woman's pulse. Many of these paintings, however, appear to be slightly ribald. In these cases, the doctors are trying to discover whether or not the patient's irregular behaviour is due to the fact that she is in love. It was a widely held belief that the 'pulse of love' or *minne-pols* could be caused by the presence or absence of a loved one.

4 · THE EXPERIENCE OF TIME

247 **Anna and Richard Wagner**
A German Christmas, 1900–1945
Photographs, 24 December 1900 to
24 December 1945
Berlin, Heimatmuseum
Charlottenburg

Whereas artists and writers have
sought for ages to describe the
effects of time's passing, it really is
only with the invention of
photography that we are able to
document the most minute external
effects of time on man and his
environment. Owing to its largely
non-selective eye, photography
captures the pervasiveness of
change in a way that the writer or
painter, with their more discerning
vision, cannot. This is particularly
true in amateur photography, where
the intrusive hand of the editor is
absent. For 45 years, Anna and
Richard Wagner sent Christmas
cards to their friends that featured a
photograph of the couple, seated in
their *gutbürgerlich* front room, with a
Christmas tree and a display of
assorted presents. The first card
shows the young and newly married
couple setting out on life's great
adventure. Richard sports a silver-
topped cane and Anna lifts up Meitz,
her cat, to show him the various
gifts they have received. The house
is sparsely decorated, but
comfortable. By 1912, their material
wealth has increased – as have their
waistlines. Richard now sits at his
desk as she tidies around him. In
1914 and 1915, a map is included in
the picture, recording the advances
of the German troops during the
early years of the First World War.

By 1917, however, the fortunes of War have been reversed and Anna and Richard pose for their picture in overcoats, signalling the lack of winter fuel. During the early 1920s, prosperity returns. In 1927, Richard begins to wear glasses and his hair is showing signs of grey. Anna's Christmas present for that year is a new vacuum-cleaner – electricity has been installed. 1934 and 1935 are frugal years and Richard, in particular, begins to look older. Both have gone completely grey by 1942, and, again, the winter coats have come out. The final picture in the series shows Anna all alone – no tree, no presents. Richard had died in August of that year. And even though Anna lived another five years, the series of photographs stops here.

Also

• Jean Dampt, *The grandmother's kiss*, 1892, Paris, Musée d'Orsay [RF 962]

Weihnachten 1917 bei Kohlenmangel

Weihnachten 1927

Weihnachten 1935

Weihnachten 1942

248 **Japanese**
The Lord, his Lady and the members of the Court
Mixed media, late 19th century
London, The Horniman Museum and Gardens [3.3.48/1]

On Girls' Day, celebrated on the 3rd day of the 3rd month (or on 3 March by modern reckoning), Japanese families traditionally set up grand displays of dolls representing the Lord, his Lady and their court. The dolls are carefully arranged on a series of raised daises. They are elaborately dressed, with each layer of clothing highly significant to their role and status. The dolls are also accompanied by a miniature banquet, complete with the appropriate foods, lanterns and other regalia. The blueprint for this display is the form and conventions for an aristocratic wedding during the flamboyant Heian period (749–1185). The purpose of this ceremony, and why it might be specifically important for Girls' Day, seems to be twofold. First, it provides a history lesson for the girls. Second, and perhaps more importantly, it prepares young Japanese girls for the elaborately ritualized conventions surrounding marriage, the act of dressing, and the way in which different kinds of foods and decorations are associated with particular dates, festivals and rituals.

Also
• North American (Plains Indian), *Girl's doll*, 20th century, Cambridge, University Museum of Archaeology and Anthropology [1957.16]

Like many other species on Earth, the human lifespan is naturally divided into a number of stages. During childhood, the young learn the skills necessary to survive. In some societies, these might be directly related to securing and preparing food or providing shelter. In others, they have components that are defined solely by religious or social conventions. The big divide between infancy and maturity is biological. With the passage through puberty and the activation of the reproductive system, a child becomes an adult. In most cultures, however, this change is also ritualized. The transformation from child to adult is formalized through a process that marks the transition from dependency to responsibility.

The final stage is also defined by biological changes. The ageing of the body signals a reverse. Many cultures characterize old age as a second childhood, with the individual's role within the society gradually redefined in anticipation of reduced involvement with its daily concerns.

249 Irvala (Gunwinggu group)

Mimi with two initiates for the Dua ceremony
Pigments on bark, 20th century
Paris, Musée National des Arts d'Afrique et d'Oceanie [MNAO 64–9–146]

One of the ways in which people are brought closer together is through the rituals that reaffirm their common past. With puberty, young Aboriginal initiates are given a series of marks that recall the marks associated with one or more of the ancestors. They then paint these marks on their bodies for all important ritual ceremonies. This painting depicts two initiates with their bodies painted. They are accompanied by a very ancient spirit-god, or *mimi*. Mimi are generally represented as being extremely thin, sometimes with their internal organs and skeleton visible as if they were being X-rayed. The designs on the bodies of the young men show that they claim ancestry from the Yam ancestor and, perhaps also, from the Bustard (wild duck) ancestor.

Also shown

• Irvala (Gunwinggu), *Ritual body markings*, 20th century, Paris, Musée National des Arts d'Afrique et d'Oceanie [MNAO 64–9–147]

250 Austria

Wimpel (binder for circumcision)
Embroidered linen, 1762
London, The Jewish Museum [J.M.533]

Circumcision of males is fundamental to Jewish faith. It is usually performed when the infant is eight days old, by a qualified *Mohel* or circumciser. Records of circumcisions are kept by the fraternity (*Hebrah*) of the circumcisers, in which the name of the infant and date of the event are noted. The circumcision ceremony also doubles as a naming ceremony. In continental Europe, it was often the practice for the mother of a newborn son to embroider a swaddling band or binder (also known as a *wimpel*) that was presented to the synagogue and used as a wrapper for the Scroll of the Law. In some cases, the *wimpel* would be re-used – again to be wrapped around the Scroll of the Law – during the child's *Bar Mitzvah*. The *wimpel* would bear the name of the infant and his date of birth, along with scenes of future good luck – such as a happy marriage and children of his own. Sadly, this *wimpel* is incomplete and does not contain the name of the boy, but the inscription tells us that he was 'born on the Holy Sabbath, 5 Tammuz 5522'.

251 **Puarmapits**

Figure of a young man with a decapitated skull
Wood with pigments, cassowary quills and seeds, 1973
Leiden, Rijksmuseum voor Volkenkunde [RMV 4928–5]

The Asmat live in south-west New Guinea. Until the middle years of the 20th century, they enjoyed a well deserved reputation as fierce headhunters and cannibals. The Asmat system of beliefs is based on a creator-god who brought the world into being by killing. Death was a prerequisite for life and headhunting was a sacred duty. Moreover, the Asmat believed in reincarnation and in the transmigration of souls into living beings. Before a young man could be initiated into the tribe, his maternal uncle had to present him with the skull of a newly decapitated victim. The young man would then meditate on this skull, holding it in front of his genitals, for three days. Afterwards, the young man would be given a 'decapitation name' – the name that had been used by the victim – and took over part of that man's identity. The initiation rite then continued with two voyages. In the first, he symbolically grew old and died and then travelled towards the west (towards the realm of the ancestors) with his skull, to be reborn. In the second journey, he travelled towards the east and the rising Sun, with the skull between his legs. He pretended to be a newborn baby who, during the trip, grows to manhood. After the initiation, the skull was often hung near banana trees or coconut palms to encourage them to grow.

Also

• Tunisia, *Shawl* (*bakhnug*), 20th century, London, collection of G.S. Barrass

• Bayram b. Ilyas, *Compass dial with qibla*, AH 990/1582–83 AD, London, The British Museum, Dept. of Oriental Antiquities [OA 1921.0625.1]

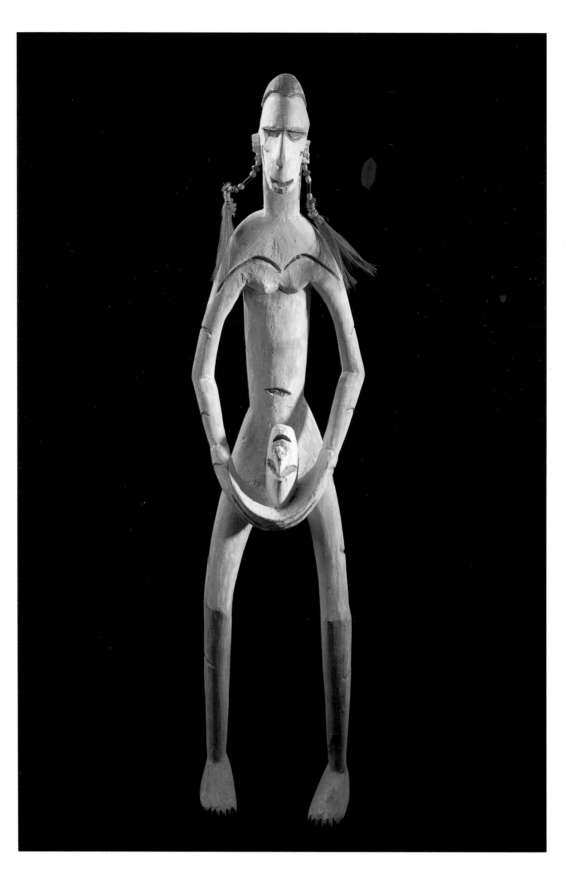

252 **Edvard Munch**
Puberty
Oil on canvas, 1902
Oslo, Munch Museet [MM M450]

In his depiction of puberty, Munch paints a young nude girl, seated on her bed with her legs tightly clenched and her arms crossed to shield her pubis. Her gaze is fixed – perhaps with fear, perhaps the result of a dreamy introspection. Behind her, there is a great shadow that seems not to be cast by the girl's slight frame but, instead, appears to be somehow welling up from her form – like smoke, or a creeping stain. Munch created one large-scale painting, several colour sketches and two prints on this theme. Of these, the painting is the easiest to understand, owing to the introduction of colour. In it, the 'shadow' is a sickeningly vivid red. The shadow – that thing that arises from the girl – is obviously the onset of her first period. Munch's depiction of the pubescent girl is intense and disturbing. And even though one understands that Munch himself had many complex and unresolved feelings about women, there is something in this image of the girl that rings true. He has, possibly despite himself, managed to capture that very odd sense of ambivalence that a young woman feels when she confronts the inevitability of her body changing. For men, puberty seems to have no drawbacks. For women, excitement, fear, disgust, anxiety, sadness and a huge nostalgia for the pleasures of girlhood are all a part of the passage.

Also

• Edvard Munch, *Puberty*, 1902, Oslo, Munch Museet [MM G.74–6]

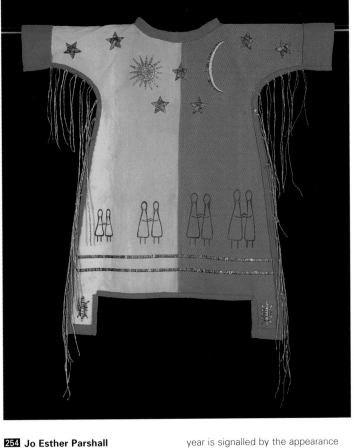

253 Attr. Giacomo Gradenigo

Dante Alighieri, *La Divina Commedia*
(*Inferno*, Canto I)
Manuscript, 1392–94
Rimini, Biblioteca Civica Gambalunga
[Sc-Ms. 1162, fol. 4 r]

Dante makes numerous references to time in his poem, *The Divine Comedy*. The best known, however, are the opening lines of the first canto of Book I of the *Inferno*:
Nel mezzo del cammin di nostra vita
mi ritrovai per una selva oscura
ché la diritta via era smaritta
('In the middle of the journey of our life, I came to myself within a dark wood, where the straight way was lost'). As the Psalms record: "The days of our life are three score years and ten" (90:10). Dante had reached that middle point of life (35 years of age) when one casts back over the past, reflects upon successes and failures and decides in which direction to move forward. Realising he has become (metaphorically) lost, Dante tries to climb to higher ground, when he encounters three beasts,

who symbolize the three main evils of the world: the leopard symbolizing lust, the lion as the vice of pride and the wolf as the sin of covetousness. In fleeing from these beasts, Dante encounters the great poet Virgil, and thus begins his mid-life journey of discovery. In the miniature, Dante is shown wearing his laurels, confronted by the three beasts. Behind them stands Virgil waiting to show him the correct path. In the sky, the constellation of Aries the Ram shines. This detail follows Dante's remark that the time of year when this journey took place was when the Sun was rising to greet "those stars that were with it when Divine Love first set in motion those sweet things". By this, Dante means that the stars matched the pattern they made when the world was created or, for the medieval mind, when the Sun was in the constellation of Aries (see cat. 2).

254 Jo Esther Parshall

Sisters of the Sundance
Canvas, quill, earth pigments, wool and rawhide, 1994
Edinburgh, National Museums of Scotland [1995.811.A]

The annual Sundance Ceremony forms a central point in the ritual year of the Cheyenne. The dance of the men has been well documented, but little has been recorded of the importance of this time of year for the women of the tribe. For the women, it is, perhaps, the one time that families are reunited. This cyclical coming together of loved ones forms the basis of the decoration of this child's dress. One year when the artist, Jo Esther Parshall, was unable to attend the Sundance because she was in her 'Moon', she received a vision, which she recorded for the other female members of the tribe. The time of

year is signalled by the appearance of seven blue stars, representing the Pleiades, or the Seven Sisters. Both the Sun and Moon are shown in the sky. In the vision, she saw the passage of time in her life as a serial progression of Sundance reunions – with the family of sisters coming together at different stages in their lives: as toddlers, as young girls, and as young women, until the present day, when they were all grown women. The blue and yellow turtles symbolize female fertility and the four directions. The horizontal quilted bands embody wishes for a long life. As the dress was never worn in a ceremonial – or, as the Cheyenne would say, it was never 'danced' – it never achieved a sacred status in its own right. For this reason, it can be exhibited as a partial witness of the important role that the Sundance plays in the life of Cheyenne women.

255 Chinese

Yun tai xian rui ('A keepsake from the cloud gallery')
Ink, gold and colour on silk mounted on blue paper, 1750
London, The British Library, Oriental and India Office Collections
[Additional Ms 22689]

In China, once a person has reached his 60th year and completed the cycle of the 'ten celestial stems' and the 'twelve earthly branches', he begins his second stage of life. From this point on, the ageing process is not considered in terms of growing older, but as growing towards becoming immortal. This is one reason why old age is venerated in China. This volume bears an inscription on the case describing it as 'The album of 1000 *li* of painting and calligraphy to wish you longevity' (the *li* is a distance equivalent to approximately one third of a mile). It was certainly intended as a birthday gift for someone who had achieved 60 years, or perhaps even his 70th, 80th or 90th year. The subject-matter concerns Daoist immortals who had achieved immortality through good deeds, following the Daoist practices, or as a gift from the gods. For example, one bifolio tells the story of Su Lin, who lived in a cavern and practised breathing exercises for 600 years in order to achieve immortality. The bifolio illustrated here tells the story of Lu Tian, who often collected medicinal herbs in the Taihang mountains. One day, he met three people who had the same name as he did. On returning home, he checked his genealogies and found out that they were very distant ancestors who had achieved immortality. The next day, he took his grandson and granddaughter to the mountains and did not return for 200 years.

Also

• Japanese, *Garments for Kanreki Iwai* (60th birthday), 20th century, Oxford, collection of Prof. Joy Hendry

○ One of the prerequisites to acceptance within a society is an understanding and commitment to the set of rules that bind that society together. One commonly agreed definition of maturity is the ability to take moral responsibility for one's own actions. In the Catholic Church, for example, the ceremony of Confirmation is an acceptance of personal responsibility to follow the teachings of the Church. The *Bar Mitzvah* also signals the moment when a young man accepts the laws that form the basis of Judaism. Although these and other 'initiation ceremonies' are often tied to specific times in a young person's life (usually to the onset of puberty), there is nothing to prevent an older adult from taking on these commitments later in life. There is, however, in all initiation processes a period of time during which the initiate must learn how to become worthy.

256 Paolo Veronese
A youth between Vice and Virtue
(*The Choice of Hercules*)
Oil on canvas, *ca.* 1580
Madrid, Museo Nacional del Prado
[POO 499, Colección Real]

Veronese's painting represents a young man caught between two women. The obvious disparity in type between them has led many scholars to identify the more luxuriant one as 'Vice' and the more restrained one as 'Virtue'. The painting belongs to a type known as 'the Choice of Hercules'. The classical source is Xenophon's *Memorabilia* which tells the tale of the young Hercules, who, when he "was passing from boyhood to youth's estate, wherein the young,

now becoming their own masters, show whether they will approach life by the path of virtue or the path of vice", was approached by two women of great stature. One, named Happiness (but nicknamed Vice), offers him the easiest road, full of pleasure and enjoyment. The other, named only Virtue, claims to know his parents well and promises that toil and effort will win him the greatest honours. Xenophon's text was widely used during the Renaissance and later as a source of ideas for the education of young princes. This painting, for example, was probably commissioned from Veronese by a Venetain father specifically concerned for the welfare and education of his young sons.

257 **After Christofero Bertelli**

The Ten Ages of Man and *The Ten Ages of Woman*
Engraving, 17th century
London, The British Museum, Dept. of Prints and Drawings
[1871.8–12.788 and 1912.1–31.1]

Traditionally, the stages through which one passed between birth and death ranged from three to twelve. In these prints, the span of human life is neatly visualized as divided into ten steps. The woman's version starts with a swaddled infant in its cradle and ends with an old woman, sitting on the closed lid of a coffin, weeping. In both versions, there are niches containing either an animal or a bird, each of which is somehow symbolic of the age that is presented on the step above it. This convention seems to have been drawn from popular folklore of the period and is relatively straightforward. For example, the young man of twenty is like a rabbit, the man of thirty like an ox and the man of forty – at the apex of his life – is like a lion. The young mother is like a hen with her chicks and the old woman is a silly old goose. Beneath both sets of stairs is a cavern in which Death waits, sharpening his scythe. Those who lead a virtuous life are led away from Death by a winged angel towards an image of Christ enthroned in glory; those who lead wicked lives are dragged by a devil into the torments of Hell.

Also

• John Goddard, *The Tree of Man's Life*, 1634, London, The British Museum, Dept. of Prints and Drawings [1847.7–23.10]

258 **Erhard Schön**

Tabula Cebetis (The tablet of Cebes)
Coloured engraving, 1532
London, The British Museum, Dept. of Prints and Drawings [E.8–5]

When it was rediscovered in the early years of the 16th century, the text of the *Tabula Cebetis* or the 'Tablet of Cebes' was believed to be a work by one of Socrates's closest circle, a certain Cebes from Thebes. As such, it was treated with great reverence and became one of the most widely translated, cited and illustrated texts of the period. It described the discovery of a picture inside a temple of Kronos by a group of pilgrims. The picture represents a walled citadel, which seems to be composed of a number of concentric circles. An old man approaches the curious pilgrims and explains that the picture represents Human Life. The gate to the first circle is guarded by an old man called Genius. Individuals wishing to enter into Life – the newborn or the spiritually reborn – must first drink from the Cup of Ignorance, held by Dame Deceit. Upon entering the first circle, they are met by beautiful women (Opinions, Desires and Deceit) and by a nude, blindfolded woman (Fortuna), who hands good fortune to some and misery to others. This is the circle of carnal pleasures and their less pleasurable consequences. If they are lucky, the individuals will encounter False Education at the next gate and are allowed to pass on to the second circle. This is the circle of the well intentioned. Most of the inhabitants of this circle, such as scholars, philosophers, scientists and poets, live contented but limited lives. Only those who find their way up a rocky and little-used path to a narrow door, flanked by Self-Control and Perseverance, manage to meet True Education and pass into the inner circle of Happiness. Happiness then crowns each of her subjects and bids them to return to the second level, where they might live as examples of virtues to those still ensnared by the limitations of False Education.

259 **Tibet**

The Transmigration of the Six Paths
Gouache on cotton, 19th century
London, The Victoria and Albert
Museum [IM. 262–1916]

This *thangka*, or temple painting, represents *Srid pa' l'khorlo* – the 'The Transmigration of the Six Paths' (sometimes also known as the 'Wheel of Becoming'). The Tibetan lamas say that Buddha drew the diagram himself in order to explain the Buddhist concept of life and rebirth according to the doctrine of ethical retribution. The diagram illustrates the path towards enlightenment and warns of the perils of leading an impious life. In the centre, there are three animals, caught up in an endless dance of vice. The cock represents lust, the snake is hatred and the pig is ignorance. They are all bound to each other, mouth to tail, and it is their dance that keeps the Wheel of Becoming moving in an anti-clockwise direction. In the next layer, the figures on the right are seen descending on a black road towards a demon-filled hell. On the left, the people are ascending on a white road towards a superior rebirth, led by a naked *yogi*. The next layer is divided into six segments, representing the six paths of existence. The superior existences include the region of the gods, of human beings and of the *asuras* (Tibetan demigods). The inferior existences are those of the animals, of hungry ghosts and of hell. Finally, the whole circle is embraced by a hideous monster, Anityata, who is the personified form of 'impermanence' and, as such, is also known as 'the devourer of life'. According to their *karma*, all sentient beings are subject to repeat this cycle of transmigration. It was the rule to hang one of these *thangka* at the monastery gate to instruct all passers-by about the nature of their cyclical existence.

261 Indian (probably Sufic)

Snakes and Ladders
Blackwood and mother of pearl, 19th century
Cambridge, University Museum of Archaeology and Anthropology
[1951.995]

The game of Snakes and Ladders is derived from a much older type of game that originated in India, Nepal and Tibet. The Nepalese version is called *nāgāpasha*, or 'to fall into the trap of the serpent'. Based on Hindu influences, the aim of the *nāgāpasha* is to move through degrees of progressive consciousness – from the level of human existence through the realm of fantasy, the realm of *karma*, the realm of balance, the realm of human consciousness, of knowledge, of reality and, finally, to the realm of the gods. The board is decorated with red snakes and black snakes, the former are benevolent and lead upwards towards enlightenment and the latter are malevolent, leading downwards towards bestiality. In Tibet, the game was called 'the game of liberation' and it tends to frame enlightenment in more Buddhist terms, with the ultimate goal being *nirvana*. In India, the game was called *jñāna cahupār* and its board was marked with snakes and arrows. This Snakes and Ladders game preserves an Islamic variant of the *nāgāpasha* idea. Each square is filled with a Persian inscription describing one of a hundred different states of man. Along the bottom, representing the lowest estate of man, one finds such words as: Want, Birth, Consent, Labour, Dishonesty, Folly, Hatred, Insult and Desire. At the top of the game, the squares read Satan, Azrael, Michael, Father Abraham, Bukabala, the Gate of the Citadel, Mohammed, Gabriel, the Angel of Death and Pride – showing that, even once one has risen through the ranks, there are still sufficient dangers to send you skidding back downwards again. At the very top of the game there is *Diswilah*, 'the Throne of God'.

260 J. Wallis & E. Newberry

The New Game of Human Life
Hand-coloured engraving on paper, mounted on canvas, 1790
London, The Bethnal Green Museum of Childhood [E.156–1933]

The primary purpose of this game is the moral education of youth. The advertisement promises that it is "the most agreeable and rational recreation ever invented for youth of both sexes". As the instructions explain: "If parents who take upon themselves the pleasing task of instructing their children (or others to whom that important trust may be delegated) will cause them to stop at each character and request their attention to a few moral and judicious observations, explanatory of each character as they proceed, and contrast the happiness of a virtuous and well spent life with the fatal consequences arising from vicious and immoral pursuits ...".

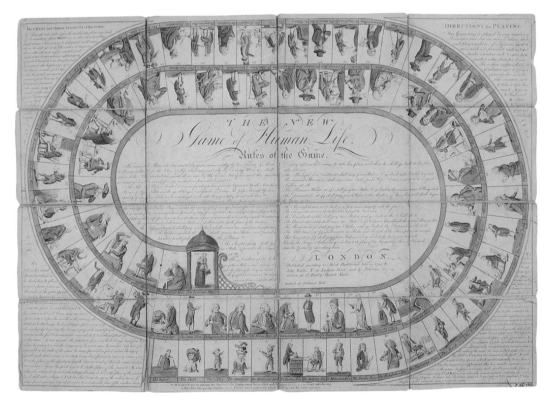

When Nature's thread, that filament never-ending,
Is nonchalantly on the distaff wound,
When unrelated things that know no blending
Send forth their vexed, uneasy jarring sound –
Who then bestows the rhythmic line euphonious,
The ordered pulse to stir or soothe the soul?
Who marshals fragments to a ceremonious
And splendid music, universal, whole?...
The might of man, in poets manifest.

Goethe, *Faust I* (*'Prelude on the Stage'*)[1]

The History of Anniversaries: Time, Number and Sign

E.H. Gombrich

Majolica dish with figure of *Chronico* (see **214**)

The German title I had chosen for the original version of this paper was '*Zeit, Zahl und Zeichen*' ('Time, Number and Sign').[2] I had chosen this alliterating form because I knew that Nelson Goodman (another speaker at the conference) had called his contribution 'Words, Worlds and Works' and I had the ambition to match his title. Had I originally written the paper in English, I might have called it 'Nature, Norms and Numbers'.

Of course, playing with language like this, one cannot make any claim towards writing poetry. But it does share with poetry the property of being created from the *donnés* of language, that is, from a prevailing system of signs. When we do something like this, the sign itself affects the signified, or the intended thought. Here, as always, language not only reflects an original thought, it also stimulates new thought. I can hardly deny, then, that the alliterative title that I wrested from the German language also influenced my plans and intentions. It was this title that gave me the idea of not speaking more generally about Cassirer's relationship to art, but of choosing the occasion of the centenary of his birth as my subject. It transpired that this subject also offers a bridge to art. For, according to Jakob Burckhardt, festive pageants present the transition from life into art.

If the occasion for celebration is, let us say, the hundredth birthday of Ernst Cassirer, then we are measuring the time passed by the number of years that have elapsed and we are giving this number the sign '100'. Obviously, the individual years – those units we have counted – are natural facts and their quantity also describes an objective fact. The sign, however, and the system of symbols that we use for our own comprehension, derives from our language and our culture. The Greek thinkers ascribed the former to *physis*, or 'nature', and the latter to *thesis*, or 'convention'. Without nature's periodicity or without the human ability to perceive repetition in nature, we could not grasp the passage of time. The choice of period – whether it is heartbeats, days, phases of the moon, or seasons – remains our own and is no less convention than the designation of number is. And yet, however fundamental this difference between a natural event and human creation may be, we should not exaggerate it. After all, convention is also chiefly rooted in natural facts – in the nature of man. Our system of counting is no exception. It not only follows from our having ten fingers, which are convenient for counting on. Even the fact that 100 is 10 times 10 ultimately derives from the limitations of our minds. A creature with unlimited powers of invention and total recall could probably dispense with such a system. It could allot a name or sign to every number in a series that was as long as it liked, and then append a category for the numbers

Without nature's periodicity, or without the

We can only make sense of the world by trying to trap it within a hierarchical system

not counted. For human beings such an unsystematic, unstructured series of numbers would be as useless as a purely nominalistic language in which every individual object was given its own sound. As humans, we must have concepts that lump things together. We have to order things and numbers into sets and sub-sets. We can only make sense of the world by trying to trap it within a hierarchical system. All numbers and systems of measurement are hierarchically constructed in this manner. The reason that the number 100 sticks out in the numerical series is that it is 10 times 10 units. In the twelve-based, duodecimal system, of course, the number 144 would have a similar psychological and cultural significance.

It is not surprising, therefore, that in people's minds the distinction between units of duration in nature and units of duration in the system we have created for ourselves has become blurred. We speak, for instance, of a 'milestone birthday' without necessarily realising that it is not the birthday (*i.e.*: the number of years elapsed) that is the 'milestone', but merely the conventional sign for the 'round number' in our system of numbers. The fuss surrounding the impending 'millennium' is a good illustration of the same fallacy. The status that we learn to accord numbers appears to be no less real than the natural cycle of the year. Whether we are thinking of the cause of our celebration or the nature of celebration generally, the tendency to merge *physis* and *thesis* can be seen everywhere.

Man's experience of nature and organic processes leads him to expect recurrence, and even to think of time as cyclical (this is especially true of primitive and prehistoric cultures).[3] In his trilogy, *Joseph and His Brothers*, Thomas Mann described this dreamlike conception with great understanding. In the section entitled 'A Journey to Hell', he says:

"What we are concerned with, is not numerical time but the abrogation of numerical time in the mystical alternation of tradition and prophecy, which means that the phrase 'once upon a time' applies to both the past and the future and acquires a charge of potential presentness. This is where the idea of reincarnation has its roots."

The extreme version of this idea can be found in Nietzsche's "eternal recurrence of the same"; but it was also expounded in ancient times by the Stoics. Chrysippus is reported to have said: "There will be another Socrates and another Plato, and they will each have the same friends and fellow citizens and this second coming will happen

not once ... but go on forever and ever".

"Forever and ever" in this context means 'countless times'. One cannot help wondering to what extent the ability to count (*i.e.* the concept of an unlimited series of numbers) was determined by the invention of the relevant signs, and whether signs and numerals themselves have not shaped the prevailing idea of time. Obviously one should not be too dogmatic here, but it is possible that marking an event in a durable way, that is, as a 'sign', also led to the ability to compare the cycles of nature with one another and fix their duration. In the first instance, of course, I am thinking of observations of celestial bodies. These observations were among the earliest achievements of human culture. What these cultures bequeathed to us are the various forms of the calendar which still accompany us in our daily lives. It was observations of celestial bodies that enabled us to count the number of days between the solstices, or the phases of the Moon within these periods. It may have been difficult to bring these natural cycles into harmony, but from that moment on the passage of time was, so to speak, disciplined and fixed for the future as well. Without this creative achievement there would be no calendar and, of course, no anniversaries in the festive calendar of every community.[4]

Nevertheless, the concept of an anniversary is not the same as that of a festive celebration. Many societies and religions stipulate commemoration days for individuals, such as the first anniversary of a father's death (an anniversary that plays a crucial role in Japanese life), or even the anniversary of a crime, for which the Austrian criminal code may prescribe the special punishment of a "hard bed, bread and water". Generally, though, anniversaries are structured so that the community participates in the commemoration day and celebrates with the individual. And this is where anniversaries merge almost seamlessly into celebration.

Psychologically speaking, these and similar religious festivals are strongly associated with a cyclical view of time. The prescribed ritual and its artistic variations are intended to stimulate believers to experience afresh the event being celebrated and ignore the time that has elapsed in between. Yet there are signs that our idea of a linear progression of years can still play a part. In the Gospel of St Luke, Jesus says at the Last Supper: "Do this in remembrance of me" – a request which, in the ritual of the sacrifice of the Mass, turns from being a commemoration into a repetition. In Shakespeare, the event itself is coupled with a prophecy of festive commemoration:

human ability to perceive repetition, we could not grasp the passage of time

"He that shall live this day, and see old age,
Will yearly on the vigil feast his neighbours,
And say, 'Tomorrow is Saint Crispian':
Then will he strip his sleeve and show his scars
And say, 'These wounds I had on Crispin's day'.
Old men forget, yet shall all be forgot,
But he'll remember with advantages
What feats he did that day: then shall our names,
Familiar in his mouth as household words ...
Be in their flowing cups freshly remember'd.
This story shall the good man teach his son;
And Crispin Crispian shall ne'er go by,
From this day to the ending of the world
But we in it shall be remembered."

King Henry V, Act II, Scene III

In this case, it is quite clear that the anniversary is rooted in the calendar. Even more meaningful – and sinister – is the scene in Act III of *Julius Caesar* immediately after the assassination, in which Brutus predicts a future festive ritual that did not, however, come to pass:

"Stoop, Romans, stoop,
And let us bathe our hands in Caesar's blood
Up to the elbows, and besmear our swords:
Then walk we forth, even to the market place,
And waving our red weapons over our heads,
Let's all cry Peace! Freedom! and Liberty!

CASSIUS:
Stoop then, and wash. How many ages hence
Shall this our lofty scene be acted over
In states unborn and accents yet unknown!"

As we know, we commemorate Julius Caesar today not on a particular day but for a whole month, since July is named after him. This decision ought certainly to make him immortal.

The same yearning for immortality inspires Horace's ode proclaiming that 'Neither the innumerable succession of years nor the flight of time' could harm what he has created:

"*Exegi monumentum aere perennius
regalique situ pyramidum altius
quod non imber edax, non aquilo inpotens
possit diruere aut innumerabilis
annorum series et fuga temporum.*

*non omnis moriar, multaque pars mei
vitabit Libitinam: usque ego postera
crescam laude recens, dum Capitolium
scandet cum tacita virgine pontifex.*"

('I have raised up a monument more lasting than bronze and loftier than the pyramids of kings, which neither the greedy rain nor the headstrong wind can destroy, neither the innumerable succession of years nor the flight of time. Not all of me shall die, a large part of me will survive death: I shall be renewed and flourish in further praise as long as the Pontifex and Vestals ascend the Capitol Hill.')

The Chinese poet Li Taipeh shared this confidence,[5] which, despite appearances, is not at variance with the cyclical view of time. The linear and cyclical models are not as incompatible in the human mind as they may be from a purely logical point of view.

At the most basic level, people have always thought in terms of the yearly cycle. Almost inevitably this led to the idea of an open-ended sequence of cycles, of recurrences and renewals that are embodied in the myths of many advanced civilizations. The boldest speculation about the length of these eras is to be found in ancient India. It is closely linked to the decimal system of numbers. The *mahayuga*, which itself has four unequal segments, lasts 12,000 years, and this is just one divine year. Three hundred and sixty of these divine years make a cosmic cycle, which therefore lasts 4,320,000 years. A thousand *mahayuga*s make a *kalpa*, and this is equivalent to a day in the life of Brahmā. Brahmā's life lasts one hundred years composed of these *kalpa*-days – after which time comes to an end and a new creation begins.[6]

In *Timaeus*, Plato also speaks of the 'Great Year' – the cosmic epoch which will end when all the planets return to their original starting point. There is a passage in *The Republic* in which he appears to suggest that the Great Year is equal to a cycle of 360 years. The concept reappears in Virgil's Fourth Eclogue, which was interpreted during the Middle Ages as prophesying the birth of the Saviour:

"*ultima Cumaei venit iam carminis aetas;
magnus ab integro saeclorum nascitur ordo.
iam redit et virgo, redeunt Saturnia regna.*"

('We have reached the last Era in Sybilline song. Time has conceived and the great Sequence of the Ages starts afresh. The Virgin comes back to dwell with us, and the kingdom of Saturn is restored.')

In these examples, however, we are probably dealing with esoteric knowledge and expectations that hardly touched the lives of ordinary people. On the other hand, in the culture of ancient Mexico there seems to have been a much shorter cycle which did impinge on the life of the community. The Mexicans had the concept of a 'bundle' of 52 years, at the end of which all fires had to be put out. The cruel rituals of this civilization insisted that the new flame was lit in the breast of a human sacrifice and from there distributed all over the country.

It was the overarching cycles of the ancient Jews that had by far the greatest impact on Western culture. The key passage here is Leviticus, 25, which is concerned with regulating work in the fields. Just as the seventh day of the week is a day of rest, so it is laid down that the seventh year must be a year of rest for the fields, which must lie fallow. After seven cycles of seven years have elapsed (a total of forty-nine years), the Old Testament prescribes a year of rejoicing, which, therefore, occurs every fifty years. In the King James version, this is expressed as follows:

"And thou shalt number seven sabbaths of years unto thee, seven times seven years; and the space of the seven sabbaths of years shall be unto thee forty and nine years. Then shalt thou cause the trumpet of the jubile to sound on the tenth day of the seventh month, in the day of atonement shall ye make the trumpet sound throughout all your land. And ye shall hallow the fiftieth year, and proclaim liberty throughout all the land unto all the inhabitants thereof: it shall be a jubile unto you; and ye shall return every man unto his possession, and ye shall return every man unto his family."

One is bound to ask whether it was ever possible to observe these rules to the letter.[7] Regardless of practicality, though, these rules continued to have an influence in the Christian era, because of the thought that underpins them. The confident hope that life will return after a long interval has been prompted by nature and the process of the dying down and growing back of plants. What has not been prompted by nature, of course, is the number of years and days that the rules stipulate and which in this instance is taken from the cycle of the week.

In itself, this way of measuring time works against the cyclical notion, but it would be idle to speculate how and when the counting of years became a cultural convention. It was probably not the number of years a person lives that produced this habit, because even today many people in primitive cultures do not know how old they are. It is a different matter, of course, for those cultures that remember lines of ancestors stretching into the dim and distant past, or the hope of descendants reaching into the remote future – a hope that is vouchsafed to us in the Bible and implies a linear conception of time.

Annus ('Year') from a Missal at St Florian (see **025**)

In some of the more advanced civilizations, larger historical time scales are provided for the community by the life spans of rulers and dynasties. These also accentuate the feeling of distance from the past. The king lists from ancient Egypt mostly add the number of the years of each king's reign to his name, while the Hellenized Egyptian historian Manetho also numbered the dynasties – a convention that is still used by Egyptologists today. The idea of using a single event as a fixed point from which to count the passage of years, in order to define an 'era', is of more recent origin. It is true that Buddhists reckoned the number of years separating them from the birth or death of Buddha, but different local traditions produced different figures. There were even variations when reckoning time since the founding of Rome, *ab urbe condita*, as there have been with the Christian and Jewish eras.

As long as people's conception of time and the length of historical eras varied, celebration had to be marked by 'commemoration days', which could be read from an *ad hoc*, public calendar. This was probably true of the centenary of the founding of Rome, for which Horace wrote his *Carmen Saeculare*, and of other founding celebrations.[8] It was only when the recognition of eras led to a universally binding framework of numbers that it became possible to fix predictable celebrations of the kind that we call anniversaries. It may come as no surprise to learn that the first such celebrations of which we know are associated with the biblical injunction about the year of rejoicing, which, in turn, was the result of a mistranslation from the original Hebrew. As will be remembered, the Old Testament speaks of the "trumpet of the jubilee" proclaiming the start of the celebrations. The Hebrew word for a trumpet made from a ram's horn is *yobel*. This, in turn, was rendered as *iubilatio* in Latin and 'jubilee' in English.

In 1300, Pope Boniface VIII announced the Catholic Church's first 'year of rejoicing' – partly because he had heard that the same thing had been done a century earlier. The influx of pilgrims in this and subsequent 'years of rejoicing' suggested that this profitable celebration should be held more often. At first it was held every 50 years, then every 33, until Pius II decreed that every 25 years would be a holy year. The first real centenary celebration, however, seems to have been held by Protestants – possibly in imitation of the papal custom. This was the centenary of the Reformation of 1517 held in Germany in 1617, as the coins struck in its honour testify. Catholics soon followed suit. The Jesuits published a sumptuous book with emblems called *Imago Primi Saeculi Societatis Iesu* to mark their centenary in 1640. What all these examples confirm is that such celebrations were originally linked to the life of a particular community. It is one's own history that is to be commemorated. Thus, in fifteenth-century Florence, the 200th return of Dante's (probable) birthday was celebrated because it was especially important for the Florentines to honour publicly their great poet,

number. I am thinking of the great 'Shakespeare Jubilee' held at Stratford-upon-Avon in 1769 and organized by the actor David Garrick. Many of the less salubrious features of such jubilees, such as their razzmatazz, kitsch and commercialism, were in evidence even then. As I have said, the date was a random one: the idea for the jubilee developed out of a project to erect a statue to Shakespeare, and in return for his efforts Garrick was given the freedom of Stratford. Handel was probably the first composer to have his 100th birthday celebrated – in England in 1785 – but this followed on naturally from the fact that his works were still performed after his death.[12]

According to the *Oxford English Dictionary*, the word 'centenary' was first used in the modern sense in 1788, when it was applied to the political celebration commemorating the "Glorious Revolution of 1688". Celebrations of this kind naturally offered the heightened national consciousnesses of Italy and Germany a welcome opportunity to enhance their community spirit. When the Academy of Arts in Berlin organized a public, secular celebration of Raphael in 1820, it was first decided that the anniversary of the death of Dürer should be celebrated no less brilliantly. Indeed, the Dürer Jubilee of 1828 turned into a national celebration of German Romanticism. Yet even this was overshadowed by the great Schiller centenary of 1859, which produced veritable orgies of patriotic rhetoric.

Only an anniversary gives a community certainty that achievements can defy mortality

whom they had banished. It was decided in 1465 that Domenico Michelino should be commissioned to replace the earlier paintings in the cathedral with the portrait of Dante that hangs there to this day.[9]

The growing popularity of counting in terms of centuries seems to have come from the teaching of history in schools. By about 1700, it was universal.[10] From then onwards reports of centenary celebrations abound. At any rate, in 1706 the University of Frankfurt an der Oder celebrated its 100th anniversary in the presence of the King of Prussia.[11] The first centenary celebration that I know of for a great philosopher is the 1746 commemorative address given in Latin by Johann Christian Gottsched at Leipzig University to mark the 100th birthday of Leibnitz. In 1728, a commemorative volume was published in Goslar in honour of the artist Albrecht Dürer, "at exactly the time that he left this world 200 years ago". This was the forerunner of a festive procession that was to know no end.

Surprisingly, the first large-scale celebration of a famous person that was specifically called a 'jubilee' is not connected with a round

It goes without saying that the ruling houses also used their hereditary rights to bind their subjects closer to them through family celebrations and jubilees.[13] My mother, who was born in 1873, remembered even in old age the magnificent procession staged in Vienna by the painter Makart to mark the silver wedding anniversary of the emperor and empress in 1879. Even Queen Victoria's 50th jubilee of 1887 lives on in folk memory. With a fine sense of irony, Robert Musil made his novel *The Man Without Qualities* revolve around futile attempts to organize a dynastic celebration of this kind – namely the hypothetical 70th anniversary in 1918 of the accession of the 'Peace Emperor' Franz Joseph, which was meant to eclipse the 30th anniversary celebrations in the same year of Kaiser Wilhelm II's accession.

There would seem little point in continuing the lines that lead from there to our present situation, in which calendars and almanacs for the coming year ensure that we will not miss the opportunity to celebrate anniversaries.[14] Publishers and exhibition organizers, radio and television producers, not to mention the tourist industry, are grateful for these aids to planning their programmes. I recently received an invitation from the Polish Cultural Institute in London to an exhibition marking the centenary of the first Polish film poster. The socio-economic factors that led to this inflation are obvious. But there are certainly deeper reasons as well for the constant increase in the number of anniversaries. In these fast-moving times of technological progress the past is all too easily forgotten. What used to hold this dangerous tendency to 'memorialize' things in check has turned out not to be very effective. We walk or drive past the statues of eminent men and women without reading or even noticing the inscriptions on them. Many of the shrines commemorating the occurrence of mythical or religious events have fared a little better and still attract swarms of pilgrims. In secular terms, these have become the well established 'sights' of tourists, who love to hear that the particular house, or at least space, is exactly as it was when the famous person left – that is, that time has stood still. Only an anniversary is capable of giving a community of like-minded people the certainty that achievements and events can defy mortality, as Horace had so rightly hoped. For its handiwork belongs to civilization, to the 'universal culture' that is aware of having its roots in the past. Consequently, the anniversary, unlike a ritual celebration, in no way denies the linear passage of time. It can also make conscious for us the distance separating us from the event that is being celebrated but must not be lost. Even a hundred years after his death, Cassirer's philosophy remains of concern to us. So I should like to close with the English expression, which is untranslatable because it blithely combines the cyclical conception of time with the linear one: 'Many happy returns of the day'.

Joseph Naylor, Astronomical calendar clock (see **061**)

1. Goethe, *Faust (Part I)*, English transl. by P. Wayne, London 1949, p. 34.

2. The original version of this paper was delivered in 1974 as part of the celebrations marking the centenary of Ernst Cassirer's birth. It was subsequently published in a volume of essays in honour of Dieter Henrich, in thanks for his having presented the original paper for me at the Cassirer centenary and for having put forward my name to receive the Hegel prize from the city of Stuttgart. See 'Zeit, Zahl und Zeichen. Zur Geschichte des Gedenktages', *Philosophie in synthetischer Absicht*, ed. M. Stamm, Stuttgart 1998, pp. 583–97.

3. See, for example, the arguments in Frank E. Manuel, *Shapes of Philosophical History*, Stanford 1965.

4. See the texts collected in *Festivals in World Religions*, ed. Alan Brown, Burnt Mill, Harlow, Essex, 1985.

5. A German paraphrase of the poem appears in H. Bethge, *Die Chinesische Flöte*, Leipzig 1920, p. 39.

6. Taken from Mircea Eliade, *Le Mythe de l'éternel retour*, Paris 1949, pp. 169–71. See also R. Gombrich, 'Ancient Indian Cosmology' in *Ancient Cosmologies*, eds. C. Blacker and M. Loewe, London 1975, pp. 110–42.

7. For the subsequent practice, see B.Z. Wacholder, 'The Calendar of Sabbatical Cycles During the Second Temple and the Early Rabbinic Period', *Hebrew Union College Annual* (Cincinnati), XLIV, 1973, pp. 153–96.

8. See M. Bernhardt, *Handbuch zur römischen Münzkunde*, Halle 1926, pp. 75 and 76ff.

9. For a reproduction and information, see C. Marchisio, *Monumento pittorico a Dante in Santa Maria del Fiore*, Rome 1956.

10. J. Burckhardt, *Die Entstehung der modernen Jahrhundertrechnung. Ursprung und Ausbildung einer historiographischen Technik von Flaccius bis Ranke*, Göppingen 1971. See also A. Witschi-Benz, review of R. Landfester, *Historia magistra vitae*, and J. Burckhardt, *Die Entstehung der modernen Jahrhundertrechnung ...*', in *History and Theory*, XIII, no. 2, 1974, pp. 181–89.

11. H.H. Monk, *The Life of Richard Benteley, D.D.*, London 1833, p. 191.

12. Dr Burney, *An Account of the Musical Performances ... in Commemoration of Handel*, [s.l.] 1785.

13. See the collection by E. Brix and H. Steckl, *Der Kampf um das Gedächtnis. Öffentliche Gedenktage in Mitteleuropa*, Vienna 1997.

14. For example, the publisher Deike in Kreuzlingen promised to include over 1600 birthdays, death-days and other commemorative dates in their almanacs.

'POSITIVIST CALENDAR' (1849) : HUMANITY · MARRIAGE · PATERNITY · FILIATION · FRATERNITY · DOMESTICITY

Niccolò Pellipario, *Apollo in his chariot* (see 022)

Time and History

Felipe Fernández-Armesto

For historians, time is the past. The future is just the past we have not yet experienced. If there is such a thing as an absolute future, it can exist only in the imagination; and imagination, as soon as we are aware of it, is already part of memory. History and time therefore seem made for each other: mutually nourishing. Time is history's subject matter and history is the diet of time.

In practice, as the concept of time has changed, the way history is written has changed with it. At the risk of over-simplification, we can summarize the state of our knowledge by saying that the history of the concept of time has gone through three phases: a cyclical phase, which encouraged cyclical notions of history; a linear phase, which encouraged teleological history-writing; and a chaotic phase, which has encouraged chaotic history. After a glance at the conceptual problems, we can review these briefly in turn.

Time is change. No change, no time. Or, in common usage, time is a means of relating different or distinguishable changes to each other. When, for example, we distinguish what appear to us to be successive states in a single process, we are exhibiting a notion of time. As I write these lines, I can see puffs of smoke rise above a hedge from a bonfire in my neighbour's garden. The fact that they appear to me successively means that I have a time-scale along which to place them. I also have a crude scheme of measurement – in terms of numbers of puffs. I can, in effect, relate one process of change to another: the appearance of smoke-puffs to my own counting. In the light of such a comparison, change appears to become measurable and to acquire a dimension.

Time, in this basic sense – as far as we know – is universal. Every human being is aware of time. Every culture has practices for 'sharing' time, or for establishing some standard and communicable scale of reference. The origins of this concept are beyond history.

Our earliest evidence precedes literacy. It consists in rudimentary calendars notched on to sticks or bones or shells, scores of thousands of years old; or in the detritus of divinatory technologies,

Time is history's subject matter and history is the

from which writing techniques emerged, like the oracle-bones of Shang China; or in the monumental calendrical devices which in some civilizations preceded or outlasted other forms of record-keeping, like the henges of megalithic Europe. The antiquity of almost all evidence of this sort is a matter of dispute. Even when it can be dated securely, it can only be speculatively interpreted because of the want of other evidence from periods at which

Chu Jun, *Jin shi tu* (see 272)

In a world of perpetual repetition there is little point in keeping chronological records

calendrical documents first appear. The Druidical divinatory calendar from Coligny, for instance, is the earliest surviving document in what is known to be a Celtic language. Its sixteen columns cover a five-year cycle of sixty-two lunar and two intercalary months, divided into propitious and unpropitious periods. Similarly, without possessing anything normally classified as a writing system, the Mixtec of Mesoamerica had almanacs listing days of good and evil import, with elaborate royal genealogies, in which portraits of kings and depictions of key events were aligned with dates recorded in a scheme of numerical annotation.

Still, enough evidence of early timekeeping survives to suggest that a more or less universal phase occurs early in the history of the concept of time, in which the standard of reference is supplied by celestial bodies. Thus, for example, one could measure one's own sense of ageing, or that of another person, or the duration of a war or hunt or journey, in terms of so many Suns or so many Moons.

Annals were a form of historical record-keeping suited to societies where astrology commanded assent and where human life – like so much of the rest of nature – was thought to reflect the solar cycle. Traditional historiography adopted the solar year, with astonishing near-uniformity, as its basic unit of periodization. The inertia of the tradition is such that, even today, in industrial or post-industrial societies, where the seasons make little difference to people's lives, historians still depend on it.

Societies with a strongly cyclical notion of time tend to invest heavily in it. They invest emotion and expectation; and they willingly carry the burdens of priests and prophets who proselytize its message. In partial consequence, historiography reflects the cyclical vision: you get cyclical history or, in deference to the ultimate changelessness the cycles encompass, no history at all. In a world of perpetual repetition there is little point in keeping historical records with chronological precision. Astonishingly – at first glance – this sort of indifference to historical chronology can persist in cultures otherwise

diet of time – they are made for each other

obsessed by time. In ancient Egypt, for instance, where people were so concerned with accurate timekeeping that they buried their dead with stellar calendars to enable them to keep abreast of the date during their underworld journey, the compilers of histories evinced no interest in when the events happened, except as far as 'such-and-such a year of such-and-such-a-pharaoh' can be said to provide a chronological framework. The order of pharaohs and dynasties –

which we are still unable to correlate satisfactorily with the chronologies of other peoples who left overlapping records – had eventually to be worked out by a Hellenized scribe who picked up a taste for chronological exactitude from the Greeks. The similar indifference of ancient Indian historiography to questions of chronology is often remarked, yet it developed in a culture with amazingly elaborate practices of cosmic timekeeping. Though Jain mathematicians reckoned the age of the Earth back through cycles revolving over hundreds or thousands of millions of years, none of their fellow-sages ever wrote a history book with a date in it. Compared with the stasis of the cosmos, in the context of such a vast scheme, the observable cycles of the heavens seem trivial and the histories of human change seem puny.

At the other extreme among cultures in which time is cyclically conceived, history fuses with prophecy. The most striking case is that of the ancient Maya. There has never been a culture so obsessed with time. The epigraphy of the Maya 'classic age' from the third to the tenth centuries AD is dominated by dates which often take up so much space in surviving inscriptions that there is hardly any room left for the text.

The surviving historical books, compiled, in their present form, in the colonial era and written in Roman characters, are arranged like annals, with entries under various dates in the sacred calendar, implying that they can apply equally to different points in the cycle – both those past and those still to come. Invasions by Spaniards blend with recollections of earlier invasions from central Mexico; accounts of disturbances and revolts merge with memories of internecine wars in the distant past; references to early bloodshed become prophecies of a successful insurrection against Spanish rule.

Celestial or natural cycles evidently have had a profound and widespread appeal for people seeking standard ways of measuring time. But another method has always been available: instead of relating linear changes to cyclical ones, the timekeeper can limit his comparison to two or more sequences of linear change. A linear concept of time, and a linear method for measuring it, is illustrated among the Nuer of the Sudan, who use the growth-rate of cattle, or other, similar points of reference, when recalling the time at which an event occurred. Thus a famine, war, flood or pestilence might be remembered as occurring 'when my calf was so high' or 'when such-and-such a generation was initiated into manhood'.

Cyclical and linear techniques are therefore probably both of immense antiquity and have cohabited for most of history.

Gradually, however, the linear method has asserted its superiority, as the imperfections of astral cycles have been revealed, and scientific enquiry has cast doubt on the existence of any true cycles. This development is often said to have started with the Jews of the period of the formulation of the Genesis creation-myth: in reality, an even older tradition of habit of mind may be responsible, but the writers of the Old Testament were of enormous importance, in this as in so many other respects, as moulders of later opinions. Against the beauty of a cycle, which has no beginning and no ending, Jewish sages were among those who proposed the argument that time began with a unique act of creation that inaugurated change within changelessness and time within eternity. This did not mean that time had to be consistently linear in character: it might have started like a loosed arrow or like released clockwork; it could exhibit some properties of both. From other cultures with a cyclical notion of time, the Jews incorporated themes of recurrence and – after the diaspora – of the restoration of ancient Israel. Nevertheless, they adhered to a predominantly linear model. Particular events might be repeated or echoed; but history as a whole was unique. In compiling historical records, the Old Testament writers emphatically avoided cyclical formulations. The historical books of the Old Testament use human generations as units of periodization. Except when the life-span of particular individuals are in question, reckonings are hardly ever made in years and never in terms of other astronomical cycles. Dates are virtually unknown. The difficulties of relating the Old Testament histories to the chronological frameworks of other peoples who have left records from the same period have defeated every attempt made so far.

The prestige of Jewish scriptures in Christendom and Islam ensured that the linear model of time was inherited and re-emphasized by the most dynamic and widest-spreading civilizations of the modern world. Indeed, a Christian could not have a cyclical notion of time without lapsing into heresy. The incarnation occurred once and the sacrifice Christ made at his death was sufficient for all people, everywhere, for ever. His second coming will not be a repeat performance but a final curtain-call at the end of time.

The results include the teleological schemes of history which have been so popular in Western writing until our own times. Typical of these is the providential scheme, by which God leads the world towards a redemptive climax. A popular form of this reading of history was devised in twelfth-century Sicily by a practitioner of biblical divination, Joachim of Fiore. In his version, the past was encompassed by two ages – of the Father and, since the Incarnation, of the Son. A future or imminent Age of the Holy Spirit would precede the end of the world, after a cosmic struggle of Antichrist against the 'Last World Emperor'.

Sometimes, since then, traditions linger from eras with cyclical notions and infect what are essentially teleological schemes with cyclical episodes. Along with a lot of other ancient wisdom, the Renaissance went some way towards rehabilitating cyclical notions. For example, Aristotle's three-part scheme for the constitutional history of city-states was devised in a period when cyclical notions prevailed, but it was adapted in the Renaissance within a linear framework. Aristotle envisaged a perpetual cycle in which monarchy, having degenerated into tyranny, was replaced by aristocratic government, which would in turn degenerate into oligarchy before replacement by democracy; the ensuing anarchy would be redeemed by a monarchy and the cycle would then start again. Machiavelli borrowed this scheme for his own reading of history, but he saw the cycle reaching a climax and a conclusion with the perfection of the republic.

The linear concept of time has been undermined

The providential model can be detected in many secular variants adopted more recently by Western historians who substitute for providence one or other form of progress: the purgative and supposedly improving effects, for instance, of class struggle or racial purification or constitutional conflict or the survival of the fittest in a universal struggle to adapt. In each of these traditions, and others like them, history was expected to stop when the process reached perfection – whether that might be the dictatorship of the proletariat, or the classless, stateless society, or the German 'race', or the British constitution, or white world supremacy, or the robotic perfection of the 'machine age'.

Indeed, until very recently we have gone on getting more and more linear in our understanding of time. Western science in the nineteenth century proposed interpretations of the past of the planet in terms of development in a single direction from a finite starting-point or towards a final end. The theory of evolution encompassed the whole history of creation in a single, progressive story. The second law of thermodynamics, developed around the middle of the century, suggested that the universe was heading for entropic immolation and that its entire history could be described as a single process of energy loss. The discovery of radioactive emissions, constant and unilinear, in 1896, seemed at once to support the linear conception of time and to provide a means of measuring time independently of sidereal bodies. The discovery of the apparent expansion of the universe in the 1920s impressed most interpreters in a similar way: it does not of itself rule out the possibility of cyclical phases in which, for example, the universe might start to shrink and rebound from infinitesimal compression to resume expansion – but the observations we can make disclose only one, unidirectional phenomenon. Meanwhile, evidence accumulated that all the apparently cyclical phenomena which tricked the senses of people in past ages are really in the grip of linear change. The rotation of the earth is slowing down. The sun and stars are burning up. The apparent revolutions of the celestial bodies conceal a common story of degeneracy.

The linear concept of time is practically useful; but its influence on historians has long been undermined, and all the teleological constructions of the past, which it formerly encouraged, have been abandoned. Current fashions in historical writing reflect instead a concept of time which has no direction at all – neither linear nor cyclical. It is imagined in a state of chaotic, directionless flux; or it is classified as a mental construct which can safely be omitted from any attempted account of an objective world. Two mutually

The subject matter of history seemed to have vanished. History had lost the past. How could it have a future?

Even more influential on historians' work, in the long run, than the ideas of philosophers were those of scientists, because, for much of the twentieth century, historians have tended to aspire to quasi-scientific status and to pretend that their discipline resembles that of scientists. The great upheaval in the way modern science has represented time began in 1905 when Einstein emerged, like a burrower from a mine, to detonate a terrible charge. The realisation that time is not objective or absolute, but that every observer has his own time, which varies according to his speed and vantage point, has no practical implications for the study of history, which is confined to such a tiny part of the cosmos that the effects of relativity can safely be ignored. Nevertheless, Einstein's intervention

for historians: current fashions reflect instead a concept of time that has no direction at all

contradictory theories of time, both popular in the late nineteenth and early twentieth centuries, have helped to induce this bewildered state of mind. The first – which has a long history in Western philosophies of time – might be called the atomic theory of time. This denies duration, acknowledges only discontinuous instants and therefore questions whether the existence of the world at any moment implies its continued or previous existence at another. In a world matched to this theory, memories are an encumbrance and history is a causeless, unconnected cascade of events – which is just how many contemporary historians write it up.

The second theory is particularly associated with the French philosopher Henri Bergson, who claimed to have invented it and called it 'duration'. All definitions of it are baffling. His own is best: "Pure duration is the shape taken by the succession of our states of consciousness when our inner self lets itself live, when it abstains from establishing a separation between its present states and the preceding states." He regarded 'the past' as a virtually meaningless concept, perceptible only in the part of imagination we call memory, where it took purely subjective forms, and where it was actually part of the present – never achieving completed action but only ever 'becoming'. This was close to literal nonsense, but like so much nonsense it was immensely influential. It helped to inspire the tradition of prose fiction known as the stream of consciousness. Transmitted via a complex chain of philosophers and linguists, it has come to have a similar effect on the work of historians, many of whom have never heard of Bergson but who, in suppressing chronology and abandoning objectivity, show the long-term effects of the idea he implanted in the tradition they have inherited.

had a belittling effect on the kind of chronological researches in which historians traditionally engaged. The fashion for experimentation with chronology – or perilously neglecting it – is among its liberating or confusing effects. These have been augmented by work on time in Einstein's wake. The 'arrow of time' could turn out to be a boomerang. Time could be reversed – for instance, by the contraction of the universe. The order in which we perceive events – and therefore the structure of cause and effect we infer from this order – is negotiable.

The shake-up of the concept of time is only one of many reasons why contemporary historical writing has shied away from traditional objectives and become more like imaginative literature: historians' work, even in the more conservative reaches of the profession, has embraced the counter-factual, the self-reflexive, the representational, the random, the causeless, the unverifiable, the liminal and the implicit. As long as historians do not lose their nerve, these novelties can enrich the discipline. The past is a name for everything in our collective experience: through all the changes in the way we understand it, its interest abides. Even our imaginings and speculations become part of it as soon as they occur. We shall go on studying it, because, in effect, we have nothing else to study.

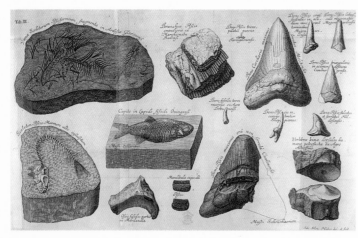

Nicolaus Steno, *De Solido intra Solidum Naturaliter …* (illustrations of fossils) (see 275)

Geologists' Time

A Brief History

*Time we may comprehend;
'tis but five days older
than ourselves.*
Sir Thomas Browne

Martin Rudwick

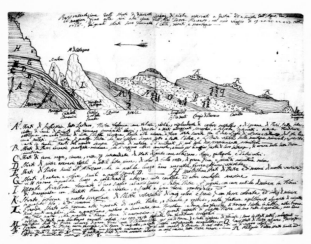

Giovanni Arduino, *Section of Val d'Agno*, manuscript, 1758

In explaining that the earth was just five days older than man, Sir Thomas Browne expressed a belief that would have been taken for granted in the seventeenth century, even by the best informed. The universe – and time itself – had, so far, lasted no more than a few thousand years; and, apart from a brief prelude to set the scene for them, human beings had been on stage throughout. James Ussher, Browne's older contemporary, claimed that the moment at which time began could be dated precisely: to a specific Saturday evening in the autumn of the year 4004 BC. Far from being scolded as an obscurantist or derided for his naivety, the scholarly Irish archbishop was admired for giving quantitative precision to the accepted account of the world's history. Other and equally learned 'chronologers' disputed Ussher's particular date, but they all shared the scholarly methods by which he had reached it. A short time scale for the universe, for the earth and for human beings was simply taken for granted.

In the modern world, a quite different time scale is equally taken for granted, at least among scientists. Astronomers and cosmologists date the 'big bang' in terms of billions of years, and do so as nonchalantly as they talk of the literally inconceivable distances to the furthest galaxies. Geologists and other earth scientists deal with time in almost equally vast quantities, but they also date a plethora of past events with a casual confidence and a sense of relative precision that tend to recall Ussher's efforts. Multicellular life first exploded into diversity in the early Cambrian seas of 540 million years ago; the mass extinction at the 'K/T boundary' (probably the result of an asteroid's impact) took place 65 million years ago; the last retreat of the Pleistocene ice-sheets happened swiftly some 10,000 years ago. Such dates, progressively refined and set with known margins of error, are the everyday currency of twentieth-century geologists.

Ussher and his colleagues practised the seventeenth-century science of chronology; modern geologists practise the twentieth-century science of geochronology. The similarity of terms points to shared concepts and even methods. Both groups have been at the forefront of intellectual life in their respective centuries. In fact, far from being diametrically opposed, what Ussher and other chronologers were trying to do was the direct lineal ancestor of what earth scientists do in the modern world. This essay describes very briefly how the conclusions of the first group evolved, by the kind of learning process that characterizes all scientific work, into those of the second.

Ussher, and many other chronologers all across Europe, aimed at correlating the calendars of all civilizations into a single universal time-line for world history. Since the correlations were often uncertain, the so called 'Julian Period' was widely used as the standard of reference. This deliberately artificial time-line, which was independent of any particular culture or religious tradition, was chosen because its starting point was further back in the past than anyone imagined Creation to have been. Ussher, for example, dated the beginning of time at 710 Julian, which meant that any earlier dates had been, as it were, in 'virtual' time. Although its use did not outlast the seventeenth century, the Julian Period was a crucial innovation, because it established the idea of a 'neutral' time-line that stood outside the flux of historical events and the diversity of human cultures.

On to this time-line, chronologers tried to plot the important events in the histories of all the ancient cultures they knew: not only Jewish, Greek and Roman, but also Egyptian, Indian and Chinese cultures

Chronology" from fossils – but he meant that they could supplement textual evidence by throwing light on the obscure earliest periods of history. Neither he nor his contemporaries had any clear notion that fossils might be witnesses of a history dating from before any written records whatever.

Naturally enough, fossil shells were initially assumed to be evidence for the one and only recorded event that seemed large enough to account for them: Noah's Flood, which seemed to be matched by similar stories in other ancient cultures. Yet the more closely fossils were studied, the less plausible this explanation became, on any literal reading of Genesis. Fossils often seemed too well preserved to have been swept hundreds of miles in a turbulent flood. Moreover, they were found in thick piles of finely layered rocks, which could hardly have been deposited in any brief event. Around the end of the

A world without a human presence seemed meaningless, indeed unthinkable

(as far as they were understood at the time). In cases of inconsistency, it is hardly surprising – given the cultural basis of Christendom – that biblical data tended to be privileged above others: Ussher, for example, dismissed as 'mythical' the ancient Egyptian claims to dynasties extending far back before his date for Creation itself. However, such debates obliged chronologers to judge all the records in terms of their relative reliability and, for the first time, to distinguish such categories as 'myth' and 'legend'. So, in effect, the practice of textual criticism undermined any naive literalism about any of the sources – even the biblical texts.

The chronologers were always frustrated by the paucity of reliable records for the earliest periods. Traced backwards, most histories ran out into the mythical times of heroes and demigods relatively quickly. Even the biblical texts became disappointingly thin, with little more than genealogical lists of who begat whom. So when, about the same time as the end of Ussher's life, naturalists began to draw attention to fossils – many of which were easy to recognize as organic in origin – these natural objects were soon recruited as evidence that could supplement the textual sources for chronology.

seventeenth century the London physician John Woodward, who later endowed a chair at Cambridge to promote the study of fossils, argued that the piles of rock, with their fossils, had been deposited out of a kind of global soup, in order of their specific gravity. But this entailed modifying the Flood story out of all recognition, and later naturalists quietly dropped the equation with Genesis. They retained, however, the idea that the pile of rocks could be read as the record of a sequence of events, or, in effect, as a history based on natural evidence.

In the early eighteenth century, naturalists came to suspect increasingly that most of the rocks and their fossils must date from long before the few millennia documented by the chronologers. At the same time orthodox beliefs in the created status of everything in the world, from atoms to human beings, were being challenged by 'deistic' or even atheistic claims that the world – including its human inhabitants – was uncreated and eternal. Such 'eternalist' views, which reached back to Aristotle, were the often covert rival to the traditional short time scale. Nevertheless, these rivals shared one crucial assumption, which distances them both from any modern

Fossil shells were initially assumed to be evidence for Noah's Flood

Fossils were referred to as 'witnesses' to events for which the textual evidence was scanty or obscure. For example, fossils that looked like living marine shellfish were found widely on dry land, far from the sea, so they were treated as nature's 'documents'. They were evidence of the greater extent of the sea at some remote period. In the 1660s, Robert Hooke, the employee of the new Royal Society in London, even suggested that it might be possible to "raise a

conception of cosmic time and history. Both took it for granted that human beings had been present throughout. A world without a human presence seemed meaningless, indeed literally unthinkable, to those on both sides of the argument.

What was novel in the eighteenth century, therefore, was the growing suspicion, among naturalists who studied rocks and fossils, that the world's time scale might be far longer than the work of the chronologers suggested – but not that it was eternal – and that most of this long history might have been non-human, with the whole of human history crammed into just its final phase. These hunches – at first they were no more than that – were, above all, the product of an increasing focus towards studying rocks and fossils in the field, and not merely indoors in museums. However, the very long but finite time scale involved could not be quantified with any confidence. The leading French naturalist Georges Leclerc, comte de Buffon, tried to put figures on it, by extrapolating the results of his experiments on the cooling of model globes. In the 1770s, he calculated that some 74,000 years had elapsed since the Earth originated as an incandescent body thrown off from the Sun. But any such figure depended, of course, on the validity of that theory of the Earth's origin, which, at that point, was deemed both speculative and controversial. In any event, Buffon himself suspected his figure was much too low to account for the piles of rock formations and, privately, he thought some three million years was more likely.

Although such a figure may seem absurdly inadequate to modern geologists, even a million years entailed a huge stretching of the imagination, to spans of time that were as inconceivable – literally – as the cosmic distances estimated by astronomers around the same time. When, in the 1780s, the Scottish philosopher James Hutton claimed that the Earth had "no vestige of a beginning, no prospect of an end", it was his blatant eternalism that drew criticism. His implicitly vast sense of time was, by then, almost a commonplace among naturalists, even if it was still unfamiliar to the wider public.

Ironically, it was the story in Genesis that provided the conceptual model for enlarging human history into a far longer 'geohistory', to use the modern term. In the 1770s, Buffon defined seven successive 'epochs' or significant moments in the Earth's history. In doing this, he offered, in effect, an updated and secularized version of the seven 'days' of the Creation story. By defining his last epoch as the first appearance of human beings – and no longer as God's Sabbath rest! – he made explicit what other naturalists already suspected: that most of geohistory had been pre-human history. This conclusion was reinforced by the continuing failure to find any signs of human life, either bones or artefacts, in any but the most recent deposits.

More specifically, and again ironically, the idea of calibrating nature's history with a quantified chronometry arose from a concern to defend the historicity of the biblical Flood, by using natural evidence to confirm the conclusions of the chronologers. The Genevan naturalist Jean-André de Luc claimed on the basis of his fieldwork that there had been a drastic change in geography in the infancy of the human race – the present continents had risen from

the ocean floor and the former continents had sunk below the waves. He matched this event with a rather loose interpretation of the story in Genesis, and he tried to date it by extrapolating the known rates of various observable natural processes back into the past. By the 1790s, he was calling these processes "nature's chronometers", deliberately recalling John Harrison's recent invention, the great 'high-tech' achievement of the century. River deltas, for example, were growing at rates that could be estimated from historical records. As they were of finite size, they could be used to calculate the approximate date at which they had originated. De Luc concluded, from several independent 'chronometers', that nature's great 'revolution' had happened only a few thousand years ago. The physical evidence was, therefore, compatible with the textual evidence assembled by chronologers. (Modern geologists would recognize much of de Luc's evidence as marking the end of the last glacial period, indeed only a few thousand years ago.)

The spirit of Ussher and his seventeenth-

In the early nineteenth century, geologists – as they may now be called without anachronism – assumed, like de Luc, that the Earth's still earlier history was far longer, but unquantifiable. Leaving the magnitude of the time scale aside, therefore, they concentrated on clarifying the sequences of rock formations. The discovery by the English mineral surveyor William Smith that some formations had what he called "characteristic" fossils, by which they could be traced across wide tracts of country, greatly aided this new study of the sequences of rock formation – or stratigraphy, as it was later called. But it was only gradually that geologists, unlike Smith himself, began routinely to treat such sequences of formations as evidence for geohistory, and it was still longer before they revived earlier attempts to quantify its time scale.

In the 1830s, for example, the London geologist Charles Lyell tried to quantify the dates of each of the more recent sets of rock formations by calculating the percentages of still living molluscan species found among their respective fossil shells. But this method depended on the validity of his theory that faunas and floras have changed continuously over time, by the appearance of new species and the extinction of old ones at a statistically uniform rate. Lyell's contemporary John Phillips pursued a more fruitful approach, exploiting the new results of stratigraphical research in many parts of the world. In the 1840s, he tried to estimate the maximum thickness of sediments deposited in each of what were now defined as successive 'periods' of geohistory, and then to match them with current rates of deposition, as far as those could be measured. Although, like Lyell's method, this depended on an assumption of uniform rates through time, it yielded figures that most geologists found much more plausible: a total of some 100 million years for the whole known fossil record, starting in what had recently been defined as the 'Cambrian' period.

In the 1860s, the Scottish physicist William Thomson (much later ennobled as Lord Kelvin) asserted, on the quite independent grounds of the Sun's supposed rate of cooling, that the Earth's solid crust was no more than 98 million years old. He put this forward with all the arrogance of a physicist lording it over mere geologists (the precision of his figure was spurious, since he conceded a margin of error ranging from 20 to 400). In fact, however, most geologists were initially content with Kelvin's estimate, which matched their own quite well. But it was incompatible with the far longer time scale demanded both by Lyell's extreme version of uniformity and by Charles Darwin's new theory of evolution by natural selection. By the end of the nineteenth century, however, Kelvin and his

breed of scientist who, unlike Kelvin, understood geology as well as physics. The Englishman Arthur Holmes, for example, summarized the research on radioactivity and the age of the earth during the second decade of this century and, after the end of the First World War, Holmes became a leading advocate of the new synthesis. By the late 1930s, most scientists were agreed that the Earth must be at least a couple of billion years old.

More unexpectedly, it also became apparent that the entire fossil record, from the Cambrian period to the present, was only a small fraction of geohistory. The 'Precambrian' period, in which almost no fossils had been found, was no longer treated tacitly as a relatively brief prelude, but became instead the bulk of the story. This new perspective was just the first sign of a radical change in the character of what was now called 'geochronology'. Instead of merely trying to estimate the total age of the Earth, geologists were beginning to use

century colleagues is alive and well among modern geologists

followers had refined their calculations and tightened the screws on the geologists. Their new estimates, allowing only some 20 or even merely 10 million years for the whole record of the rocks, were resisted not only by the followers of Lyell and Darwin, but by a much broader range of geologists. They argued that the physicists must have made some mistake, because the geological evidence, however hard it might be to quantify, seemed to demand substantially more time. They were reassured, for example, when the Irish physicist John Joly offered an estimate of about 90 million years – not far from Phillips's figure – based on the inferred rate of accumulation of salt in the world's oceans.

The discovery of radioactivity at the turn of the century upset both apple carts. Since this strange phenomenon occurred in rocks, its potential as a new kind of 'natural chronometer' was quickly appreciated: once the rates of radioactive decay had been measured, analysis of its products could, in principle, yield quantitative ages for rocks. Even the early results, uncertain though they were, suggested a time scale far in excess of what geologists had expected and, of course, still further in excess of what Kelvin and his colleagues had allowed. The geologists, having been bitten once by the physicists, declared themselves twice shy: if the physicists could swing suddenly from one extreme to another, the geologists thought it prudent to be sceptical about both.

After the First World War, however, geologists gradually came to appreciate that their own distinctive evidence did, in fact, support the longer time scale now offered by the physicists. This reconciliation was greatly facilitated by the emergence of a new

the new 'radiometric' methods for truly geochronological purposes: to put dates (however approximate they might be) on many successive events in geohistory and to calibrate the stratigraphical sequence into quantified periods of time. During the years following the Second World War, a crucial technical development made this far more reliable: the mass spectrometer greatly accelerated the conversion of rock samples into accurate and consistent dates.

By the end of the twentieth century, the radiometric time scale had come to be taken for granted by geologists and other Earth scientists. It provided a dimension of time that lay, as it were, outside the flux of terrestrial events, thereby allowing them to be dated with a steadily increasing confidence and precision. This geochronology offered a framework for the more interpretative work of understanding the causal relations of events and their roles in geohistory. Geohistory had turned out to be unimaginably lengthy and complex. The history of our species had been reduced to a brief final phase, while the history recorded in textual records had become an even smaller sliver of time. Yet the achievements of modern geochronology should not blind us to the continuity that links it back to the more modest history offered three or four centuries ago by the science of chronology. The time scale is now almost inconceivably greater in magnitude, but the ideal of a quantified history and the underlying passion for precision remain the same. At the start of a third Christian millennium – a date first established by the old chronologers – it is right and proper to acknowledge that the spirit of Ussher and his colleagues is alive and well among modern geologists.

262 Jacob de Wit

Truth and Wisdom help History to record Memories
Oil on canvas, 1754
Amsterdam, Amsterdams Historisch
Museum [SA 7439]

Whereas many cultures have, at one stage or another in their development, relied on an oral tradition to preserve their histories, written histories provide the surest format for survival. Indeed, the earliest allegorical personifications of History show her writing. The uppermost figure in this painting personifies Truth, nude, but modestly covered with white drapery – to show that she has nothing to hide, but is never totally revealed. Wisdom is depicted as Pallas Athene, the patroness of the arts and sciences. The fact that the nude female figure seated with her back to the viewer is writing and sits upon a ruined classical column suggests that she is a personification of History. The large terrestrial globe upon which she leans serves to underline how History must record all the events in terms of both time and place. The overall meaning of the painting is that both Truth and Wisdom are needed to construct a proper reckoning of memories, but that History – in writing these memories down – preserves them for posterity.

263 Rarotonga, Cook Islands (Polynesia)

Staff gods
Toa (ironwood), 19th century
Cambridge, Museum of Archaeology and Anthropology [1895.158 (Z 6099)]

The staff gods seem to have some connection with ancestral memory. They have been seen as 'national idols', as genealogical records, or as successive male generations of a tribe (all connected to one another by an elongated symbolic phallus). The most recent interpretation has been that the largest head, at the top of the staff, represents the principal historical ancestor of Rarotonga.

Once man has ordered his present, he begins to re-create his past. The present is often a rather precarious place to be and the future is always unknown. The only means left towards establishing some sort of meaning to one's existence is to anchor it in the past. The past gains its authority from having survived the present. It is the bedrock from which we derive our identity.

The earliest forms of a chronicled past are the genealogies and dynastic lists of ancient Babylon and Egypt. Here, the very act of listing provided the sense of continuity that man needed. One might be tempted to relegate this activity to antiquity were it not apparent that genealogies have continued to play an important role up to the modern day. Searching the archives to secure one's past has become an international pastime. Even a number of preliterate cultures form ancestries leading backwards towards the dawn of time and their society's mythical origins. Indeed, one of the recurring preoccupations of this kind of history is the search for an unbroken list of names or events that will lead back to the creation of mankind.

Western notions of history are based on a combination of Hebrew and Greco-Roman writings, both of which are inclined towards systematization, interpretation, selectivity and the creation of patterns. The past can be ordered according to any number of models: evolutionists propose that things get better; others have suggested a quasi-biological pattern of development – societies grow, become vigorous, then decay and die. As we rely on the past to provide models of behaviour for the present, however, each generation tends to write the kind of history in which the present world view is supported. Regardless of the exact form it takes, history is expected to tell us how and why we arrived at where we are today.

264 Egyptian
Pedigree of Priests
Limestone, 10th century BC
Berlin, PKB, Ägyptische Museum und Papyrussammlung [23673]

Following the ancient Egyptians themselves, modern Egyptologists tend to divide the past according to the successive reigns of the Egyptian kings. Individual dates were and are recorded in terms of the day, month, and year of the current king's reign. The Berlin 'Pedigree of Priests' portrays history in terms of the succession of priests at the Temple of Ptah in the city of Memphis from the equivalent of 2055 BC to 945 BC.

265 Old Babylonian
The Ur-Isin king list
Baked clay, 1813–1812 BC
The Schøyen Collection [Ms 1686]

The most basic building blocks of history are those documents offering some sort of record of what happened when. Today, we can find 'historical documents' almost anywhere – in the rings of trees, in the sediment of riverbanks, in the tusks of an elephant. There is a great difference between these traces of time left to us by the processes of the natural world and the self-conscious creation of a documented chronology, however. One of the certain signs of an emerging culture is when it begins to document its own history. For most of the early civilizations, this history was the story of its kings. By creating and preserving a list of the successive kings, a community provided a bedrock of legitimacy for the state. The rulers, the priests and the people all drew strength from the knowledge that their history was bound together by a chain of names that stretched from the beginning of time to the present day. This list, probably copied out from a larger list, details the kings and their regnal years for the Third Dynasty of Ur (which ended in 2004 BC) and the First Dynasty of Isin. The last entry concerns year 4 or 5 of Damiq-ilishu, who reigned until 1794 BC. It is the oldest known king list to have survived from ancient Babylon. For historians, these lists provide a crucial time matrix for any study of ancient Mesopotamia. As most inscriptions in this part of the world are dated according to the name and regnal year of a presiding ruler, Assyriologists use king lists to date and interpret other finds.

Also
• France, *Hallmark Plaque of the Goldsmiths of Rouen*, 1408, Paris, Musée des Thermes et de L'Hôtel de Cluny [CL 3451]

266 Northern Italy (Venice?)

Bible

Manuscript, early 14th century
London, The British Library, Dept. of
Manuscripts [Additional Ms 18720]

The use of the metaphor of a tree to
describe one's genealogy appears in
a number of different cultures. The
image of the so called Tree of Jesse
represents the genealogy of Christ. It
is based on a combination of texts:
the prophecy in Isaiah 1:3 "And
there shall come forth a rod out of
the stem of Jesse, a branch shall
grow out of his roots; and the spirit
of the Lord shall rest upon him" and
the genealogy of Christ related in
Matthew 1:1. A large tree sprouts
from Jesse's body and, nestled in
the tree, there is a series of portraits
of Christ's ancestors.

Also

• England (East Anglia), *The
Gorleston Psalter*, ca. 1310–20,
The British Library, Dept. of
Manuscripts [Additional Ms 49622]

267 Northern French

Genealogical and chronicle roll

Vellum with pen, watercoloured
miniatures and burnished gold,
ca. 1470–75
Leeds, University of Leeds Library,
Brotherton Collection [Ms 100]

The kings and queens of
Renaissance Europe were anxious to
legitimize their right to rule by calling
upon the authority of a long and
distinguished lineage. Genealogical
manuscripts and rolls, such as this
one, fulfil a dual role. On the one
hand, they provide a historical
overview of all the major events and
personages from the dawn of
Creation. Reading forwards, one is
provided with a simply drawn
chronicle of events or, basically, a
'history lesson'. On the other hand –
working backwards from the present
to the past – these manuscripts
provide a series of genealogical
avenues through which any member
of a royal or aristocratic household
might trace his or her ancestry.

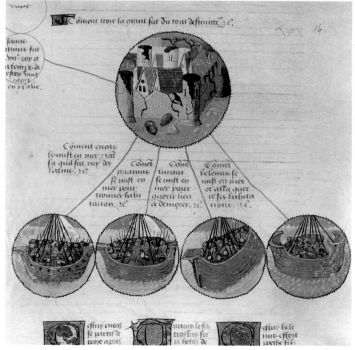

268 Miss Seagar

The Seagar Family genealogical sampler
Embroidered cloth, 1881
Courtesy of Whitney Antiques, Oxfordshire

In some societies, the duty of recording ancestral history falls to the girls and young women. Samplers were generally embroidered by young women before they were married. Most contain verses from the Bible or edifying maxims; many contain specific genealogical information – the same sort of thing you might find entered on to the flyleaf of a family Bible. The Chapman Golding Family Sampler records the names and birth dates (and, in one case, the date of death) of all the siblings of one particular family. It was embroidered by the eldest sister of the family. The Seagar family sampler, however, is a more talkative piece. It lists the date and time of birth for each of the five girls of the Seagar family, who hailed from Brierly Hill (Kingswinford) in the West Midlands. The accompanying doggerel outlines the humble aspirations of a young woman in the early 19th century: "Woman born to dignity retreat/ Unknown to flourish and unseen by great./ To give domestic life its sweetest charms/ With softness polish and with virtue warm./ Fearful of Fame, unwilling to be known,/ No censure dread but those which crimes impart/ The censures of a self-condemning heart./ Heavens ministering Angel shall she seek the cell/ Where modest want and silent anguish dwell./ Sustain the languid head, the feeble knees/ Cheere the cold heart and chase the dire disease./ The splendid deeds which only seek a name/ Are paid their just reward in present fame."

Also

● Louisa C. Golding, *Chapman Golding Family genealogical sampler*, 1812, courtesy of Whitney Antiques, Oxfordshire

269 Chinese

Ancestor scroll of the Li family
Tempera on canvas, 1941
London, collection of G.S. Barrass

The Chinese veneration of ancestors has a past as long as that of China herself. The practice helps to establish a link between the past, the present and the future. Sons draw strength from the unbroken lineage of their forebears; fathers gain comfort from knowing that their spirits will be well tended by later generations. In this regard, the teachings of Confucius (551–479) were important. His stress on mutual obligation, in which the role of filial piety is paramount, helped to strengthen well tested concepts and existing practices. Prior to the Communist Revolution, it was not uncommon to find ancestral shrines within most well-to-do Chinese houses. Here, on special occasions – such as the New Year, on birthdays or on any occasion that one would expect the 'whole family' to be present – formalized portraits of one or more generations of the ancestors would be displayed above makeshift altars containing food and incense. The names and birth and death dates of each of the ancestors would be preserved on wooden tablets. In some cases, though, the portraits themselves have blank spaces to serve as two-dimensional ancestor tablets. As each member of the family passed away, his or her memory would be preserved in a Chinese version of 'the family tree'.

Also

● Chinese, *Ancestor tablet*, Qing dynasty, Edinburgh, National Museums of Scotland [L.209.1]

270 James Ussher

The Annals of the World deduced from the Origin of Time ...
Printed book, London 1658
London, The British Library, Dept. of Printed Books

An interest in history often generates curiosity about the nature of pre-history. There were very few documents, however, to help the learned chronologer judge how much time had elapsed between the creation of the universe and the first historical records. Up until the late 19th century, the most widely used historical 'document' was the Bible. Apocalyptic passages from Daniel (which suggested that the world would end when it had aged 1290 'days') or from Revelations (which foretold that the world would end after "seven ages" had passed – or in its 6000th year), tended to turn men's minds towards speculation about how old the Earth itself might be. One of the catalysts towards the Venerable Bede's study of the calendar for example, was an attempt to find out how old the world was. In the 9th century, the rabbinate calculated that the world had been created at 3760 BC and have dated the Jewish year-counts accordingly ever since (the year 2000 will be *anno mundi* 5760). But Christians have always been slightly at odds about the exact age of the Earth. James Ussher, Archbishop of Armagh, having read all the available chronological sources in their original languages, conflated, cogitated and deduced that the world had been created just before nightfall on 23 October 4004 BC. Using this date as a starting point, he created his *Annals of the World*, in which all the major events from all the great cultures of the world have been organized in sequence. For providing a clear linear matrix for all history, Ussher was greatly championed and his work remained influential well into the succeeding century.

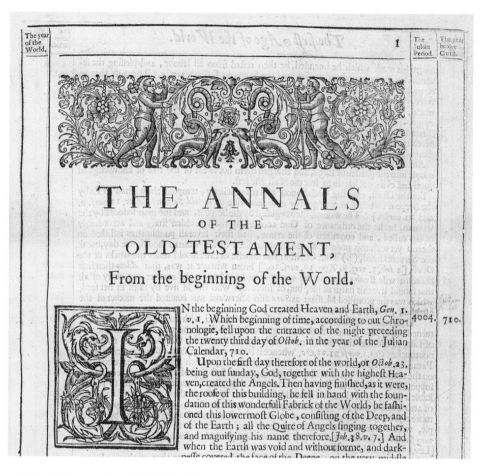

271 English (London)

Wallis's New Game of Universal History and Chronology
Hand-coloured engraving mounted on linen, 1814
London, The Bethnal Green Museum of Childhood [E.218–1944]

From the 15th century onwards, educational games were often used to teach children and adults the rudiments of history. Games such as this underline the notion that history is both linear and progressive. In this game, the player moves forward by reading sequential histories aloud. History, in this case, is composed of 75 stages (or squares). The culmination of the game is to arrive in the square containing a portrait of the Prince Regent (later George IV).

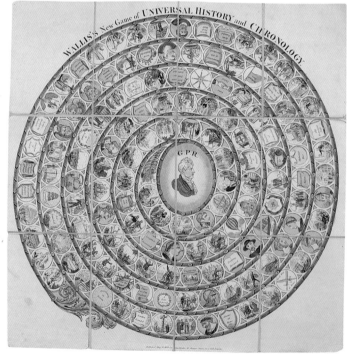

272 Chinese

Ritual vessel (he)
Bronze, 11th century BC (Shang or
Western Zhou dynasty)
London, The British Museum, Dept.
of Oriental Antiquities
[OA 1953.5-1.1]

Inscriptions first begin to appear on
cast bronze ritual vessels about 1250
BC, during the Shang dynasty. The
earliest examples are limited to the
name of the clan or the individual
who owned the vessel. On this *he*,
for example, there is a clan mark set
within a rectangular border and
characters that read 'to father Ding'.
The character *ding* is one of the so
called ten stems, which were used
by the Chinese as time-markers
analogous to the days of the week. It
was often the tradition to name an
ancestor after one of the ten stems
and to keep that day set aside for
sacrifices. Ritual bronzes became
one of the gifts that a king or
provincial ruler might bestow on his
favoured subjects along with land
and honorific titles. The vessels
would often be inscribed with
detailed accounts of these gifts or
might describe the exact nature of
the fealty that bound the subject to
his lord. As the bestowing of gifts
tended to coincide with major
political events, the inscriptions on
these bronzes often serve as
valuable historical documents. They
also served the purpose of acting as
an intermediary between ancestors
and heirs. A remembrance of one's
ancestors would be part of the ritual
process; and the deeds of an
illustrious family member would be
kept alive for future generations
through the inscriptions themselves.
The translation of these early
historical records occupied the same
sort of effort that one sees in Europe
during the Renaissance and
afterwards, when classicists strove
to record all the extant Greek and
Roman inscriptions – save that, in
China, the study began much earlier.
In order to aid these studies,
rubbings or 'ink squeezes' were
taken from the bronzes and these
paper transfers were circulated
amongst the scholarly community.
The second stage in this
dissemination of inscriptions is the
Jin shi tu, in which reduced,
woodblock versions were produced
by the engraver Chu Jun and were
complemented by a commentary
offered by the scholar Niu Yunzhen.

Also

• Chu Jun, *Jin shi tu* (with
commentary by Niu Yunzhen),
1743, London, The British Library,
Dept. of Oriental Manuscripts
[15299.d.14]

273 Attr. Swift Dog (Ta-sunka-duza)

High Dog Winter Count
Muslin with pigments, *ca.* 1912
Bismarck (ND), State Historical
Society of North Dakota [00.00791]

Many tribes of the North American
Plains kept a record of their tribal
history through the medium of
'winter counts'. The winter count
was a series of pictograms, each of
which encapsulated the most
important event of a given year. After
consulting with the elders of the
tribe, the 'Keeper of the Count' would
decide upon which event had been
most memorable for the tribe that
year and what would separate it out
from all the previous and succeeding
years. The position of Keeper of the
Count was passed on from
generation to generation. When this
particular winter count was 'closed',
it recorded 113 years of Lakota
history, from 1798/99 to 1911/12.
The cycle of the High Dog Winter
Count starts in the upper right-hand
corner and proceeds in a clockwise
direction. From contemporary
descriptions, we know that the
pictogram for the first year (1798)
records the initiation of a ruling class
in the tribe, all of whom wore blue
feathers. The fifth year (1802) notes a
battle in which some crinkly-haired
horses were taken; the following year
(1803) marks the acquisition of some
shod horses. The square in the upper
right-hand corner of the count (1810)
shows that the tribe had suffered a
smallpox epidemic. Year 14 (1839)
recalls the year in which a woman
hanged herself (the written legend
tells us that 'there was a love-
romance behind the act') and, in year
44 (1841), snowshoes had to be
worn. The last years of the winter
count are marked with a number of
deaths. In year 111 (1919), there was
a comet and, in the final year of the
count (1912), the children were
covered with spots because they had
the measles and a star disappeared
from the sky having burned up.

274 **William Dyce**

Pegwell Bay, Kent: A Recollection of October 5th, 1858
Oil on canvas, 1858–60
London, Tate Gallery [NO 1407]

If one were to offer another title for this painting, it might be 'Variations on the theme of time'. The first kind of time depicted is the 'now' of human time. Dyce portrays the shore at Pegwell Bay near Ramsgate, with the tide out, on a cool autumn day. In the foreground, the artist has included figures of his wife and his wife's two sisters and one of their sons. A male figure, perhaps Dyce himself, is shown carrying artist's materials at the far right of the canvas. Each of the figures is caught as if in a photograph. 'Astronomical time' is represented by the appearance of Donati's comet, which was first observed on 2 June 1858. The *London Illustrated News* had advertised that, on 5 October, the comet would be at its brightest and easily observable in the sky. To the Victorian imagination – still struggling with the nostalgia for a smaller and more comprehensible cosmos – recent astronomical discoveries presented a universe composed of unimaginably large distances and incomprehensibly immense time-spans. Comets, in particular, were seen as witnesses of epic portions of time. As the *Illustrated News* reporter pointed out, this comet would not be seen again for another 2100 years. Now, with the scales recalibrated, the great temporal patterns drawn by the elliptical orbit of a comet must have seemed rather awesome. 'Geological time' is represented by the women collecting shells and by the huge chalk cliffs that loom behind the holiday-makers. In the 19th century, people were also having to come to terms with the long history of their own planet. The humble seashell that Dyce's wife is stooping to collect might have fallen from the adjoining cliff and been used as evidence that molluscs existed on Earth 250 million years ago. The recognition that fossils were petrified examples of life-forms that had existed hundreds of thousands of years before man was something that had become apparent only recently. It was not until the 1840s, for example, that John Phillips defined the 'Cambrian period' of 100 million years ago. Seen from this perspective, the title of Dyce's painting takes on a new significance. He is remembering a specific time in his life when the vastness of geological time and astronomical time came together for one brief moment – which he saved for posterity by painting it.

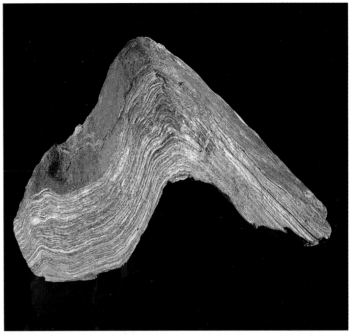

275 English

Carved ammonite fossil in the form of a snake, 185 million BC
York, The Yorkshire Museum
[YORYM:1993.315]

One of the topics that occupied the mind of many scholars was where stones came from and why peculiarly formed stones – shaped like teeth or like seashells – might turn up on the top of a mountain or at the bottom of a deep cave. Some suggested that stones grew in the bodies of animals and were excreted as part of their natural bodily functions. Others thought that they grew on plants. Still others thought that stones were generated by the heavens and fell to Earth, especially during storms. Ammonites, for example, were believed to be remnants of some stage in the evolution of the snake. In order to make this point more apparent – perhaps to a non-believer in the miracles of medieval lore – this ammonite has been cleverly carved to make the resemblance more apparent. One particular stone that seems to have caught the attention of ancient experts was the so called *glossopetrus*, or 'stone tongue'. Pliny,

for example, believed that the *glossopetri* fell from the sky during a lunar eclipse. By 1565, however, scholars had noticed that these triangular stones greatly resembled shark's teeth. Similarly, it was recognized that the so called 'Devil's toenails' were actually fossilized mollusc shells.

Also

• English (Blackheath, London), *Fossilized shark's teeth*, Eocene era, London, The Natural History Museum [BMNH 19895–19905]

• English, '*Devil's toenails*' *(Gryphaea arcuata Lamarck)*, Jurassic era, London, The Natural History Museum [BMNH.28488,L.28489]

• Nicolaus Steno, *De Solido intra Solidum Naturaliter Contento Dissertationis Prodromus*, Florence 1669, London, The British Library, Dept. of Printed Books

276 Switzerland (Jura)

Gneiss fold
Stratified rock, 250 million BC
Basle, Naturhistorisches Museum
[NMBa Nr.43953]

Between the 16th and 18th centuries, geologists made great progress in understanding the ways in which rocks and mountains are formed and began to come to terms with the tremendous time scales that must have been involved. Lazarro Moro (1687–1740) seems to have been the first geologist to suggest that the complicated folding and contortions that one often sees in stratified rocks was due to tremendous subterranean forces that had been released through the action of the central fires of the Earth. Gneiss folds are extremely evocative documents of the way in which stratified layers of rock can be manipulated. This example, from the serpentine complex in the Binn Valley, in the Canton of Valais in Switzerland, was transformed by Alpine metamorphism, with temperatures of approximately 500–550°C and pressures of up to 5 kb. Under this pressure, the

originally granite-like composition of the rock was transformed. What was once white mica in the rock was changed into a greenish mica mineral called phengite. As the crystals of mica were being re-formed, the pressure flattened each crystal. The resulting structure is not unlike *mille-feuille* pastry, composed of dozens of layers of thin and elongated crystals. The undulating shape of this fold records the varying directions from which the external pressure was exerted.

CHARLES DARWIN
YOU KNOW WE ALL SPRANG FROM MONKEYS?"

RICHARD OWEN
WHAT A PITY YOU DIDNT SPRING A LITTLE FURTHER?"

277 English

Two cartoons from Vanity Fair: 'Men of the Day, No. 33 – Charles Darwin' and 'Men of the Day, No. 57 – Richard Owen'
Lithographs, 30 September 1871 and 1 March 1873
English Heritage (The Darwin Museum at Down House, Downe) [88202629]

Darwin's *Origin of Species* transgressed a number of long and passionately held cultural beliefs. Amongst these was the idea that God had not created all living forms at once, but that the varied species roaming the Earth were the result of a long process of adaptation and change. Another challenge was the concept that one species could evolve into another. The Christian Church held that God the Father had created the Earth over a period of days. During that process, He created each one of the animals in their perfected form. The idea that He might have created a world filled with amoebae, which – by chance – had evolved into mankind, was as threatening an idea as Copernicus's charge that the Earth might revolve around the Sun. Both proposals undermined the teachings of the Bible and the special relationship between man and his god. By the time Darwin published his text in 1859, the premise of evolution had become widely accepted among scientists. Nevertheless, those outside the scientific community – or those with a personal axe to grind – portrayed Darwin and his proposals as a dangerous anathema to Christian beliefs. One opponent, Richard Owen, was particularly spiteful in his attacks, the tenor of which is recorded in this pair of cartoons from *Vanity Fair*.

278 David Regan

'The Evolution of the Species' tureen
Salt-glazed porcelain with sgraffito black liquid slip, 1993
Manchester, Manchester City Art Galleries [1994.149]

On one side of the tureen, the planet Earth is depicted; on the other, there is a bust-portrait of the naturalist Charles Darwin. The oversized finial of the cover is shaped like a brain. Five monkeys are placed on the rim of the tureen and large lizards lounge on each handle. There are two inscriptions: *Survival of the fittest* and *Charles Darwin, Origin of Species*. Using pottery as a basis for decorative or narrative scenes has a long history, dating back thousands of years. The confrontation of an effusive, almost rococo style with a 19th-century subject matter is just the sort of juxtaposition, however, that leads the viewer to believe that the artist must have invested some complex, 'hidden' meaning in the form and decoration of this tureen.

Did he know that Darwin had married the daughter of Josiah Wedgwood? Is the tureen itself a metaphor for the great 'primordial soup' out of which all life on Earth had evolved? Is the brain-shaped finial an allusion to the most highly developed organ of the most highly evolved species on Earth – or is the brain just a convenient handle, the use of which allows the most fit to survive?

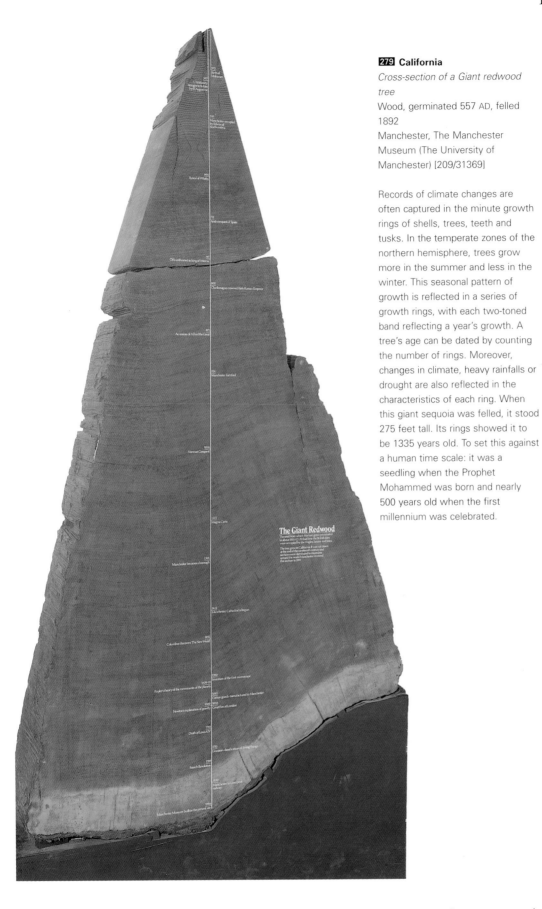

279 California

Cross-section of a Giant redwood tree
Wood, germinated 557 AD, felled 1892
Manchester, The Manchester Museum (The University of Manchester) [209/31369]

Records of climate changes are often captured in the minute growth rings of shells, trees, teeth and tusks. In the temperate zones of the northern hemisphere, trees grow more in the summer and less in the winter. This seasonal pattern of growth is reflected in a series of growth rings, with each two-toned band reflecting a year's growth. A tree's age can be dated by counting the number of rings. Moreover, changes in climate, heavy rainfalls or drought are also reflected in the characteristics of each ring. When this giant sequoia was felled, it stood 275 feet tall. Its rings showed it to be 1335 years old. To set this against a human time scale: it was a seedling when the Prophet Mohammed was born and nearly 500 years old when the first millennium was celebrated.

280 New York State (Albany County)

Fossilized coral (Zaphrentis prolifica)
Fossil, Middle Devonian period (350 million BC)
Hamilton (NY), Colgate University Dept. of Geology [CU Spec. 467]

Increases in light or food supply usually lead to increased growth. For example, the variation in the growth rings of this fossilized coral, show that, during the Devonian period, the phases of the Moon were of a shorter duration and that thirteen lunations occurred every solar year, instead of twelve. This was due to the fact that the Moon was then closer to the Earth and it took only about 28 days to circle it. The thicker bands of fossilized coral show the heavy summer growth and the thinner bands record the waxing and waning of the Moon.

Mary Dixon (Nungurray) Yuendumu

Australian Aboriginal Concepts of Time

Howard Morphy

The experience of time is a significant part of the individual's relationship with the world. It influences how they feel and how they act. In a life of hunting and gathering, the time of the day and the time of the year affects what people do. There is a time for hunting: when the kangaroo are resting from the heat of the day and can be approached stealthily and caught unawares. There is a time for fishing: along the beach in the late afternoon when the sun is less intense and the fish run inshore. Depending on the season, the yams will be ready to dig, the native apples will be ripe, the turtles will lay their eggs high on the beach or the kingfish will have migrated close to the shore. Time has its social side: there are times for people to be in small groups and times when they gather together for commemorative and ritual events. Time also has its physical and metaphysical aspects: the comforts and discomforts of personal ageing; the sense of where a person fits in the scheme of the universe. All these senses of time have their own scales, durations and frequencies.

Some events, like eating food, happen each day. Some changes, like the Sun rising above the horizon, are visible in minutes. Other events, such as major ceremonial gatherings, occur after months or years of preparation and some changes, such as human ageing, occur over a lifetime. All these can be nested within cycles of repetition in which the uniqueness of each individual event is lost. Human beings have a tendency to place finite events into infinite sequences through a process of generalization, which, in turn, allows infinity to enter into the singular or the momentary event. With repetition, an episode of hunting becomes a habitual activity, and such activities may come to stand metaphorically for universal processes.

The process of transmitting knowledge can be seen as a matter of timing the transfer of skills from one generation to the next: any

Tim Leura Tjapaltjarri with Clifford Possum Tjapaltjarri, *Napperby death spirit Dreaming*, synthetic
polymer paint on canvas, 1980, 207.7 × 670.8 cm, Melbourne National Gallery of Victoria, Felton Bequest

'Dreamtime' is as much a part of the present and the future as it is of the past

individual event – the killing of an animal or the singing of a song – is the result of accumulated knowledge. Aboriginal cultures have the concept of 'readiness' in much the same way as others do – it comes, in part, from viewing the world and the life of people within it as a process. There is a right time for boys to be circumcised, and for middle-aged men to stop hunting. There is even a right time to die.

Time is constituted observationally and relationally. The changes of ageing are measured against the bodies of friends and relatives before they are experienced in personal memory. The yams will be ripe after the rains have stopped and before the creek beds are dry. People have a capacity to measure one event by another and to take one thing as a sign for another. Pragmatically, this enables them to be ready for particular events – to move where a food supply is, to plan for seasonal abundance or to anticipate the arrival of the kingfish by making ready the fish spears. It allows the scheduling of events by ordering them into possible sequences and it allows communication about the relative duration of events and the placing of people and events in time and space. The process of temporal sequencing, when linked to the emotions and events of the world, becomes a rich source of both analogy and metaphor that builds on meaningful coincidences and transferable emotions. For example, the tide that daily moves up the beach and sweeps it clean of the debris of life becomes an ideal metaphor in Northern Australia for the transient nature of human lives.

While different events have their own durations and are caught up in independent rhythms associated with seasonality, with cycles of reproduction and growth, and with celestial movements, the very fact that the one event can be used as a sign of the other (or as a metaphor or analogy for the other) brings them together. The need to co-ordinate leads to the development of implicit scales which allow one set of activities to be related to another. The ceremonial cycle must be made to fit the lunar cycle and the seasonal cycle. Many ceremonies, for instance, are timed to end with a Full Moon. Moreover, they must be held at a time of seasonal abundance to support the people gathered together. They are also linked with the life cycles of the individuals who participate in them since the elders must be ready to pass on their knowledge and the initiates be of the right age to receive it.

The activities with which the time is filled affect the way time itself is experienced – whether it moves slowly or fast, whether its duration is painful or pleasurable. However, how time is experienced is also profoundly influenced by theories or beliefs about the nature of the world. Such theories affect perceptions of the temporal distance

between events and the way they are thought to affect each other. Aboriginal people's experience of time is influenced by the existence of an encompassing cosmological schema – or theory about the world – that is generally referred to in English as the 'Dreaming' or 'Dreamtime'. The effect of the Dreamtime is to displace events in time so that things that happened in the past often have an immediacy that makes them part of the present. The Dreamtime is lived. Cosmological time has this characteristic in many societies since it is infinite in its duration and can be as much a part of the present and the future as it is of the past.

The terms 'dreamtime' and 'dreaming' were first used in the late nineteenth century. They arose both out of attempts by early anthropologists to translate Aboriginal concepts into English and out of Aboriginal attempts to explain their religious ideas and values to European colonists. Aboriginal people talking among themselves also began increasingly to use them. 'Dreamtime' corresponds to a word or set of words in many Aboriginal languages: the Yolngu *wangarr*, the Warlpiri *tjukurrpa*, and the Arrernte (Aranda) *altyerrenge* are often given as examples. Indeed, it was in translating the Arrernte term that Baldwin Spencer and Frank Gillen first used the word 'dreamtime' in print, in 1896. They argued that the word *altyerrenge* (*alchuringa* in their orthography) was applied to events associated with ancestral beings in mythic times and to representations of those times; it thus represented a past period of a vague and dreamy nature. The word *altyerra* was also used for 'dream' and the suffix -*enge* signifies possession or belonging to. A literal translation might have been 'belonging to dreams' or 'of the dreams'. But, in order to differentiate the concept from everyday dreams and to signify the connection to the ancestral past, Spencer and Gillen used (and may have coined) the phrase 'dream times'. It would be wrong to see the word 'dreamtime' as a literal translation of an equivalent term in all Aboriginal languages. The word *wangarr*, for example, used by the Yolngu-speaking peoples of Eastern Arnhem Land, cannot be translated literally as 'dreamtime', and indeed some Yolngu feel that the connotation of 'dream' is inappropriate: *wangarr* is not a dream, but a reality. So the words 'dreaming' and 'dreamtime' should not be understood in their ordinary English sense but, rather, as terms for a unique and complex religious concept.

The Dreaming exists independently of the linear time of everyday life and the temporal sequence of historical events. Indeed, it is as much a dimension of reality as a period of time. It gains temporality because it was there in the beginning, underlies the present and is a determinant of the future. It is time in the sense that once there was *only* Dreamtime. But the Dreamtime has never ceased to exist and, from the viewpoint of the present, it is as much a feature of the future as it is of the past. And, as we shall see, the Dreamtime has as much to do with space as with time – it refers to origins and powers that are located in places and things.

In most if not all Aboriginal belief systems, there was a time of world creation before humans existed on Earth. Often it is said that, in the beginning, the surface of the Earth was a flat featureless plain. Ancestral beings emerged from below and began to give shape to the world. These beings were complex forms capable of transforming their own bodies. Many formed in the likeness of animals, such as kangaroo, emu and possum, or caterpillars or witchetty grubs. Others were inanimate objects, such as rocks and trees. And others still comprised whole complexes of existence, such as bushfires or bees with their beehives and honey. Whatever their form, however, they were not subject to the constraints of everyday life. If they were boulders they could run, if they were trees they could walk, if they were fish they could move on land or dive beneath the surface of the Earth. Frequently during its travels an ancestral being might transform from animal to human to inanimate form, swimming like a fish or jumping like a kangaroo, walking like a person, singing songs or performing ceremonies, and transforming into a feature of the landscape.

Every action of the ancestral beings had a consequence on the form of the landscape. The places where they emerged from the Earth became waterholes or the entrances to caves; where they walked, watercourses flowed; and trees grew where they thrust their digging sticks in the ground. They lived as humans do today but on a grander scale, and their actions had grander consequences. Great battles occurred between groups of ancestral beings and where they died hills arose in the shape of their bodies, or lakes formed from pools of their blood. Over time, the features of the Earth as they now exist began to take shape, and as long as the ancestral beings lived on the surface of the Earth they modified its form little by little. They left permanent marks here in a hillside by cutting down trees; they left behind a rocky bar where they crossed this river; when they threw their boomerangs they pierced a hole through that cliff.

Although their actions often appear to be violent, capricious, or amoral, ancestral beings also instituted many of the rules by which humans subsequently lived. They invented ritual practices such as circumcision, originated the form of ceremonies, established marriage rules, and lived according to the social divisions that characterize present human groups. They also created material culture objects such as stone spearheads, boomerangs and string bags. Plants and animals that occurred in particular areas were named by the ancestral beings and they spoke in the languages of the

groups who eventually took over guardianship of the land. They invented the songs, dances and paintings that commemorated the great acts of their lives and their journeys. The mythological beings of the Dreamtime also created the human beings who were to succeed them on the earth. Gradually, after the creation of humans, the ancestral beings removed themselves from the surface of the earth. They did not, however, end their engagement with human life.

The ancestral dimension continues to be connected to the present in many different ways. The ancestors ensure the fertility of the land and, through ceremonies, their powers can be summoned for different purposes: to return the souls of the dead back to their ancestral lands, to ensure the fertility of people and animals or to endow youthful initiates with spiritual strength. The most direct link between people and the ancestral dimension is through the process of spirit conception. Each individual is believed to be in part the product of a conception spirit associated with one of the ancestral beings who created the land. Conception spirits are often associated with major features created by an ancestral being – in particular, waterholes or ochre deposits – though they can be present more generally in the body of an animal of the same species as the ancestor, or may be conjured up by the power of ceremonial performance. The conception spirit initiates the pregnancy and provides the individual thus animated with a direct link to the ancestral world and to particular places associated with it. As a person grows older and takes in more and more ceremonies, their spiritual power increases and eventually they become identified with the ancestors themselves. On death, ceremonies are performed to ensure that their spirit returns to its own country to rejoin the reservoirs of spiritual power infused in the land. The Dreaming thus involves a recycling of spiritual power between the world of the living and the ancestral dimension. Life is thought to be conditional on maintaining contact with the ancestral past, on following the rules that the ancestral beings instituted and on keeping alive their memory. The ancestral beings provided a precedent for the living: people follow in their footsteps, guided by the traces that they left in the landscape.

The landscape is redolent with memories of other known human beings. The ancestral beings, fixed in the land, become a timeless referent – beyond the politics of daily life – to which the emotions of the living can be permanently attached. Ancestral essence is, in effect, frozen forever at a particular event or action in the continuing event of the ancestral journey, so that this part of the action becomes timeless. Place has precedence over time in Aboriginal ontogeny. Time was created through the transformation of ancestral beings into place – the place being forever the mnemonic of the action or event. They 'sat down' and, however briefly they stayed, they became part of the place forever. In Yolngu terms they 'turned into' the place. Whatever events happened at the place, in whatever sequence they occurred and whatever intervals existed between them – all this becomes subordinate to their representation in space. Sequences in time are represented only if they were spatially segregated and occurred at separate places in association with separate features, and even then synchronicity or perhaps timelessness is built into the way they are presented. Transformed into features of the landscape, mythological events are represented simultaneously even if they could be said to have occurred at different points in time. What remains is the distance between places rather than the temporal distance between events. The time it took ancestral beings to complete their journey, the precise interval between events, is seldom part of the recounting of myths. There is no 'and many years later' in Yolngu myth-telling, though there are certain identifiable strata of time or bands of synchronicity. The Dreaming encompasses within it other temporalities: it

of the Macassans. These references are not to the historical Macassans but to Dreamtime Macassans. This was brought home to me graphically when I was talking to a Groote Eylandt man, Jack Markarrakarra, about a sculpture in the shape of a Macassan boat made in sand as part of a mortuary ritual. I suggested that Yolngu had learnt the form of the design from observing the Macassans. He strongly denied this and said that the designs were from the *wangarr* and that Yolngu had always reproduced them. Then, as a concession to my line of questioning, he said that maybe it was only when the historical Macassans arrived that Aborigines knew that the ships were made from wood and the sails of cloth.

Everything is prefigured in the Dreaming – everything that was, is, and will be. When I began my research at Yirrkala in Eastern Arnhem Land on the Gulf of Carpentaria, it was two years after the establishment of the nearby mining township of Nhulunbuy, a year after the introduction of alcohol to the community and, thus, a time

In Aboriginal Australia, infinity is more readily graspable as part of the present

integrates processes, personal memories and historical events within its framework. The seasonal cycle and its relationship to the environment are understood in the minutest detail and are precisely delineated, but it is seen to be intimately involved with the action of ancestral beings – for example, the lightning snakes, who bring about the wet season and cause the flood waters to rise. The memories of members of previous generations are kept alive and provide much of the emotional energy for ceremonial performance, but, eventually, through those very performances, the recently dead are reabsorbed into the ancestral dimension and lose their individuality. Major events – such as the annual visits of the Macassans to the coasts of Arnhem Land and the Kimberley (until they were banned by Europeans in 1907), the arrival of Europeans themselves, the impact of European settlement and the two World Wars – are remembered as part of oral history, but they, in turn, are partly processed through myth.

People talk of the 'Macassan times' or the 'killing times' or 'mission times'. Such time periods are only relatively discrete since they too can be connected to the Dreaming and can be seen to have a Dreamtime precedent. Throughout Arnhem Land, for example, the Macassans are an important theme of myth. For a period of at least 300 years the 'historical' Macassans used to come at the same time every year to collect *trepang* (bêche de mer). Macassan themes are an important part of the ritual knowledge of clans from the region. Macassans are transformed into features of the landscape, sacred designs represent Macassan boats, houses, and material culture objects, and ceremonial actions represent the daily life and activities

of great difficulty and uncertainty. The artist Narritjin Maymuru would often tell me that we were returning to the time of Bamabama. Bamabama was a trickster ancestral being who broke all society's rules, who committed incest, behaved violently and disregarded the authority of his elders. In the end, a great flood came and he and the people he lived with changed back into the form of their totemic animals and returned to their own habitats so that life could begin anew. Narritjin looked at the world around him, at the increase in violence, at the deaths of young people and at the challenges to the traditional system of authority. We might say he drew an analogy with the Bamabama story, but it would be more accurate to say that he saw current events prefigured there.

The structure of the Dreamtime is just one of the frameworks through which time is experienced in Aboriginal Australia. Time is 'lived and apprehended concretely' through people's relationships with others and the activities of their lives. The proximity of the Dreaming, however, has its effect on the way temporal events are experienced and the ways in which different temporalities are related to each other. Whereas Western conceptualizations of time are heavily influenced by ideas of 'progression' and 'development', Aboriginal concepts of time are essentially atelic (purposeless). Or, rather, purpose lies in the Dreaming, which is in many respects infinite and timeless. In Aboriginal Australia, there is a shorter distance between historical time and infinity. As a consequence, infinity is more readily graspable as part of the present.

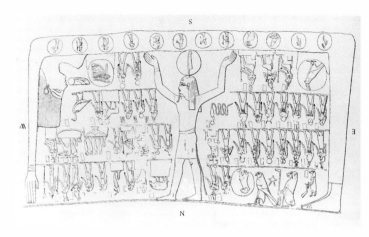

Shu supports the bending figure of Nut, the Egyptian sky goddess,
Nagᶜ Hamad B Tombs, Ptolemaic or Roman period (drawing from Neugebauer
and Parker 1969, III, pl. 39)

Egyptian Time

Anthony Spalinger

Before one can enter into a discussion of early Egyptian timekeeping and calendrics, it is important to remember that our understanding of ancient Egypt is, perhaps, somewhat skewed owing to the arbitrary nature of what artefacts and buildings have managed to survive. The survival rate of religious artefacts and buildings far outweighs that of what one might consider secular monuments. In addition, the survival of material specifically related to funerary practices dominates what we think we know about the Egyptian people and their concerns. The vast majority of the so called 'clocks', astronomical instruments and astronomical scenes that are still preserved, such as the various depictions of the Goddess Nut encompassing the hours and constellations of the night sky and the various decan-clocks one finds engraved on temple walls and painted on the inner lids of coffins, have been salvaged primarily from temple precincts or tombs. Therefore, one must be cautiously circumspect when one begins to consider the 'why' and the 'how' of Egyptian timekeeping.

Having said that, however, it does seem that systematic timekeeping in Egypt originated within the temple institutions. In the beginning, the need to know the time had a religious imperative. In particular, the earliest information relating to Egyptian timekeeping is found on material relating to funerary practices. On a number of coffins dating from the late First Intermediate Period and the Middle Kingdom (*ca.* 2100–1850 BC), there are illustrations of a set series of stars connected to ten-day periods of time. These stars, now known as 'decans' after the name the Greeks later gave them (*deka* is the Greek for 'ten'), were first used as a means for marking the time of the rising Sun. Owing to the difference between the solar and sidereal (or stellar) year, the Sun appears to slip backwards by about four minutes every night. The star that rises with the Sun on one day will rise four minutes before the Sun on the following night. The Egyptians set up a system of 36 important stars or groupings of stars that they could use as relatively accurate heralds of the dawn. After ten days, when the accuracy of the first decanal star had slipped by 40 minutes (what was considered a decanal hour), the next star was used as the herald and so on. In this way, the Egyptians developed an equal-hour system for measuring the 1440 minutes of the day (36 decans × 40 minutes 'decan hours'). When a 30-day month was introduced, the decans were modified. Then, 24 bright stars were chosen to cover intervals of 15 days. This system is, ultimately, the source of our 24-hour day, because a 1440-minute day divided by intervals marked by 24 bright stars will yield hours of 60 minutes each. Also, as records from the Twelfth Dynasty attest (*ca.* 1990–1785 BC), the Egyptians switched from marking the stars as they rose on the eastern horizon to using

In the beginning, the need to know

Regularized timekeeping, scholars believe, was the outcome of political centralization

a meridian-based system for these stellar observations in order to regularize better the timing of their nocturnal events. Again, it seems to have been the need to perform certain religious rites during the night that provided an impetus towards this decanal system of night hours.

How such 'night clocks' were actually used is not clear. In earlier periods, such as during the Old Kingdom (*ca.* 2700–2200 BC), there is no evidence that the Egyptians used any system based on decan-stars. The major religious corpus of the period, the so called *Pyramid Texts*, indicate that astronomical knowledge was somewhat simple at this time. The hymns do not refer to any specific hours, much less to calendars. And it is only from a few non-religious references concerning 'watches' at night that we can hypothesize some sort of nocturnal timekeeping arrangements.

Another means for keeping the time at night employed the human body. We have records dating from the Eighteenth Dynasty (*ca.* 1500 BC) that describe two men sitting on a temple roof, one of whom observes the position of a key star against the template of the body of a man who is seated in a specific location. At certain key points during the night, specific stars might be found hovering over his shoulder or above his head. By the Middle Kingdom, it seems that part of this exercise involved both men sitting along a prescribed north–south meridian line. This means that, by watching the position of the star relative to the subject's body, one was also actually observing what was more or less meridian transit of that star. The step from using a human body as a sighting device to the employment of rudimentary instruments was not a large one. And, not surprisingly, references to transit instruments and (slightly later) to water clocks first appear during the Eighteenth Dynasty. From the reign of Thutmose I, there is an intriguing, private inscription that describes the invention of the water clock. Whether or not this individual actually invented this timepiece seems unlikely, though he may have invented one variant of the type. What seems clear, however, is that water clocks were originally developed as an adjunct to or, possibly, as an anticipated replacement for the laborious business of naked-eye observation. By having an interval-timekeeper, one could easily measure a night's worth of equal hours.

The earliest Egyptian calendars appear to have been used primarily for religious purposes and to have been exclusively lunar in their construction. Nevertheless, the main system of cylindrical reckoning throughout pharaonic history was the so called civil calendar, which

time had a religious imperative

seems to have been developed some time around the turn of the third millennium BC (either during the Predynastic period, *ca.* 3100 BC, or three to four centuries later at around 2780 BC). Scholars believe that the regularized system of timekeeping was the natural outcome of the political centralization of the Nile Valley. The emerging bureaucracy required an effective dating system and the growing treasury needed an easy method of bookkeeping. The old lunar-based calendar, which had been in use before the creation of a civil calendar, was then discarded so that the state could preserve and date its records more efficiently.

The civil calendar was made up of twelve months, each of which was named after a significant feast that fell within that month. By and large, most of the names that are used in the civil calendar can be traced back to a predominantly lunar festival year. Panyi (month nine), for example, was named after the famous Valley Feast, which was celebrated during the ninth Moon. Weprenpet (month one) referred to the *wep* ('opening') of the *renpet* ('year'), which took place on New Year's Day. To this extent, there was a degree of continuity from the original lunar system.

In the civil calender, however, the months were fixed at 30 days, each of which comprised three 10-day 'weeks'. To this total of 360 days, five extra or epagomenal days were added to make up the civil year of 365 days. The schema of the Egyptian civil year is somewhat artificial. Theories abound, however, concerning the reasons why the figure 365 was chosen. Was it based on astronomy or on nature? Or, to put it more crudely, was the civil year based on man looking upwards towards the Sun and the stars or was it based on him looking downwards towards the Earth and its seasonal changes? The difference is important because it marks a fundamental difference in the ways that the Egyptians may have thought about the cosmos and their place within it. Although the question is far from being 'solved', the general trend amongst modern scholars is to posit a solar origin for the civil calendar. The true solar year, however, is 365¼ days long. Astronomical records suggest that the Egyptians were aware of this fact from a very early stage, but they never seem to have considered introducing leap days to bring the solar calendar back into line with the stars. The decision not to include leap days meant that, in practice, the calendar lost one day against the pattern of the stars every year. In 1500 years – not a long time period for a civilization as stable as Egypt – the calendar had fallen behind by a whole year. This may be one reason why the Latin term for the Egyptian civil year is *annus vagus* or 'the wandering year'. For this reason, one suspects that the 365-day year was symbolically important to the Egyptians at an extremely early date. In fact, the pharaoh had to swear an oath never to tamper with the calendar at the occasion of his accession to the throne of Egypt.

The question of the origin of the civil calendar may be more complex than it first appears owing to the importance and portentousness of the brightest star in the night sky, Sirius (α CMa) – or Sothis as the Egyptians called it. Originally, the Egyptian year began with the morning heliacal rising of Sothis following a period of 70 days of invisibility. Therefore, it is possible that the civil calendar was derived from this astronomical event. Nevertheless, one still faces the quarter day problem. Once the civil calendar was established, however, the heliacal risings of Sothis ceased to be a yearly marker. In later times, whereas there still may be notations of heliacal risings of Sothis within the civil calendar, it does not serve to mark the beginning of a new administrative year. Whether or not there was ever an official 'Sothic calendar' (one based on the star's return to the same date in the civil calendar – an event that would happen once every 1460 years) is another matter. Such cycles are known in the Greco-Roman world and it is not impossible that it was a time-reckoning system that was native to the Nile Valley. Nonetheless, although the religious significance of Sothis was paramount, the heliacal rising of the star was not pivotal to the structure of the civil calendar. Again, the evidence seems to suggest

Egyptian, Vertical sun dial, 1st–2nd century AD (see **102**)

that once a significant form, such as a 365-day year, was established and accepted, variants from this form were either 'tidied up' or conveniently ignored.

Generally speaking, as is the case with so many civilizations, the Egyptian state appears to have moved from a lunar-based system to a predominantly solar one. The religious feasts and festivals, which had originally been calculated according to a strictly lunar calendar, eventually became subsumed by the civil one. From as early as the New Kingdom (ca. 1555–1079 BC), for example, those religious festivals of which the celebration may have been determined by the lunar day (that is, by the appropriate phase of the Moon), were placed within a specific civil month. That is, a lunar festival would be planned in accordance with the proper phase of the Moon, but there was the additional requirement that a naked-eye sighting of the Moon occurred within the appropriate civil month. It was possible – as evidence from the Middle Kingdom site of Illahun bears out – that a temple might schedule its festivals according to a purely lunar calendar. Indeed, it appears to have done exactly that for more than a millennium (from ca. 3100 BC to 1800 BC). But, still, the temples were forced to date these festivals in its records according to the accepted civil calendar.

Temple walls were often carved with calendars that described the official religious activities of the residing deity or deities. Generally, though, they are tersely worded and mainly reflect the purely religious side of the cult. Nevertheless, their non-narrative rendering of the key events of the Egyptian civil year provide us with the major portion of our knowledge of Egyptian calendrics. In fact, they are often crucial for reconstruction of the calendrical variations between different temples and different periods and for understanding the complex economic affairs of the priests and their community.

The most ancient festival calendar that survives dates from the Old Kingdom. It is written on two sides of the doorway in the funerary Sun temple at Niuserre. The lengthy account preserved here is a paradigm for the basic arrangement of this type of calendar. Generally, there is a preamble, covering the construction of the temple or addition that has been made to an existing one. Often descriptions of the donations that were made by the pharaoh, which are regularly dated, and the purpose of these donations are provided. Niuserre's text then lists all the festival celebrations themselves. Exact dates within the civil year are listed in conjunction with precisely described foods (*i.e.* 'one haunch of beef' or 'five bundles of vegetables'). Even when the festival is related to

The brightest star in the night sky, Sirius, was of

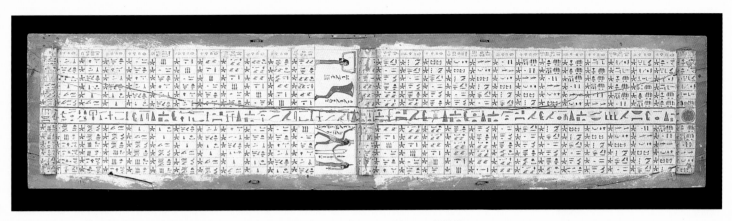

Egyptian, Sarcophagus lid with decan imagery, *ca.* 2154–1783 BC, Hildesheim, Peltzaeus-Museum [5999]

the Moon, it is listed according to the 365-day civil year.

Lunar-based 'feasts of heaven' were expressly separated from the 'seasonal festivals' that occurred only once a year and were set within an exclusively solar calendar. Accompanying a list of endowments made by Thutmose III (Eighteenth Dynasty) at Karnak, there is a complete calendar drawn up on a grid pattern. The left-hand column contains only dates. The right-hand columns have numbers that refer to separate headings, such as oxen, bulls and fowl. This feature of mathematically ruled horizontal and vertical boxes is a typically New Kingdom device. In fact, there are at least five similar extant calendars like this dating from the reign of Thutmose III alone. In the calendar at the temple in Elephantine, there is an important reference to an 'ideal New Year's day', when the heliacal rising of Sothis coincides with a specific day in the Egyptian civil calendar. Any astronomer would have known, however, that there could be no set date for the heliacal rising of Sothis when a 365-day calendar was in use.

Often, we find evidence of a pharaoh renewing the offerings of past monarchs without altering the earlier or original religious calendars. But, almost as frequently, they will expand or revise old calendars. By juxtaposing these different calendrical forms, one can begin to reconstruct the development of calendrics in Egypt. Alongside the contemporary, late New Kingdom calendar from the reign of Rameses II in Abydos, we have the great Medinet Habu exemplar, which can be dated from the reign of Rameses III. The latter can be recognized as a copy of the Abydos calendar with a series of minor additions, such as alterations in the lists of daily offerings. In one case, a key festival of victory has been added later as a palimpsest over the original account.

With the arrival of the Greeks and Romans in Egypt, the original purpose of the calendar appears to have changed. They no longer mention provisioning the temples and the endowments of the feasts are ignored. In their place, there are elaborate lists detailing official processional feasts. It is as if the texts move from substance to pure form. Any sense of the social and economic substratum upon which the old temple complexes had operated disappears. Events are described, but they are always somewhat lacklustre. The daily travels of a specific deity or his or her particular rituals are recounted, but any sense that these calendars reflect the lifeblood of an important cult centre has died. Everything that is anecdotal, such as the marginal references to the surplus that the priests would receive from unused food offerings, disappears. From this point on, it is impossible really to think of there still being an 'Egyptian' time.

great importance for the Egyptian calendar

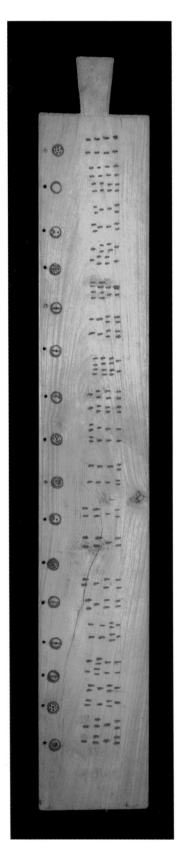

281 Benin (Abomey tribe)

Divination calendar for the king and the country
Wood, 20th century
Paris, Musée d'Histoire Naturelle
[MH 36.21.108]

This calendar was owned by Bokono Gedegbe, one of the influential members of the royal court of Abomey. It is called a calendar, even though its primary purpose does not seem to be the calculation of any specific date. Instead, it is used to divine the future for the king of Abomey and for his people. It is fire-engraved on both sides. On the verso, there are 16 holes punched at regular intervals, below which is a series of 32 punched designs. The marks on this side are used to predict the future for the country. On the recto, there is a similar set of marks to which a series of circular motifs have been added. The circular marks seems to bear some relation to the seven-day week, which was not an indigenous concept in Benin and must reflect contact with Islamic traders. This face is used to predict the future for the king.

Western philosophers have agreed that time is divided into three parts: the past, the present and the future. Each of these three faces of time is seen as inherently different, but, somehow, they each partake of or, perhaps, manifest the same essence: 'the medium of time'. Because we have conceptualized time and given it certain qualities, time itself is seen as being able to impart a structure – some form of inherent rationality – on the events which occur within it. From a historical point of view, we might argue that something is 'true' because it occurred within the medium of time. For the present, something that exists does so 'in time'. The natural and rational extension to these understandings is that the future also exists in time. It, too, is as 'real' as the past or the present and is subject to the same natural laws as everything else that exists 'in time'.

One might be rightly sceptical about notions concerning the existence of the future. Nevertheless, there is scarcely a culture that exists or has existed that does not believe that it is possible to see into the future. Beyond this, there are very few cultures that do not have elaborate means for predicting and, often, claiming to be able to influence forthcoming events. The belief is that, since time is all one substance, one is able to use the experience of the past and the observation of the present to decipher the language in which information about the future is conveyed. Fortune-telling, astrology, geomancy, chiromancy, clairvoyance, divination, omens and portents all provide ways in which men believe they can navigate through the medium of time to visit the future and reveal its secrets.

282 Neo-Babylonian (Sippar)

Model of a sheep's liver
Baked clay, 7th–6th century BC
London, The British Museum, Dept. of Western Asiatic Antiquities [WAA 50494]

A number of cultures share the belief that the physical world is filled with numerous 'clues'. The patterns found in nature reflect the gods' desire to communicate their divine intent to man. This belief is the basis of astrology, geomancy and extispicy – the examination of internal organs of a sacrificial victim – in the ancient world. The general shape of the extispicial ritual involved the priest whispering a specific question into the ear of the sacrificial victim whilst killing it. The organs of the dead animal would then reveal the answer to the question posed. The liver was one of the internal organs that was credited with being the 'seat of life' in both man and animal. The ancient Babylonians, for example, developed the art of hepatoscopy – examination of the liver – in order to answer certain specific questions about the health of the state: to forecast success or failure of a proposed military campaign, to determine the allegiance of well placed officials or the relative health of members of the royal family. The idea was that a healthy liver forecast a healthy state – deviations from this norm could be favourable or unfavourable.

283 Chinese

Feng shui compass
Lacquered wood, 18th century
Cambridge, The Whipple Museum of
the History of Science [Wh. 1203]

From at least 200 BC Chinese diviners were using instruments that enabled them to predict the outcome of an event by noting how its timing would correspond with the cycles of the heavens and the Earth. By the 11th century the principle had been extended to foretell the outcome of a proposed enterprise or activity in view of the position that it would take on Earth; and, shortly, a magnetic needle was fixed to the centre of the instrument, so as to ensure that the chosen position or direction for the action – building a house, siting a grave or, more recently, erecting a multi-storey bank – would be satisfactory. Such a choice was intimately connected with relations between time and space. The diviner's compass includes a large number of circular bands, each one of which is carefully graticulated, allowing the operator to count the intervals of time and the movements of the natural world of the heavens and Earth. By co-ordinating these with spatial factors, he could advise whether a chosen site would invite the entry of life-giving energy (*qi*) and repel the advance of destructive energy (*sha*). The outermost rim accommodates the 28 divisions of the zodiac, of unequal size, but together accounting for the 365¼ degrees of the circle.

284 Edward John Poynter

The Ides of March
Oil on canvas, 1883
Manchester, Manchester City Art
Galleries [1883.18]

In Shakespeare's *Julius Caesar*, the world of omen and prophecy, the wilfulness of a vain leader and the unrelenting march of history all collide. In Act I, Scene III, there is a terrible storm, described by Casca as a sure sign of impending doom: "Either there is civil strife in heaven,/ Or else the world, too saucy with the gods,/ Incenses them to send destruction ..../ For I believe, they are portentous things,/ Unto the climate that they point upon." In Act II, Scene II, Caesar's wife Calphurnia relates a horrific dream, in which she has seen blood spurting "like a fountain with a hundred spouts" from Caesar's statue. She also mentions the recent appearance of a malefic comet in the sky. Astrologically, as Calphurnia knows, comets signal the death of kings. She also consults the augurs, who, having examined the entrails of a slaughtered beast, claim that they could not find its heart – another sure sign of trouble. But all visions of the future are subject to interpretation. The sly Decius offers another, suggesting that the image of the statue spurting blood is a metaphor for Caesar's gift of life to Rome. Caesar, willingly seduced by this imagery, ignores the warnings raging around him and heads off to the Capitol – to fulfil his destiny.

One of the fundamental components of astrological belief is that the character and fortune of a person is set by the configuration of the stars and planets at the moment of his or her birth. The orientation of certain zodiacal signs to the local horizon the 'native' or subject, the placement of the planets within each sign and how that planet is then positioned relative to the other planets, the other signs and the local horizon – all these heavenly signifiers set a blueprint for the future. A person with the planet Jupiter 'well aspected' will go on to become a leader; one with Mercury well aspected will become a scholar. The casting and interpretation of horoscopes was (and, in many cases, still is) a major component of virtually every society since at least 1000 BC. Babylonians, Egyptians, Greeks, Romans, Arabs, Indians, Chinese, Japanese – astrology plays or has played a vital role in almost every major civilization on Earth.

285 Greek and Coptic
Horoscope for 13 April 95 AD
Papyrus, 95 AD
London, The British Library, Dept. of Western Manuscripts [Pap. 98]

286 Egyptian
Horoscope for 1 April 81 AD
Papyrus, 81 AD
London, The British Library, Dept. of Western Manuscripts [Pap. 130–1]

These two hybrid horoscopes help to illustrate the pan-Mediterranean nature of astrological prediction during the early years of the first millennium. The first papyrus was written in Greek and then copied in Coptic, with Greek headings to the paragraphs. On the recto, there is a very detailed horoscope that seems to have been cast for 13 April 95 AD.

It lists the positions of all the planets and describes their influences on the native or subject. It seems less a specific prediction than a slightly personalized almanac entry – the same sort of thing as one might find in newspapers or magazines today. The second was written in Greek and compiled by a certain Titus Pitenius at Hermopolis in Egypt. In it, he describes the latitude of Hermopolis as having 'the ratio 7:5'. – contrasting the length of the longest day of the year with the shortest day of the year for that specific latitude (in this case 30°N). The horoscope itself begins with an attempt to record the date as 'the time of the [equinoctial] tropic of the third year of the divine emperor Titus; 6th of the month Pharmouth; at the third hour of the night; on the Kalends of April, Roman style; ancient Egyptian style, first to second day at the month of Pachon'. This corresponds, in modern terminology, to 1 April 81 AD. The positions of all the planets and the *horoscopos* ('rising sign') are provided, alongside notations of the influences of the Egyptian decans, the Greek *dodekamatoria* and the Roman 'terms' (each of which provides further subdivisions of the degrees of each sign of the zodiac).

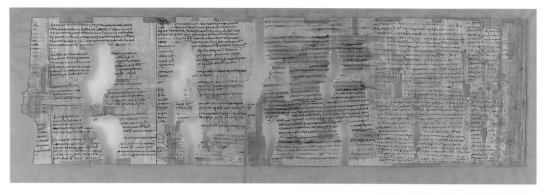

287 **Attrib. Jost Amman**
Plate from Jacob Rueff, *De Conceptionis Generationis Hominis*
Engraving in a printed book, Frankfurt 1580
London, The British Museum, Dept. of Prints and Drawings
[1867.10–12.663]

Owing to the importance of the natal or birth horoscope, it was not uncommon for astrologers to be called in to advise not only the appropriate time for conception, but the most auspicious time for delivery. Here we see two astrologers plotting the stars at the moment of a child's birth. A pair of dividers and a blank horoscopic chart lie waiting for the right moment. Though this is a largely fanciful rendition, there are accounts – especially connected with royal births – where the astrologer is known to have advocated hurrying or delaying the birthing process in order to achieve a more harmonious celestial aspect for the newborn child.

288 **Antonio Minelli**
Mercury
Marble inlaid with bronze, 1527
London, The Victoria and Albert Museum [A.44–1951]

Horoscopes were not only cast to record the astrological configuration of an event – such as a birth or a marriage – they were also cast in anticipation of events in order to ascertain the most auspicious moment to carry out a particular deed. This statue of Mercury was commissioned by the Venetian patrician and humanist Marcantonio Michiel. The Latin inscription of the left side of the altar can be translated as: 'A statue of Mercury, dedicated to Marcus Antonius Michiel, patrician of Venice, in the 1106th year of the city of Venice. The sculptor Antonio Minelli of Padua started this work on 14 February and finished it on 15 June 1527.' (The city of Venice was believed to have been founded in 421 AD, so 1106 *urbis venetorum* equals 1527 AD.) On Mercury's right side, there is an engraved bronze disc that represents a horoscopic diagram. Opinions differ as to the exact time – either 8:00 AM or 11:46 AM – but the date it depicts is certainly 15 June 1527. The most important element in the horoscope is the placement of the planet Mercury at the mid heaven, conjunct with the Sun and Jupiter. For anyone who saw himself as a child of Mercury, this could have been interpreted as a good omen.

One of the basic beliefs of Christianity is that God created the world in all its possibilities and for all time within the first seven days of the Genesis cycle. In creating the world, He created all its future histories and even its end. One of the goals of Christian life is to understand how God's original plan makes itself manifest in the unfolding of daily events. The Church Fathers were constantly looking for clues – in the Old Testament, in the Scriptures and in nature itself. One of the teachings, for example, was that the events of the Old Testament prefigured events in the New Testament. One extremely popular iconographic device in Christian art is the 'type' and 'antitype'. Stemming from the writing of the early Church Fathers, such as Tertullian and Augustine, 'typology' was based on the idea that significant events from the Old Testament,

classical history and even pagan mythology ('types') prefigured the events of the New Testament ('antitypes'). For example, God speaking to Moses from the Burning Bush was seen as a 'type' for the Annunciation to the Virgin; the creation of Eve from the rib of Adam was the 'type' for the creation of the Church (*ecclesia*) from the wound in Christ's side.

The most popular form for the presentation of 'type' and 'antitype' are the late medieval manuscript copies of the *Speculum Humanae Salvationis* ('The Mirror of Human Salvation') and the *Bible moralisée* (the 'moralized Bible'). With the coming of the Renaissance and its artists' penchant for creating single, spatially unified pictorial fields, the format was developed so that the 'type' was incorporated as a decorative element, such as a fictive painted or sculpted relief, in the architecture surrounding or supporting the main subject of the picture. The early fifteenth-century *Virgin and Child with saints* (see **289**) represents a transitional phase between medieval and Renaissance forms of the type/antitype formula.

289 The Master of the Strauss Madonna

The Virgin and Child enthroned with angels and saints
Tempera and gold on panel, *ca.* 1410
Stalybridge, Astley Cheetham Art Gallery (Tameside MBC Leisure Services) [ASTAC: 1923.43]

In this Florentine painting, the Virgin and Child are depicted enthroned with St James the Great and St John the Baptist on the left and with St John the Hospitaller and St Dorothy on the right. At the Virgin's feet, there is a semi-nude, reclining female figure with a halo. Odd as it may seem, this figure has been identified as Eve. The plant she is holding is a miniature version of the Tree of Knowledge. Eve and the Virgin Mary are depicted as 'type' and 'antitype'. As the Old Testament figure of Eve brought man into a state of sin by encouraging Adam to eat fruit from the Tree of Knowledge, so Mary, acting as the vehicle for Christ's birth, offered mankind redemption from sin.

290 **William Holman Hunt**

The Shadow of Death
Oil on canvas, *ca.* 1870
Leeds, Leeds Museums and Galleries
(City Art Gallery) [196/03]

One part of Christian teachings maintains that every aspect of Christ's life can be interpreted as a sign pointing towards the events surrounding the final Crucifixion and the centrality of Christ's role as the Redeemer of Mankind. For his large-scale painting of *The Shadow of Death* (now in the Manchester City Art Galleries), Holman Hunt executed a number of smaller-scale studies. A youthful Christ is shown with his arms outstretched, enjoying the late afternoon Sun after a hard day's work in Joseph's carpenter's shop. A female figure, identified by Holman Hunt as the Virgin Mary, stops her inspection of the gifts of the Magi to look at Christ's shadow on the wall. In it, she sees a prefiguration of his Crucifixion – the board on which the carpenter's tools are hung forms the cross-bar of the Cross and the tools provide premonitions of the nails that will be driven into his hands. Other features that Holman Hunt placed in the picture as 'signs' include the reeds which allude to the sceptre thrust upon Him during the Mockery of Christ, the sawhorse as an image of the 'scapegoat', the scarlet fillet as the Crown of Thorns, the pomegranates as symbols of Christ's Descent into Hell, and so on.

291 **Gaetano Previati**

The opium smokers
Oil on canvas, 1887
Piacenza, Galleria d'Arte Moderna
Ricci-Oddi [1887]

Psychologists often refer to mind-altering drugs as 'psychotomimetic' (the effects of these drugs seem to mimic psychotic behaviour in the user). One of the symptoms shared between psychotic behaviour and the effects of drug use, for example, is a disassociation from the natural rhythms of time. In particular, most of the users of psychotomimetic drugs – from alcohol and tobacco to opium and hallucinogens – experience chronosystole, or the sense that time has become contracted. Under the influence, a few hours of elapsed time often seems to be much longer. To quote one trial report from a study of "hemp intoxication" of the 1930s: "... during the intoxication, time appears to be remarkably lengthened .... The dense experience that crowds the observer causes him to realise that what seemed like several days during it was but a few hours". In addition, many users describe the sensation that time seems to be moving in waves. This appears to be due to the way in which the brain alternates between phases in which it can focus on the outside world and those in which it detaches from it. As Thomas de Quincy describes it in his *Confessions of an Opium Eater*, "The ocean, in everlasting but gentle agitation, and brooded over by a dove-like calm, might not unfitly typify the mind and the mood which then swayed it. For it seemed to be that I stood at a distance, and aloof from the uproar of life, as if the tumult, the fever, the strife, were suspended .... A tranquillity that seemed no product of inertia, but as if resulting from mighty and equal antagonisms; infinite activities, infinite repose."

292 Roman (Cyrene, Libya)

Hypnos or *Morpheus*
Marble, 1st–2nd century AD
London, The British Museum, Dept.
of Greek and Roman Antiquities [GR
1861.11–27.74, Sculpture 1426]

This fragment once formed part of a
larger funerary grouping connected
with the Temple of Aphrodite in
Cyrene. Owing to its location, the
figure has usually been identified as
Eros or Cupid. The presence of dried
poppy-heads in his right hand,
however, suggests that an
identification as Hypnos or even
Morpheus might be more
appropriate. Hypnos was the god of
sleep, son of Nyx (the Night) and
brother of Death (Thanatos).
According to Hesiod (*Theogony*,
211), he lives in the underworld and
never sees the Sun. He is the god of
sweet death. His poppies symbolize
a death that is as easy as falling
asleep. Morpheus is the son of
Hypnos. He sends dream visions to
those who sleep or who enter into a
drug-induced state. Poppies are the
usual attribute of Morpheus,
signalling the dreamy effects of the
drug that bears his name –
morphine.

293 Tiki Paningajak

Human-animal transformation mask
Stone, ca. 1982
Toronto, The Royal Ontario Museum
[994.197.4]

Among the Inuit, many celestial
objects are given a mythical
terrestrial origin. For example, the
Sun and Moon were sister and
brother who were banished to the
sky after having murdered their
parents and embarked upon an
incestuous relationship. In addition,
though, apparently inexplicable
things that appear on earth are
credited as being part of the natural
functions of the stars. Certain kinds
of mushrooms that seem to spring
from nowhere are called 'star
faeces', and a special kind of red
lichen is called 'the Sun's urine'. The
interlocutors between the terrestrial
and celestial worlds are the
shamans, who, in their spiritual
flights, take on the body of another
being and visit other worlds in the
sky, below the seas and beneath the
surface of the Earth. The shaman
travels in order to atone for broken
taboo or to gather more information
to help the tribe. Many shamans
claim to have been reincarnated from
one or more species of animal – the
greatest ones several times.

294 Northern Sweden (Lapland)

Sámi shaman drum
Animal skin and wood, 19th century
Cambridge, University Museum of
Archaeology and Anthropology
[E.1887.75]

This drum hails from Lule Sámi or
northern Sweden. It was used to
help the shaman attain a trance-like
state in which he could make contact
with the spirit world, travel outside
his body and voyage into the future.
The drum was beaten with a
hammer, usually made of reindeer
antler. The shaman would sit on his
heels, with his knees touching the
ground. He held the drum in his left
hand, resting his elbow directly on
the artery of his inside left thigh.

Drumming with his right hand, he
would sway to and fro until the
rhythms allowed him to travel.
The pictograms drawn on the skin of
the drum seem to be a schematized
view of Sámi cosmology, depicting
the topography of the journey that a
shaman's soul takes between the
three levels of the universe – the
celestial, the terrestrial and the sub-
terrestrial. Sámi drums were also
used for divining. A small triangular
'indicator' (*vei'ke*) was placed on the
skin. As the drum was beaten with
the stick, the vibrations caused the
marker to skitter across its surface.
The 'journey' made by the marker, as
well as the place upon which it
landed, indicated the good and bad
fortunes due to befall the subject.

295 Francesco Bartolozzi after John Webber
The Death of Captain Cook
Engraving, 1786
London, The National Maritime Museum

296 Mathias Kauage
Captain Cook discovering Australia
Pen and ink on paper, 1998
London, The Rebecca Hossack Gallery

One of the most harrowing stories resulting from a misunderstanding of how different cultures mark and divide time concerns the death of Captain James Cook on the big island of Hawaii on 14 February 1779. When he and his men arrived in Kealakekua Bay, Cook was received as a god. This was because, according to Hawaiian legend, the god Lono was expected to come to the island during that season of the year. The myth stipulated that Lono would arrive not by canoe, but on a floating island, covered with trees. When the islanders beheld the *Resolution*, with its tall masts, it seemed as though the legend had come true. As Cook was clearly the leader, he must be Lono incarnate. When the *Resolution* finally set sail again, they ran into a storm that damaged the foremast. Cook knew that he and his men had probably

outstayed their welcome – their long landing had seriously drained the island's resources – but he had no choice but to return. Indeed, the welcome the second time was much more muted. From the islanders' point of view, Cook's return at the incorrect season signalled either a greedy god or, perhaps, an imposter. Despite attempted appeasement by the chief, the islanders turned ugly. To them, it was now clear that this Cook was not Lono at all and his continued stay on the island was an omen of evil times ahead. As Cook and his men retreated down the beach, the islanders became quite menacing. Cook's men fired to gain time by dispersing the crowd. As they reloaded, however, the islanders attacked. Cook stumbled at the water's edge and was clubbed to death by the angry mob.

Most people understand the term 'relative time' from its use by Einstein in the formulation of his theories of relativity. By proposing that mass is a form of energy (the well-know formula of 'e = mc²'), Einstein redefined the laws of physics that had been dominant since the late seventeenth century. Time and space were no longer conceived as physical absolutes, but were subject to external forces, such as gravity. But the term 'relative time' has leaked into a number of other disciplines. Psychologists, for example, might use the term 'relative time' to describe how and why different people seem to perceive certain phenomena, such as the passage of time, in different ways; and anthropologists often use the phrase to describe how disparate cultures structure and experience time uniquely.

We assume that the time we carry on our wrists is the only time that is 'real'. But, as one has seen, time is infinitely relative. Each person's understanding of time is both psychologically and culturally conditioned. For, despite the acts of the International Meridian Conference of 1884, time itself cannot be zoned. The face of a clock may proclaim one truth, but an hour in New York is not the same as an hour in Tahiti and an hour in purgatory is not the same as an hour in paradise.

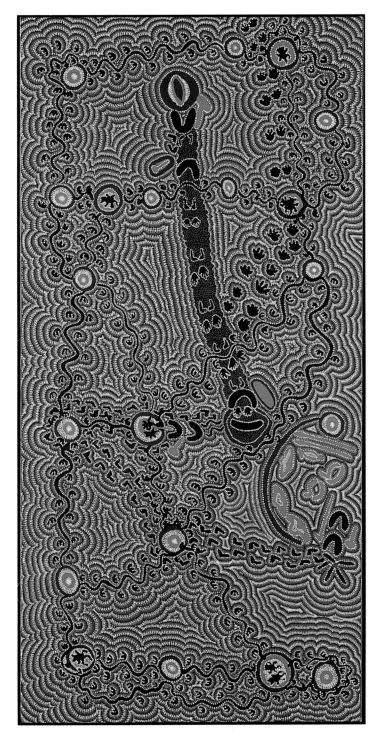

297 **Jeanie Nungarrayi Egan and Thomas Jangala Rice**
Jangganou Jukurrpa (Possum dreaming)
Acrylic on canvas, 1990
Paris, Musée National des Arts d'Afrique et d'Oceanie [Ap.91–67]

'Dreaming' or 'Dreamtime' paintings reflect a world view in which the concept of time is inextricably linked with a sense of place. The paintings are like maps that contain all of a people's history – past, present and future. And, in the same way that a map showing the city of Rome depicts a city that is simultaneously ancient Rome, medieval Rome and modern-day Rome, so the Dreamtime paintings reflect a topography that synthesizes the past, the present and the future. Most often, Dreamtime paintings depict one or more episodes in which certain features of the landscape were created. The episodes themselves, however, are not ordered sequentially. They are ordered by the space upon which each episode left its mark. One might therefore say that the landscape was created by events that existed in the 'past'. But an Aboriginal would never describe a soak hole or mountain ridge as having been 'made' by an ancestor. Instead, the feature would be described as the place where the ancestor lives. The feature is continually in the process of 'being made' by the ancestor because it continues to exist. By revisiting particular landscapes associated with particular stories or by recounting the stories connected with them either visually or verbally, the tangible presence – the 'here and now' – of the land brings the 'past' into the 'present', fuses the two and ensures that it continues to be vital. This painting represents one part of *Possum dreaming*. The possums, marked by their E-shaped footprints, are shown travelling north from Jarrawarri (on the bottom right) looking for gum leaves.

298 **Albert Einstein**
Autograph manuscript notes on the Special Theory of Relativity
Manuscript on paper, *ca.* 1912
Jerusalem, The Israel Museum [Gift of the Jacob E. Safra Foundation]

Until Einstein published his Special (or Restricted) Theory of Relativity, it was generally believed that time was an absolute: it passed at the same rate for everyone in every part of the universe. Newton, for example, stated that time intervals are independent of the motion of the observer. In his theory, though, Einstein postulated that not only was all motion relative, but that the velocity of light is always constant relative to an observer. The ramifications of Einstein's theories were widespread. As the speed of light is constant, the 'time' at which an event occurs will change according to how far an observer is from that event and whether he is stationary, moving towards or moving away from that event. A timekeeper carried on a fast-moving vehicle will run slower than one that is stationary. And, if two vehicles are moving at a constant velocity relative to each other, it will appear to each that the other's time processes are slowed down.

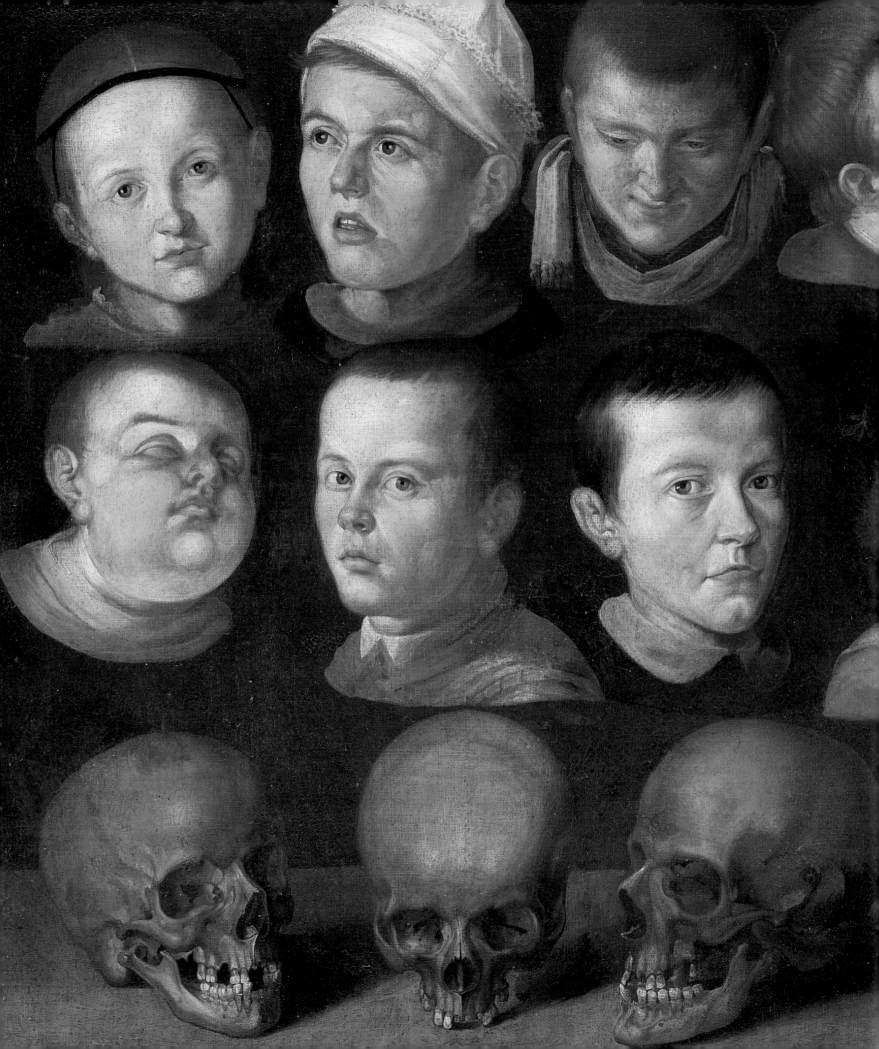

Conrad Meyer, *A father with his sons*
(see 304)

THE End OF Time

One discovers something touchingly human when exploring different cultures' thinking about the end of time. Apocalyptic visions, for example, are shared by almost all cultures that have a linear sense of time. But, surprisingly, the end of time rarely seems to signal the end of everything. Regardless of how strict or deterministic a belief system might be, there seems to be a shared conviction that God will save a small band of the righteous elect, who will survive the end of time and dwell in His glory for all eternity. This tendency to believe that something will survive beyond the end of time seems to be a fundamentally human trait. Human beings find it difficult to believe that all aspects of their existence might disappear. In fact, of all of the great cultures and religions of the world, the Maya and Aztecs, who had apocalyptic visions that were both linear and cyclical (believing that the Earth had been destroyed and re-created by the gods several times) appear to be unique in believing that the next apocalypse will be the final one. In their minds the gods had given the world four chances already; this creation was based, literally, on 'do or die'.

The end of time is much simpler for those who have a cyclical vision of time. In Hinduism, for example, the temporal deities, such as the Sun, the Moon and all of the weather-related gods and goddesses, are seen as relatively minor deities, who 'die' with the rest of the universe at the end of each great age, only to be reborn when the world is fashioned again by the great gods. The end of one world is celebrated because it signals the beginning of the next.

On a personal scale, of course, every creature on Earth experiences an individual 'end of time'. As birth begins the human clock, death ends it. Nevertheless, and even though there is copious evidence to the contrary, most people still hold on to some sort of core belief that they will not really die. This belief reappears in a number of forms: the body may die, but not the spirit; death is just a journey to the other side; one lives on through one's children; or, for the artist, he or she will live forever through his or her work.

StScl-PRC96-10 – Gravitational Lens Galaxy Cluster 0024+1654
(Courtesy of W.N. Colley, R. Turner (Princeton University), J.A. Tyson
(AT&T Bell Labs) and NASA)

Understanding the Beginning and the End

Martin Rees

Our ancestors could weave cosmological theories almost unencumbered by any facts. But technical advances, especially in the last decade, have transformed cosmology from speculation to serious science, vastly expanding our horizons in space and time.

Centuries ago, terrestrial maps had blurred boundaries, where the cartographers could write "there be dragons". But, after the pioneer navigators had traversed the globe, delineating the main land masses, there was no expectation of a new continent, nor that we would ever drastically revise our estimates of the Earth's size and shape.

At the end of the twentieth century, we have now, remarkably, reached the same crucial stage in mapping our Universe – its spatial extent, its large-scale structure, its main constituents and its vast span in time. For the first time, the broad cosmic picture is coming into focus. This story – a collective achievement of thousands of astronomers, physicists and engineers, using many different techniques – can now be presented (at least in outline) with compelling conviction.

Let us start with our own star, the Sun: how did it begin, and how will it end? The proto-Sun condensed from a cloud of dusty gas in our Milky Way. Gravity pulled it together until its centre got squeezed hot enough to trigger nuclear fusion – the same process that makes hydrogen bombs explode. The temperature within the Sun is continually adjusted by the force of gravity so that fusion supplies power at just the exact rate needed to balance the heat shining from the Sun's surface. Even though it is already 4.5 billion years old, less than half the Sun's central hydrogen has so far been used up. The Sun will keep shining for a further 5 billion years. It will then swell up and become the kind of star known as a 'red giant', large and bright enough to engulf the inner planets and to vaporize all life on Earth. After this red giant phase, some of the Sun's outer layers will be blown off, leaving it to become a 'white dwarf' – a dense star no larger than the Earth, which will shine with a dull glow, no brighter than the full Moon today, on whatever remains of the Solar System.

To help in visualizing these vast time spans – of the future as well as past – a simile can help. If one represents the lifespan of the Sun as a walk across America, starting in New York when the Sun formed and ending up in California ten billion years later, when the Sun is about to die, then each footstep in the journey would represent 2000 years. Moreover, all recorded history would be just a few steps that would lie a little before the half-way stage – somewhere in Kansas, perhaps. Likewise, we are far from the culmination of evolution. The progression towards diversity has much further to go. Even if

We are far from the culmination of evolution

life is now unique to the Earth, there is still a great deal of time for it to spread from here through the entire Galaxy, and even beyond.

Astrophysicists are able to compute the life cycle of a star that is, say, half, twice, or even ten times the mass of the Sun. Smaller stars, for example, burn their fuel more slowly. Stars heavier than the Sun – such as the four blue Trapezium stars in the constellation of Orion – shine much more brightly and consume their fuel more quickly. This theory can be tested by observing other stars, which are at different stages in their evolution.

Not everything in the cosmos happens slowly, though. Stars more than ten times as heavy as our Sun expire in a more violent way, by exploding as a supernova. The nearest supernova of modern times was seen in 1987. On February 23–24 of that year, a new bright 'star' appeared in the Southern Sky that had not been visible the previous night. Astronomers have studied this particular supernova, especially how it fades and decays. They have even found, on images taken before the explosion, that the precursor was a blue star equal to about 20 solar masses. When a heavy star has consumed all of its available hydrogen, its core contracts and heats up, releasing energy via a succession of reactions involving progressively heavier nuclei, composed of helium, carbon, oxygen, silicon, and so on. The star then faces an energy crisis when it cannot draw on any further nuclear sources. The consequences are dramatic – a supernova explosion, from which the former star throws back into space a mixture of atoms. The important information we have gained from studying dying stars is that the material they expel has roughly the same proportions of the elements as exist on Earth.

Our understanding of stars explains why elements such as carbon and oxygen are so common here on Earth, but gold and uranium so

an axis in the plane of its orbit) may indicate another large collision. The craters on the Moon bear witness to the violence of Earth's early history, before the planetesimals had been depleted by impacts or coalescence. Space probes have now visited the other planets in our Solar System, showing that they (and their larger moons) are highly distinctive worlds. There is no reason to expect other solar systems to have the same configurations or same numbers of planets as ours happens to have.

Planets on which life could evolve, as it did here on Earth, must be rather special. Their gravity must pull strongly enough to prevent the atmosphere from evaporating into space. They must be neither too hot nor too cold and, therefore, they need to be exactly the right distance from a long-lived and stable star. There may be other special circumstances required. For example, it has been claimed that the existence of the planet Jupiter was essential to life on Earth, because its gravity captured or 'scoured out' a large number of the asteroids that might have otherwise caused catastrophic impacts on Earth. Also, the tides induced by our large Moon may have stimulated some phases in evolution. But even if there are extra requirements like these, planetary systems are (we believe) so common in our galaxy that Earth-like planets should be numbered in the millions.

Searches for Earth-like planets are now a main thrust of NASA's space programme. This is a long-range goal and it will require vast optical interferometers in space – but will stimulate much excellent science on the way. And once a candidate has been found, several things could be learnt about it. Suppose, for example, that an astronomer forty light years away had detected our Earth: it would appear to be, in Carl Sagan's phrase, a "pale blue dot". If Earth could be seen at all, its light could be analysed and would reveal that it had been transformed (and oxygenated) by a biosphere. The shade

For the first time, the broad cosmic picture is coming into focus

rare. Our galaxy is like a vast ecosystem, recycling gas through successive generations of stars and gradually building up the entire periodic table of elements. Before our Sun even formed, several generations of heavy stars could have been through their entire life cycles, transmuting pristine hydrogen into the basic building blocks of life – carbon, oxygen, iron and the rest. Everything that exists on Earth is, literally, the ashes of long-dead stars.

The actual layout of our Solar System is the outcome of many 'accidents'. In particular, our Moon was torn from Earth by a collision with another proto-planet. The odd spin of Uranus (around

of blue would be slightly different, depending on whether the Pacific Ocean or the Eurasian land mass was facing the observer. Distant astronomers could, therefore, by repeated observation, infer the Earth was spinning. They could learn the length of its day and even determine something of its topography and climate.

The most interesting question, of course, is – even when a planet offers a propitious environment – what is the chance of life getting started? We still do not know whether life's emergence is 'natural', or whether it involves a chain of accidents so improbable that nothing remotely like it has happened on another planet anywhere else in our galaxy. That is why it would be so crucial to detect life, even in simple and vestigial forms, elsewhere in our Solar System – on Mars, or under the ice of Europa. If 'life' had emerged twice within our solar system, this would suggest that the entire galaxy could be teeming with life.

But there is a second stage. For even when simple life exists, we do not know what the chance is that it might evolve towards intelligence. The year 2000 also marks the fourth centenary of the death of Giordano Bruno – burnt at the stake in Rome. One of his many alleged heresies included a belief in distant inhabited worlds. The belief that intelligent life exists elsewhere is widely shared, but it is based on very little evidence. Only in the last five years of this millennium have we known for sure that 'worlds' exist in orbit around other stars. But, even if innumerable planets exist, we are little closer to knowing whether any of them harbour anything alive.

We can now see vast numbers of galaxies, stretching to immense distances. One amazing picture taken with the Hubble Space Telescope shows only a small patch of sky, less than 1% of the area covered by a full Moon. It is densely covered with faint smudges of light – each a billion times fainter than any star that can be seen with the unaided eye. But each smudge is an entire galaxy, thousands of light years across, which appears so small and faint only because it is several billion light years away.

In about 1970 the Canadian cosmologist James Peebles isolated two discoveries that he called "golden moments in cosmology". The first of these was Edwin Hubble's realisation that our universe was expanding. This was based on his observation that the more distant a galaxy was from the Earth, the more stretched out the characteristic patterns of its wavelengths were. The further away they were, the more their light shifted to the longer, red end of the spectrum. Hubble suggested that this red-shift was caused by something like the Doppler effect in acoustics. It signalled that the universe was expanding at speeds proportional to their distance from an observer. The second "golden moment" was the detection by Arno Penzias and Robert Wilson of the 'afterglow of creation' – or cosmic background radiation. They discovered that intergalactic space was not entirely cold. Their antennae were picking up microwaves which seemed to arrive in equal strength from all directions and had no obvious source. This 'warmth' is a relic of the original 'fireball' phase; the microwaves are an echo, as it were, of the 'explosion' that initiated the universal expansion.

During the first millisecond of the universe, everything would have been squeezed denser than an atomic nucleus or a neutron star. Particles would have repeatedly collided with each other. The Large Hadron Collider at CERN in Geneva will soon be able to simulate the type of energies that prevailed when our universe was 10^{-14} seconds old, but it will never come near to being able to generate the sort of energy that the universe carried even slightly earlier. Even though we know a great deal about the nature of the 'big bang', its earliest phases confront us with a condition so extreme that we do not know enough physics to solve its riddles. The assumptions underpinning the two great foundations of twentieth-century physics – quantum physics and Einstein's general theory of relativity – are still disjointed. What we really need to understand is how 'quantum gravity' might work; and that is a task for today's scientists.

As it stands, our understanding of the history of the universe divides into three parts:

The beginning – a brief but eventful era when the key features of our universe were imprinted, spanning the interval from the Planck time (10^{-43}) up to the first millisecond. This is the intellectual habitat of mathematical physicists and quantum cosmologists. The relevant physics is speculative.

From the first millisecond to when the universe was about a million years old. This is an era where cautious empiricists (like me) feel more at home. There is good quantitative evidence and the relevant physics is well tested in the lab. The densities are far below nuclear density, but everything is still expanding smoothly. The relevant physics is firmly based on laboratory tests and theory is corroborated by quantitative evidence – the cosmic abundance of helium, the background radiation *etc*. This part of cosmic history, though it lies in the remote past, is the easiest of the three to understand.

The 'recent' era that initiated when the first gravitationally bounded structures condensed out – when the first stars and quasars formed and lit up. This is the era studied by traditional astronomers and the one in which we are able to witness complete manifestations of well-known basic laws. Gravity, gas dynamics, and feedback effects from early stars combine to initiate the complexities around us. This part of cosmic history is difficult because it involves ultra-complex manifestations of simple laws.

Given this state of knowledge, the educated lay-person may wonder if it might not be presumptuous for an astronomer to claim to know anything, with any level of confidence, about cosmology. In the sciences, though, it is generally complexity, and not sheer size, that makes things hard to understand – a star, for instance, is simpler than an insect. In the primordial fireball, for example, everything must have been broken down into its simplest constituents. This means that the early universe really could be less baffling, and more within our grasp, than the smallest living organism.

In about five billion years the Sun will die and the Earth with it. At about the same time (give or take a billion years), the Andromeda Galaxy, already falling towards us, will crash into our own Milky Way. Our own galaxy will surely end in a great crash. But will our universe go on expanding for ever? Or will the entire firmament eventually re-collapse to a 'big crunch'?

The answer depends on how much the cosmic expansion is being decelerated. Everything in the universe exerts a gravitational pull on everything else. The expansion of the universe could eventually be

density contrasts. If one had to summarize, in just one sentence, what has been happening since the big bang, the best answer might be to take a deep breath and say: ever since the beginning, gravity has been amplifying inhomogeneities, building up structures, and enhancing temperature contrasts – a pre-requisite for the emergence of the complexity that lies around us ten billion years later, and of which we are part.

The answers to our future lie in our past – in something remarkable that happened when our entire observable universe was compressed to a microscopic size. Right back at the beginning, the problems of the cosmos and of the microworld are interlinked. But cosmology is not only a 'fundamental' science, it is also the grandest of the

Will the entire firmament eventually re-collapse to a 'big crunch'?

reversed if there were, on average, more than about five atoms in each cubic metre. That does not sound like much; but, if all the galaxies were dismantled and their constituent stars and gas spread uniformly through space, they would make an even emptier vacuum – containing one atom in every ten cubic metres or the equivalent of one snowflake in the entire volume of the Earth. This figure is fifty times less than the 'critical density', and, at first sight, this fact would seem to imply perpetual expansion, by a wide margin.

But the real problem is not so straightforward. Astronomers have discovered that galaxies, and even entire clusters of galaxies, should fly apart unless they are held together by the gravitational pull of about ten times more material than we actually see. This means that something that we cannot see is holding things together.

There is almost certainly enough so called 'dark matter', mainly in galactic halos and clusters of galaxies, to contribute 20% of the critical invisible density. And, until recently, we could not rule out the possibility that several times this amount – comprising the full critical density – might exist in the space between clusters of galaxies. But that suggestion now looks unlikely. Sufficient density does not seem to exist within the universe and it does appear that the expansion of our universe will be never-ending – our universe will become ever-emptier and ever darker, as the galaxies recede from each other and their constituent stars exhaust their fuel, leaving dead remnants.

People sometimes wonder how our universe can have started off as a hot amorphous fireball and ended up manifestly far from equilibrium. One key component is the gravity that renders the expanding universe unstable to the growth of structure, in the sense that even very slight initial irregularities evolve into conspicuous

environmental sciences. How did a hot amorphous fireball evolve, over a period of ten to fifteen billion years, into our complex cosmos of galaxies, stars and planets? How did atoms assemble – here on Earth, and perhaps on other worlds – into living beings intricate enough to ponder their origins? Answering these questions is a task for the new millennium. Or, perhaps, answering those questions is an unending quest.

STScI-PRC95-11 – Cygnus Loop
(Courtesy of J. Hester (Arizona State University) and NASA)

299 Hellenistic Greek

Grave stele

Marble, 2nd–3rd century AD
London, The British Museum, Dept.
of Greek and Roman Antiquities
[Towneley Coll., 1805.7–5.211;
Sculpture 2391]

The transience of man's life was
certainly not a theme invented by the
early Christians. In this grave stele,
there is a skeleton depicted lying in a
catacomb (or rough-cut wall tomb).
The inscription addresses the viewer:
'He who passes by – looking at this
fleshless corpse, who can say
whether it was Hylas or Thersites?'.
The obvious point is that no one
would be able to recognize the
identity of this body in its current
state. But the average Greek would
have known that Hylas was one of
Jason's Argonauts whose beauty
earned him the companionship of
Hercules and caused him to be
abducted by the nymphs of a spring
at which he was drawing water; and
that Thersites was a common soldier
at Troy, given to heckling his
superiors, whom Homer condemned
for his physique and for his manners.
Two other references would have
occurred to a literary-minded Greek
reading this inscription. The first is an
anonymous epigram that can be
roughly translated as: 'The just and
the unjust are equal in death; so, in
the present, the fair and the ugly are
equal'. The second is Lucian's
Dialogues of the Dead (2nd century
AD) and the conversation held in
Hades amongst Nireus, Thersites
and Menippus. Nireus, described by
Homer as "the handsomest man of
all who came to Troy", is incensed
by the claim of the boorish Thersites
that he is more beautiful. But the
wise man of the group, Menippus,
has this to say: "Neither you nor
anyone else is handsome here in
Hades. All are equal and all are
alike." So the stele is not just a
reflection of the anonymity created
by death, but focuses specifically on
how both the beautiful and the ugly
will still end up as a corpse.

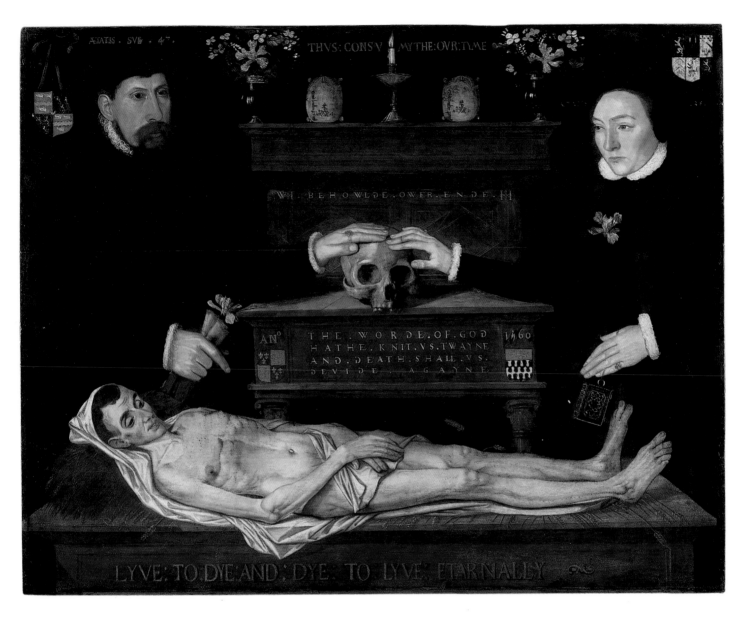

300 **English**

The Judd Memorial (also known as
The Judd Marriage)
Oil on panel, 1560
London, Dulwich Picture Gallery
[DPG 354]

Despite the fact that one might be
aware of those aspects of
Christianity which portray the body as
a despised vessel and life itself as a
type of punishment from which we
will be released only with the Second
Coming of Christ, it is quite rare to
see these attitudes expressed quite

so boldly in paint – let alone in a
format which appears to 'celebrate' a
marriage. The male figure at the left
has been identified as William Judd,
the second son of Thomas Judd of
Wyckforde, Essex. An inscription
tells us that he is 47 years old. The
female figure is his wife (either Joan
or Anne Cromwell, daughter of
Walter Cromwell). They are shown
reaching out towards a skull, over
which there is the legend *WE
BEHOWLDE OWER END*. Their
professed credo in this life is: *LYVE
TO DYE AND DYE TO LYVE*

ETERNALLY. Below the skull, there
is another inscription: *THE WORDE
OF GOD HATHE KNIT US TWAYNE
AND DEATH SHALL US DIVIDE
AGAIN*. This phrase alludes to the
notion that, with the Second Coming,
all mortal contact and relationships
will end. Again, a sentiment that
seems somewhat misplaced in the
context of a marriage portrait. Given
their rejection of the flesh, it is
curious that the Judds would have
wished to have their likenesses
painted at all. Clearly, all portraiture is
vanity. But, for all their piety, the

Judds were human and prey to
human temptations. Their
ambivalence comes through clearly in
the inscription appearing on the
original frame of the painting: *WHEN
WE ARE DEADE AND IN OWR
GRAVES,/ AND ALL OWRE BONES
ARE ROTTUN,/ BY THIS WE SHALL
REMEMBERD BE, WHEN WE
SHULDE BE FORGOTTYN.*
Nevertheless, they did commission
the portrait.

301 **Didymos the Turner**

Funerary inscription

Limestone, 1st–3rd century AD
London, The British Museum, Dept.
of Greek and Roman Antiquities [GR
1888.9–20.7]

One of the great fears of life is not
knowing how or whether one will be
remembered after death. Dread of
the anonymity of the grave was
certainly one of the major factors
towards the development of funerary
stelae, inscriptions, sculptures and
tomb decoration. In most cases, it is
the family that looks after the task of
memorializing the dead. In some
cases, though, this task has been
taken on by the incipient victim.
This limestone slab commemorates
a man known as Didymos, who lived
in Egypt sometime between the 1st
and the 3rd centuries. On it, he has
roughly carved his own epitaph. It
reads: 'Didymos, the turner. I carved
[this] while I was still alive, and now
here I lie, aged sixty-six'. Whereas

Didymos's own carving skills are not
tremendously good, there is an
obvious difference in hand between
the main body of the inscription and
the last two numbers (the Greek for
66). It seems that Didymos himself
must have carved the text, leaving
the date of his death blank for
another to fill in. When the time
came, a friend obliged and helped to
immortalize Didymos.

302 **Persian (Meshed)**

Funerary stele

Glazed terracotta,
AH 1050/1640–41 AD
Paris, Musée de l'Institut du Monde
Arabe [AI 86–03]

From very early Islamic times,
funerary stelae are often shaped in
the form of a *mihrab*, the prayer
niche in a mosque. This served as a
reminder towards prayer, but it also
fulfilled a symbolic role, presenting
the gravestone itself as associated
with sanctuary and relief from the
worldly domain. Along the sides of
this stele, there are two lone
cypresses – a universal symbol of
death. In the upper section, there is
a scene in which the turban, rings,
shield, sabre, quiver and bow and
arrows of the deceased are

displayed. The owner no longer
needs them. The inscription, written
in *nasta'liq*, contains two lines of
verse and three lines of prose. The
verse can be roughly translated as:
'Alas, the Sun in the times of his
youth had been as ephemeral as the
dawn.
Alas, suddenly, the flower that has
opened is now felled by a brisk
autumn wind.
He is dead – the one whose god
was his refuge and who so loved a
garden.'
The prose inscription tells us that the
deceased was named Hasan Khān
beg Ibn Maghfirat Panā Zulfiqār beg
Shāmlu and that he died in the
month of Sawal, AH 1050.

303 Lovis Corinth

Self-portrait with Death
Drypoint engraving, 1916
London, The British Museum, Dept.
of Prints and Drawings
[1980.12–13.27]

In this self-portrait, Corinth depicts himself drawing or, perhaps, etching the image of a skeleton. (The confusion arises from the fact that Corinth seems to have utilized the medium of drypoint almost in the same way that other artists would use a sketchpad.) It is hard to know exactly when artists began using cadavers or skeletons to improve their knowledge of human anatomy. The earliest surviving drawings that seem to record the practice date from the last years of the 14th century. Pisanello's sketchbooks, for example, include a number of drawings of executed criminals. And there is ample evidence that a number of 15th-century artists, such as Donatello and Antonio Pollaiuolo, attended disecting sessions held at local medical schools. Indeed, the drawings of Leonardo da Vinci suggest that he even dissected his own cadavers. In Corinth's engraving, however, the image is only partly concerned with a depiction of the artist improving his skills in rendering the human form. The inscription above the head of the skeleton – reading THANATOS or 'Death' – adds a *memento mori* tone to the etching. It is a double-edged theme that first appears in Corinth's art in his well known *Self-portrait with skeleton* (1896) in the Lembachhaus in Munich. On the one hand, one recognizes that Corinth is trying to come to terms with his own mortality by creating such a pairing. Particularly in the Munich painting, where he depicts himself as robust, colourful and defiant, the contrast between artist and skeleton is striking. On the other hand, Corinth must have been aware that, by creating a work of art, he was creating a memorial to himself that would live on past his own death. Artists have a peculiar slant on the *ars longa, vita brevis* dictum. Certainly, they want to make their mark on posterity through their art – but it is only the very romantic young artist who might want to die in order to test the endurance of his own greatness.

304 Conrad Meyer

A father with his ten living sons and six dead sons
Oil on canvas, *ca.*1650
Zurich, Kunsthaus (Kellersche Sammlung 201)

Without being a great artist, there are certainly other ways to leave your mark on the memory of succeeding generations. The easiest way to ensure that at least some part of you is never forgotten is to pass on your DNA. In this extraordinary family portrait, we see how the features of the father have been transmitted to each of his ten living sons. Some have his eyes, others have his nose. From a biological point of view, this father will live forever in the genetic material he has passed to his descendants. Their very number, however, is a reminder of the extent to which families of the past had to be larger in order to offset the effects of a higher infant mortality rate. In ancient Rome, for example, a woman had to have at least five children to ensure that one would live to see adulthood. In this man's experience, the odds have been slightly better, with a little more than half the children having survived.

305 Etruscan

Cremation urn of Arui Heiesa
Glazed terracotta, late 3rd–2nd
century BC
Liverpool, The National Museums of
Merseyside, Liverpool Museum
[M10463]

The Etruscans believed that the
afterlife was precisely that – a life
lived in conditions pretty much the
same as they had been before death.
In Etruscan paintings from the 5th
and 6th centuries BC, one sees the
land of the dead like a sort of
Elysium. There are scenes of
feasting, playing at sport, listening to
music and, generally, enjoying
oneself. The most common motif on
Etruscan funerary sculpture, such as
the sarcophagi and funerary urns
used to contain the ashes of the
deceased, is the symposium, or
drinking party. The deceased is
shown as a guest, reclining on an
eating couch – propped up on one
elbow or by a small pile of pillows –
with a drink in hand. In most cases,
he or she is shown alive and alert;
sometimes as if sleeping – either
because inebriated in the party or
because death itself is but an
extended sleep.

306 Chinese

Cicada
Jade, Han dynasty (206 BC–220 AD)
London, The British Museum, Dept.
of Oriental Antiquities
[OA.1937.4-16.114]

From Neolithic times onwards, jade
was regularly included as part of
burial goods in China. It seems not
to have been used primarily as a
symbol of wealth. Instead, it
functioned as a protective talisman.
Indeed, the well known 'jade suits'
excavated in Mancheng in 1968
were obviously constructed to
encase the whole body. Numerous
early Chinese sources refer to the
body-preserving properties of jade.
Ge Hong, in his *Baopuzi,* wrote that
if gold and jade "are put in the nine
orifices [of the body], the result is
that the body will not decompose".
The Chinese have regularly used
images of the cicada beetle as a
symbol for resurrection, probably
owing to the beetle's reproductive
cycle, in which it hibernates for a
period of 17 years. Carved jade
images of a cicada were often
inserted into the mouth of the
deceased as an aid towards helping
him pass over to the afterlife.

307 Myra Kukiiyaut

The dying man becomes a wolf
Stone-cut stencil on paper, 1971
Winnipeg Art Gallery
[G-76–884]

Myra Kukiiyaut derives her subject
matter from traditional Inuit songs,
legends and beliefs. In this picture
she records the magic song of her
father-in-law, Amaroq. The word
amaroq means 'wolf' and, in his
song, he expressed the wish that,
once he had died, his soul might
pass into that of a wolf, so that he
would remain happy for all eternity,
hunting the caribou. The picture
shows a dead Amaroq, lying in his
grave. But, through the power of the
magic song, which a female relative
– possibly Myra herself – is shown
singing at the top of the page, he
has been transformed into a wolf
and is, indeed, happily hunting a
caribou. The marks above the head
of the caribou are Inuit script. They
record the words of Amaroq's song
about the caribou:
'When he is eating and walking along
　Even after my death, I will stalk him
　For a warm mattress.'

308 Chinese

'TLV' mirror
Bronze, Han Dynasty (206 BC–220 AD)
London, The Victoria and Albert
Museum [M. 16–1935]

As in many other cultures, the
ancient Chinese ritually buried
artefacts with the deceased in order
to ease his or her passage towards
the hereafter. The so called TLV
mirrors appear as part of Chinese
burial furniture as early as the 1st
century BC, but they seem to be
most popular during the early years
AD. The TLV mirror is named
according to three distinctive
geometrical patterns set within the
decoration. The 'T' is generally found
flanking the central square and the
'L' and 'V' patterns appear along the
inner circular band. These mirrors
symbolize a vision of the cosmos in
which the two universal principles of
'the five' and 'the twelve' have been
resolved. At its most basic, the
mirror is composed of a circular
shape (symbolizing the heavens)
enclosing a square (symbolizing the
Earth). In this mirror, there are a
number of small animals occupying
the spaces between the central
square and the circular border. The
significance of the rest of the
decoration remains the subject of
scholarly debate, with the more
convincing arguments having based

their interpretations on the similarity
between the decoration of the TLV
mirrors and the marks found on
diviner's boards from the same
period, of which the purpose (we are
told by a contemporary Han Dynasty
source, the writer Zheng Zhong, *fl.*
50–70 AD), was for "comprehending
the appointed seasons of heaven
and apportioning good and ill
favour". The TLV mirror, then, seems
to be a stylized version of the
diviner's board in which the celestial
omens have been set permanently
into the most harmonious resolution
– in order to offer the deceased
optimal ease during his
travels after death.

310 Egyptian

Model of a boat
Painted wood, 2200 BC (6th Dynasty)
Manchester, Manchester Museum
(University of Manchester) [11266]

Boats played an important role in the
Egyptian funerary cult, symbolizing
the journey of the deceased to
Abydos, the centre of the cult of the
god Osiris. For the ancient
Egyptians, death was not an end to
life, merely a temporary cessation of
activity. With a successful transition
to the afterlife the king would join
the gods as an equal. One of the
earliest pyramid texts states that a
king would become one of the
circumpolar stars – thus existing for
ever. Later texts describe how he
spends his afterlife in the boat of the
Sun god Ra: "I am pure, I take the
oar to myself, I occupy my seat ... I
row Ra towards the west".

309 English

Chain of lockets
Gold, glass and human hair, *ca.* 1889
private collection

The use of hair in jewellery seems to
begin during the Middle Ages and
seems to be a peculiarly Christian
practice. The reason for this may be
related to prophecy concerning the
Last Judgment. In Revelations
20:11–15, John describes how all the
interred bodies of deceased saints
and sinners will rise up from their
graves to be judged. In interpreting
these passages, some of the Church
Fathers argued that, if the various
parts of the body had been dispersed

either before or after death, the
deceased was obliged to go and
collect them together before he or
she was fit to be judged. This is one
of the reasons why the Catholic
Church had strong interdictions
against cremation. Another area of
controversy was whether a husband
would recognize his wife when the
Day of Resurrection was at hand
(see 300). One romantic solution
would be to ensure that your loved
one had a piece of your body – such
as a lock of hair. In this way, you
were sure to meet again and be
reunited for the whole of eternity.

○ There are a number of major religions in which a transcendence of time is a central aim. Whether through meditation, mortification of the flesh, asceticism, trance or drugs, the goal is to exist in a realm in which time and its effects have no consequence. As Prince Siddhartha Gautama asked after witnessing the myriad tragedies of life: "Life is subject to age and death. Where is the realm in life in which there is neither age nor death?"

In Buddhism, the transcendence of time is a liberation from transience. By detaching oneself from both the pain and the pleasures of bodily existence and the 'ills of impermanence' (*anicca* or *anitya*), one can begin to transcend the illusion of time. The various schools of Buddhist thought maintain different attitudes towards time. The state of transcendence in Buddhism is called *nirvāna* or, literally, 'extinction'. There is no self or soul; therefore, there is no quest to attain perfection or union. *Nirvāna* is the absolute cessation of any attachments, interaction or involvement with the appearances of this world.

Patañjali, the author of the classical text on *yoga* written during the second or third century AD, states that the end purpose of *yoga* is 'the cessation of mental fluctuation', which leads to higher levels of consciousness or absorption (*samādhi*) and the purification of the self (*ātman*). The word *yoga* is Sanskrit for 'yoking' or 'joining'. In this respect, it differs radically from Buddhism. The ultimate aim of *yoga*, however, is still liberation (*moksa*) from both time and space. Timelessness is the ultimate reality. Time is an illusion. Moments may exist, but sequences of moments cannot be combined to form a 'thing' like an hour or a day. When the mental state of *dharmamegha* is achieved, the sequence of moments come to an end and, in *dharmamaghasamadh*, the *yogi* attains a state of zero-time experience, before joining his consciousness to timelessness (*kaivalya*).

311 **Tibetan**
Inscription of 'Om mani padwe hum'
Stone, early 20th century
London, The Hornimann Museum
and Gardens [NN 4433]

A *mantra* or 'instrument of thought' is a single- or multi-syllable phrase believed to spring from a divine origin. *Mantras* are used for both ritual and meditative purposes in a number of Indian religions. The phrase *'Om mani padwe hum'* is the *mantra* of Avalokiteshvara, one of the most important *bodhisattvas* of the power of the equally compassionate Buddha, Amithbha or Amida. The phrase is generally translated as 'hail, the jewel in the lotus, hail'. By repeating a *mantra*, the meditator's consciousness is moved towards that of the deity. In *mantra yoga*, for example, the repetition of a *mantra* is used as a means towards liberation (*moksa*). The belief is that, by chanting, one is able to 'transcend' towards the absolute, the unmanifest origin of sound.

312 **Japanese**
Seated Shaka
Wood, gold lacquer, gilding and
crystal, 18th century
London, The British Museum,
Dept. of Japanese Antiquities
[1885. 12–27. 30]

In Japan, what Westerners might call Buddhism is called Shakkyo, 'the teachings of Shak[yamuni]'. Shaka is the Japanese name for the Buddha Sakyamuni, a name describing the fact that the Buddha was descended from the Sākya tribe, who originated in a region located in present-day Nepal. The Shaka's posture is presented in a specific form. With his hands, he makes specific *mudra* or 'signs' which embody aspects of his spiritual power. With his right hand, he makes the *mudra* or gesture of reassurance, known in Japanese as *semui-in*. His left hand makes the *mudra* of wish-fulfilment, or *yogan-in*. As for the Budhha's smile, one can only quote a well known Buddhist verse:
"A thousand questions remain, but the Buddha is silent.
Others abide our questions. Thou art free.
We ask and ask; thou smilest and art still."

THE STORY OF TIME

The National Maritime Museum and the Royal Observatory, Greenwich, wish to thank the following institutions and individuals for the objects they have loaned to *The Story of Time*:

Great Britain
Her Majesty the Queen

The National Trust (Anglesey Abbey and Powys Castle)
English Heritage (Down House, Downe)
The Visitors of the Ashmolean Museum of Art and Archaeology, Oxford
Bodleian Library, Oxford University, Oxford
The British Library, London
The British Museum, London
Dorset County Museum (The Dorset Natural History and Archaeological Society), Dorchester
The Trustees of Dulwich Picture Gallery, London
The Provost and Fellows of Eton College, Eton College Library, Windsor
The Syndics of the Fitzwilliam Museum, Cambridge University, Cambridge
The Trustees of the Harris (Belmont) Charity, Faversham
The Horniman Museum and Gardens, London
Rebecca Hossack Gallery, London
The Jewish Museum, London
Leeds Museums and Galleries, Leeds
City Art Gallery, Leeds
Liverpool Museum, National Museums & Galleries on Merseyside, Liverpool
Manchester City Art Galleries, Manchester
The Manchester Museum, The University of Manchester, Manchester
The Museum of Artillery in the Rotunda, Woolwich, London
Museum of London, London
The Board of Trustees of the National Gallery, London
National Gallery of Scotland, Edinburgh
The Trustees of the National Museums of Scotland, Edinburgh
National Physical Laboratory, Teddington
Natural History Museum, London
Royal Academy of Arts, London
The John Rylands University Library, The University of Manchester, Manchester
Schott Publishing, London
The Board of Trustees of the Science Museum, London
Tameside MBC Leisure Services, Astley Cheetham Art Gallery, Stalybridge
Tate Gallery, London
The University of Leeds Library, Leeds
The University Museum of Archaeology and Anthropology, Cambridge University, Cambridge
The Chairman and Trustees of the Victoria and Albert Museum, London

The Wellcome Institute for the History of Medicine, London
The Whipple Museum of the History of Science, Cambridge University, Cambridge
The Whitworth Art Gallery, University of Manchester, Manchester
The Worshipful Company of Goldsmiths, London
Whitney Antiques, Whitney
Yorkshire Museum, York
and the private collections of
G.S. Barrass, Jonathan Betts, David Bedford, John Gleave, Peter Gosnell, P.N. Haward, Prof Joy Hendry, Helen Rees Leahy, R.H. Miles

Austria
Augustiner Chorherrenstift, St Florian
Gesellschaft der Musikfreunde in Wien, Vienna
Kunsthistorisches Museum, Vienna
Österreichisches Nationalbibliothek, Kartensammlung und Globenmuseum, Vienna
Salzburger Museum für Kunst und Kulturgeschichte, Salzburg
Steiermärkisches Landesmuseum Joanneum, Graz

Belgium
Musée de la Vie Wallonne, Liège
Musées Royaux d'Art et d'Histoire, Brussels

Canada
Royal Ontario Museum, Toronto
Winnipeg Art Gallery, Winnipeg

Czech Republic
Národní Technické Muzeum, Prague

France
Bibliothèque Nationale de France, Paris
Musée des Antiquités Nationale, Saint-Germain-en-Laye
Musée des Arts et Métiers (CNAM), Paris
Musée Carnavalet, Paris
Musée de l'Institut du Monde Arabe, Paris
Musée National des Arts d'Afrique et d'Océanie, Paris
Muséum National d'Histoire Naturelle, Paris
Musée National du Moyen Age (Musée des Thermes et de l'Hôtel de Cluny), Paris
Musée d'Orsay, Paris

Germany
Bayerisches Nationalmuseum, Munich
Bayerische Staatsbibliothek, Munich
Deutsches Museum, Munich
Deutsches Uhrenmuseum, Furtwangen
Germanisches Nationalmuseum, Nuremberg
Hamburger Kunsthalle, Hamburg

Heimatmuseum Charlottenburg, Berlin
Kunstgewerbesammlung der Stadt Bielefeld, Stiftung Huelsmann, Bielefeld
Museum für Kunst und Gewerbe, Hamburg
Staatliche Kunstsammlungen Dresden, Dresden
   Gemäldegalerie Alte Meister
   Mathematisch-Physicalischer Salon
Staatliche Museen zu Berlin - Preussischer Kulturbesitz, Berlin
   Ägyptisches Museum und Papyrussammlung
   Antikensammlung
   Gemäldegalerie
   Museum für Islamische Kunst
   Museum für Völkerkunde
   Skulpturensammlung
Staatsbibliothek zu Berlin - Preussischer Kulturbesitz, Berlin
Staatsgalerie, Stuttgart
Stadtbibliothek, Braunschweig
Universitätsbibliothek, Heidelberg
Wuppertaler Uhrenmuseum, Wuppertal

Ireland
Trustees of the Chester Beatty Library, Dublin
National Museum of Ireland, Dublin

Israel
The Israel Museum, Jerusalem

Italy
Archivio di Stato, Siena
Biblioteca Estense Universitaria, Modena
Biblioteca Gambalunga, Rimini
Biblioteca Medicea Laurenziana, Florence
Biblioteca Nazionale Centrale, Florence
Galleria Estense, Modena
Galleria d'Arte Moderna Ricci Oddi, Piacenza
Musei Civici d'Arte e Storia, Brescia
and an anonymous private collection

The Netherlands
Amsterdams Historisch Museum, Amsterdam
Koninklijke Bibliotheek, The Hague
Koninklijk Instituut voor de Tropen, Tropenmuseum, Amsterdam
Museum Boijmans Van Beuningen, Rotterdam
Rijksmuseum, Amsterdam
Rijksmuseum voor Volkenkunde, Leiden
Stedelijk Museum de Lakenhal, Leiden
and an anonymous private collection

Norway
Munch-Museet, Oslo
The Schøyen Collection

Spain
Museo Nacional del Prado, Madrid

Sweden
An anonymous private collection

Switzerland
Bernisches Historisches Museum, Berne
Château des Monts, Musée d'Horlogerie, Le Locle
Chronometrie Beyer, Zurich
Kunsthaus, Zurich
Musée d'Histoire des Sciences, Geneva
Musée de l'Horlogerie, Geneva
Museum der Kulturen, Basle
Naturhistorisches Museum, Basle
Parmigiani Fleurier, Fleurier
Uhrensammlung Kellenberger, Gewerbemuseum, Winterthur

USA
American Museum of Natural History, New York NY
Amherst College, Mead Art Museum, Amherst MA
Colgate University, Department of Geology, Hamilton NY
Cranbrook Institute of Science, Bloomfield Hills MI
Dumbarton Oaks Research Library and Collection, Washington, D.C.
Museum of Fine Arts, Boston MA
New York Public Library, New York
Philadelphia Museum of Art, Philadelphia PA
Rhode Island School of Design, Museum of Art, Providence RI
Salvador Dali Museum, Inc. St Petersburg FL
Smithsonian Institution, Washington, D.C.
   National Museum of American History
   National Museum of the American Indian
   Smithsonian Institution Libraries
State Historical Society of North Dakota, Bismarck ND
United Limited Art Editions, Inc, West Islip NY
and the private collection of Prof Silvio A. Bedini

Bibliography

General works

Aveni, A., *Empires of Time: Calendars, Clocks and Cultures*, New York 1989

Brandon, S.G.F., *History, Time and Deity: A Historical and Comparative Study of the Conception of Time in Religious Thought and Practice*, Manchester 1965

The Encyclopedia of Time, ed. S.L. Macey, New York and London 1994

The Oxford Dictionary of World Religions, ed. J. Bowker, Oxford 1997

Time [exh. catalogue, Amsterdam, Nieuw Kerk], ed. A. Turner, Amsterdam 1990

Time. A Bibliographic Guide, ed. S.L. Macey, New York and London 1991

The Study of Time, eds. J.T. Fraser *et al.*, Berlin 1970 and ongoing (Proceedings of the Conferences of the International Society for the Study of Time)

Toulmin, S., and Goodfield, J., *The Discovery of Time*, London 1965

Whitrow, G.J., *The Natural Philosophy of Time*, London and Edinburgh 1961

Whitrow, G.J., *Time In History. Views of Time from Prehistory to the Present Day*, Oxford 1989

Whitrow, G.J., *What is Time?*, London 1972

THE CREATION OF TIME

d'Alverny, M.-Th., 'Le cosmos symbolique du XIIe siècle', *Archives d'histoire doctrinale et littéraire du moyen âge*, xxvii, 1953, pp. 31–81

Bertolini, G.L., 'La cosmografia teologica del Camposanto di Pisa', *Nuova antologia*, ser. V., xvl, 1910, pp. 720–25

Blunt, A., 'Blake's "Ancient of Days": The Symbolism of the Compasses', *Journal of the Warburg and Courtauld Institutes*, ii, 1938–39, pp. 53–63

Boman, T., *Hebrew Thought Compared with Greek*, English transl. by J.L. Moreau, London 1960

Bouché-LeClercq, A., *L'Astrologie grecque*, Paris 1899

Callahan, J.F., *Four Views of Time in Ancient Philosophy*, Cambridge MA 1948

Christiansen, K., Kanter, L.B., and Strehlke, C.B., *Painting in Renaissance Siena, 1420–1500* [exh. catalogue, New York, The Metropolitan Museum of Art, 20 December 1988–19 March 1989], esp. pp. 192–98 and 246–48

Dodds, E.R., *The Ancient Concept of Progress and Other Essays on Greek Literature and Belief*, Oxford 1973

Erdman, D.V., *The Illuminated Blake*, London 1975

von Erffa, H. M., *Ikonologie der Genesis. Die christlichen Bildthemen aus dem Alten Testament und ihre Quellen*, Munich 1989–1995

Fraser, J.T., *The Genesis and Evolution of Time*, Brighton 1982

Friedman, J.B., 'The Architect's Compass in Creation Miniatures of the later Middle Ages', *Traditio*, xxx, 1974, pp. 419–29

The Intellectual Adventure of Ancient Man. An Essay on speculative thought in the ancient Near East, eds. H. Frankfort *et al.*, Chicago 1946

Jaki, S.L., *Science and Creation. From Eternal Cycles to an Oscillating Universe*, revised and enlarged edn, Edinburgh 1986

Klein, P., 'Date et scriptorium de la Bible de Roda. Etat des recherches', *Les Cahiers de Saint-Michel de Cuxa*, III, 1972, pp. 91–102

de Laborde, A., *Étude sur la Bible Moralisée illustrée*, Paris 1911–27

Leeming, D.A., and Leeming, M.A., *Encyclopedia of Creation Myths*, Santa Barbara CA 1994

Lippincott, K., 'Giovanni di Paolo's "Creation of the World" and the Tradition of the '*thema mundi*' in Late Medieval and Renaissance Art', *The Burlington Magazine*, cxxxii, 1990, pp. 460–68

Norden, E., *Die Geburt des Kindes. Geschichte einer religiösen Idee*, Leipzig 1924, esp. pp. 16ff.

Onians, R.B., *The Origins of European Thought about the Body, the Mind, the Soul, the World, Time and Fate ...*, Cambridge 1951

Panofsky, E., and Saxl, F., *Dürers 'Melencolia I': Eine quellen- und typengeschichtliche Untersuchung*, Leipzig and Berlin 1923, pp. 67ff.

Sorabji, R., *Time, Creation and the Continuum. Theories in Antiquity and the Early Middle Ages*, London 1983

Tolley, M.J., 'Europe, "to those ychain'd in sleep"', in *Blake's Visionary Forms*, eds. D.V. Erdman and J.E. Grant, Princeton 1970, pp. 115–45

Tous les savoirs du monde. Encyclopédies et bibliothèques, de Sumer au XXIe siècle [exh. catalogue, Paris, Bibliothèque National de France, 20 December 1996–6 April 1997], Paris 1996, pp. 66–67, no. 15

Weinstock, S., 'A New Greek Calendar and Festivals of the Sun', *Journal of Roman Studies*, xxxviii, 1948, pp. 37–42

Zahlten, J., *'Creatio mundi': Darstellungen der sechs Schöpfungstage und naturwissenschaftliches Weltbild im Mittelalter*, Stuttgart 1979

THE MEASUREMENT OF TIME

(Divisions of Time)

Aujac, G., 'Le Zodiaque de l'astronomie grecque', *Revue d'Histoire des Sciences*, xxxiii, 1980, pp. 3–32

Aveni, A., 'Astronomy in the Americas', in *Astronomy before the Telescope*, ed. C. Walker, London 1996, pp. 269–303

Blunt, A., *The Paintings of Nicolas Poussin: A Critical Catalogue*, London 1966, pp. 123–24, no. 172

Boll, F., *Sphaera. Neue griechische Texte und Untersuchungen zur Geschichte der Sternbilder*, Leipzig 1903

Boyancé, P., 'L'Apollon Solaire', *Mélanges d'archéologie, d'épigraphie at d'histoire offerts à Jérôme Carcopino ...*, Paris 1966, pp. 149–70

Burkert, W., *Greek Religion. Archaic and Classical*, English transl. J. Raffan, London 1985, esp. pp. 143–49

Cieri-Via, C., 'I luoghi del mito fra decorazione e collezionismo', in *Immagini degli dei. Mitologia e collezionismo tra '500 e '600* [exh. catalogue, Lecce, Fondazione Memmo, 7 dicembre 1996–31 marzo 1997], Venice 1996, pp. 29–48

Clark, V.A., 'The Illustrated *Abridged Astrological Treatise of Albumasar*: Medieval Astrological Imagery in the West', PhD thesis, University of Michigan 1979

Ford, B.J., *Images of Science. A History of Scientific Illustration*, London 1992

Forster, L., 'Die *Emblemata Horatiana* des Otho Vaenius', *Wolfenbüttler Forschungen*, xii, 1981, pp.117–28

Fowler, J., *On the Medieval Representations of the Months and Seasons*, London 1873

Gagé, J., *Apollon romain. Essai sur le culte d'Apollon et le développement du 'ritus Græcus' à Rome des origines à Auguste*, Paris 1955

Geertz, A.W., 'Pueblo Cultural History', *Photographs at the Frontier. Aby Warburg in America, 1895–96*, eds. B. Cestelli Guidi and N. Mann, London 1998, pp. 9–19, esp. pp. 16–17

Gerards-Nelissen, I., 'Otto van Veen's *Emblemata Horatiana*, *Simiolus*, v, 1971, pp. 20–63

Gingerich, O., 'The Origin of the Zodiac', *Sky and Telescope*, March 1984, pp. 218–20

Grant, E., *Planets, Stars and Orbs: Medieval Cosmology, 1200–1687*, Cambridge 1994

Gundel, W., *Dekane und Dekansternbilder. Ein Beitrag zur Geschichte der Sternbilder der Kulturvölker*, Glückstadt and Hamburg 1936

Gundel, H., *Zodiakos. Der Tierkreis in der antiken Literatur und Kunst*, Stuttgart 1972

Günther, S., 'Peter und Philip Apian. Zwei desutsche Mathematiker und Kartographen. Ein Beitrag zur Gelerten-Geschichte des XVI. Jahrhunderts', *Abhandlungen der königlich böhmischen Gesellschaft der Wissenchaften*, ser. VI, xi, 1882, pp. 1–136

Hartner, W., 'The Earliest History of the Constellations in the Near East and the Motif of the Lion-Bull Combat', *Journal of Near Eastern Studies*, xxiv, 1965, pp. 1–16

Heintze, K. , *Mythes et symboles lunaires (Chine ancienne, civilisations anciennes de l'Asie, peuples limitrophes du Pacifique)*, Antwerp 1932

Holloway, J., *Patrons and Painters: Art in Scotland, 1650–1760* [exh. catalogue, Edinburgh, The National Gallery of Scotland, 17 July–8 October 1989], Edinburgh 1989, pp. 85–92

Hollstein's Dutch and Flemish Etchings, Engravings and Woodcuts, 1450–1700, [XLIV–XLVI, Maarten de Vos], ed. D. DeHoop Scheffer, Rotterdam 1995–96

Howse, D., 'Some Early Tidal Diagrams', *The Mariner's Mirror*, lxxix, no. 1, 1993, pp. 27–43

Ivanoff, N., *Bazzani. Saggio critico e catalogo delle opere* [exh. catalogue, Casa del Mantegna, Mantua, 14 maggio–15 ottobre 1950], Mantua 1950, esp. pp. 53–54, no. 62

King, H.C., and Millburn, J.R., *Geared to the Stars: The Evolution of Planetariums, Orreries and Astronomical Clocks*, Toronto 1979

Kristeller, P., *Kupferstich und Holzschnitt in vier Jahrhunderten*, Berlin 1920, pp. 302–03

Lippincott, K., 'Gli dei-decani del Salone dei Mesi di Palazzo Schifanoia', in *Alla corte degli Estensi. Filsofia, arte e cultura a Ferrara nei secoli XV e XVI* , ed. M. Bertozzi, Ferrara 1994, pp. 181–97

Lippmann, F., *Die Sieben Planeten*, Berlin 1895; also, *The Seven Planets*, English transl. by F. Simmonds, London 1895

Lunais, S., *Recherches sur la lune. Les Auteurs latins de la fin des Guerres Punique à la fin du règne des Antonins*, Leiden 1979

McDonald, R., and Harris, S., *Stoats and Weasels*, London (The Mammal Society) 1998

Maurmann-Bronder, B., 'Tempora significant. Zur Allegorie der vier Jahrzeiten', *Verbum et signum*, i, 1975, pp. 69–101

Neugebauer, O., 'The Egyptian Decans', in *Vistas in Astronomy*, ed. A. Beer, London and New York 1995, I, pp. 47–51

Neugebauer, O., *The Exact Sciences in Antiquity*, 2nd edn, Providence RI 1957

Neugebauer, O., and Parker, R.A., *Egyptian Astronomical Texts*, London and Providence RI, 1960–69

Panofsky, E., *Albrecht Dürer*, revised 2nd edn, Princeton 1945, II, p. 42 and fig. 215

Pingree, D., 'The Indian Iconography of the Decans and Hôras', *Journal of the Warburg and Courtauld Institutes*, xxvi, 1963, pp. 223–54

Pogo, A., 'Calendars on the coffin lids from Asyut', *Isis*, xvii, 1932, pp. 6–24

Rackham, B., *Victoria and Albert Museum. Catalogue of Italian Maiolica*, London 1940, p. 192, no. 573

Reiner, E., and Pingree, D., 'Babylonian Planetary Omens, Part 2: Enuma Anu Enlil, Tablets 50–51', in *Bibliotheca Mesopotamica*, ed. G. Buccelatti, Malibu CA 1981

Rice, D.S., 'The Seasons and the Labours of the Months in Islamic Art', *Ars Orientalis*, i, 1954, pp. 1–39

Rusk Shapley, F., *Paintings from the Samuel H. Kress Collection. Italian Schools, XVI–XVIII Century*, London 1973, p. 115

Ryan, W.F., 'John Russell, R.A., and Early Lunar Mapping', *The Smithsonian Journal of History*, i, 1966, pp. 27–48

Saxl, F., *Verzeichnis astrologischer und mythologischer illustrierter Handschriften des lateinischen Mittelalter in römischen Biblotheken*, Heidelberg 1915

Saxl, F., *Verzeichnis astrologischer und mythologischer illustrierter Handschriften des lateinischen Mittelalters, II. Die Handschriften der National-Bibliothek in Wien*, Heidelberg 1926

Simon, E., 'Poussin's Gemälde "Bacchus und Midas" in München', *Jahrbuch der Hamburger Kunstsammlungen*, xviii, 1973, pp. 109–118, esp. pp. 110–12

Smith, C.H., *Catalogue of the Greek and Etruscan Vases in the British Museum, III. Vases of the Finest Period*, London 1896, pp. 284–85

Smith, A.H., *A Catalogue of the Sculpture in the Department of Greek and Roman Antiquities. British Museum*, London 1904, III, pp. 231–32

Stevenson, E.L., *Terrestrial and Celestial Globes. Their History and Construction including a Consideration of Their Value as Aids in the Study of Geography and Astronomy*, New Haven 1921

The Yavanajâtaka of Sphudjidhvaja, ed., English transl. and commentary by D. Pingree, Cambridge MA 1978

Thiele, G., *Antike Himmelsbilder. Mit Forschungen zu Hipparchos, Aratos und seinem Fortsetzern, und Beiträgen zur Kunstgeschichte des Sternhimmels*, Berlin 1898

Tuve, R., *Seasons and Months: Studies in a Tradition of Middle English Poetry*, Paris 1933

Valerio, V., 'Historiographic and Numerical Notes on the Atlante Farnese and its Celestial Sphere', *Globusfreund*, xxxv–xxxvii, 1987, pp. 97–124

van der Waerden, B.L., 'Babylonian Astronomy, II. The Thirty-six stars', *Journal of Near-Eastern Studies*, viii, 1949, pp. 6–26

van der Waerden, B.L., 'History of the Zodiac', *Archiv für Orientforschung*, xvi, 1952–53, pp. 216–30

Voss, W., 'Eine Himmelskarte vom Jahre 1503 mit Wahrzeichnen des Wiener Poetenkollegiums als Vorlage Albrecht Dürers', *Jahrbuch der preussischen Kunstsammlungen*, lxiv, 1943, pp. 80–150

Warner, D.J., *The Sky Explored. Celestial Cartography 1500–1800*, New York 1979

Wolf, W., *Der Mond im deutschen Volksglauben*, Bühl 1929

(Calendars)

Ackermann, S., 'Medal and Memory : The Secrets Divulged by Perpetual Calendars', forthcoming

Baczko, B., 'La Révolution mesure son temps' in *La Révolution dans la mesure du temps. Calendrier Républicain heure décimale, 1793–1805* , La Chaux-de-Fonds 1989, pp. 9–29

Le Biccherne. Tavole dipinte delle Magistrature Senesi (secoli XIII–XVIII), eds. L. Borgia, E. Carli, M.A. Ceppari, U. Morandi, P. Sinibaldi and C. Zarrilli, Rome 1984, esp. p. 274

Borst, A., *The Ordering of Time. From the Ancient Computus to the Modern Computer*, English transl. by A. Winnard, Cambridge 1993

Burnaby, S.B., *Elements of the Jewish and Muhammadan Calendars, with Rules and Tables and Explanatory Notes on the Julian and Gregorian Calendars*, London 1901

'The Calendar', in *Explanatory Supplement to the Astronomical Ephemeris and the American Ephemeris and Nautical Almanac*, London (HMSO) 1961, ch. 14, pp. 407–442

Capp, B., *Astrology and the Popular Press. English Almanacs, 1500–1800*, London and Boston 1979

Cappelli, A., *Cronologia, cronografia e calendario perpetuo*, 5th edn, Milan 1983

Carrigan, R.A. Jr, 'Decimal Time', *American Scientist*, lxvi, no. 3, 1978, pp. 305–13

Chabas, F., *Le Calendrier des Jours Fastes et Néfastes de l'année égyptienne. Traduction complet du papyrus Sallier IV*, Chalon-sur-Seine and Paris 1863

Corpus Inscriptionum Latinarum. Inscriptiones Latinae Antiquissimae, ed. T. Mommsen, Berlin 1863, I, pp. 294–95, 303–09 and 358–59

Dessau, H., 'Zu den Milesischen Kalenderfragmenten', *Sitzungsberichte der königlich preussischen Akademie der Wissenschaften (Philosophisch-historischen Klasse)*, xxiii, 1904, pp. 266–68

Davis, J.B., 'Some Account of Runic Calendars and "Staffordshire Clogg" Almanacs', *Archaeologia*, xli, 1867, pp. 453–78

Dershowitz, N., and Reingold, E.M., *Calendrical Calculations*, Cambridge 1997

Dicks, D.R., *Early Greek Astronomy to Aristotle*, London 1970, esp. pp. 84–85
Diels, H., and Rehm, A., 'Parapegmenfragmente aus Milet', *Sitzungsberichte der königlich preussischen Akademie der Wissenschaften (Philosophisch-historischen Klasse)*, xxiii, 1904, pp. 92–111
The Gregorian Reform of the Calendar. Proceedings of the Vatican Conference to Commemorate its 400th Anniversary, 1582–1982, eds. G. V. Coyne, M.A. Hoskin and O. Pedersen, Vatican City 1983
Les Heures Révolutionaires (= *Horlogerie Ancienne*, nos. 25 and 26), eds. Dorez, Y., and Flores, J., Paris 1989
Heitz, P., and Haebler, K., *Hundert Kalender Inkunabeln*, Strassburg 1905
Hind, A., *Early Italian Engraving: A Critical Catalogue with Complete Reproduction of all the Prints Described*, London 1938, I, p. 83, and II, pl. 129
Houzeau, J.C., and Lancaster, A., *Bibliographie générale de l'astronomie*, Paris 1887, esp. pp. 1458–76
Jenkins, I.D., 'Newly Discovered Drawings from the Museo Cartaceo in the British Museum', in *Atti del Seminario Internazionale di Studi* (Napoli,18–19 dicembre 1987), ed. F. Solinas, Rome 1989, pp. 131–76, esp. p. 154, nos. 30–31
Lippmann, F., *Die Sieben Planeten*, Berlin 1895, pp. 6–7; also, *The Seven Planets*, English transl. by F. Simmonds, London 1895, pp. 6–7
Menniger, K.W., *Zahlwart und Ziffer. Ein Kulturgeschichte der Zahl*, Göttingen 1957
Michels, A. K., *The Calendar of the Roman Republic*, Princeton 1967
Neugebauer, O., *A History of Ancient Mathematical Astronomy*, Berlin and New York 1975
The Paper Museum of Cassiano del Pozzo [exh. catalogue, London, The British Museum, 14 May–30 August 1993], esp. pp. 132–33 (g.v. = Ginette Vagenheim)
Parthey, G., *Wenzel Hollar. Beschreibendes Verzeichniss seiner Kupferstiche*, Berlin 1853, p. 95, no. 483
Poole, R., '"Give Us Back Our Eleven Days!": Calendar Reform in Eighteenth Century England', *Past and Present*, cxlix, 1995, pp. 95–139
Rehm, A., 'Parapegmastudien. Mit einem Anhang: Euctemon und das Buch De Signis', *Abhandlung der bayerischen Akademie der Wissenschaften (Philosophisch-historische Abteilung)*, N.F. xix, 1941, pp. 1–145
Rehm, A., 'Das Parapegma des Euktemon', in *Griechische Kalender, III*, ed. F. Boll, Heidelberg 1913
Richards, E.G., *Mapping Time. The Calendar and its History*, London 1998
Salman, J., *Een Handdruck van de tijd. De almanak en het dagelijks leven in de Nederlanden, 1500–1700* [travelling exh. catalogue] , Zwolle 1997
Scullard, H.H., *Festivals and Ceremonies of the Roman Republic*, London 1981
The Secular Spirit: Life and Art at the End of the Middle Ages [exh. catalogue, New York, The Metropolitan Museum of Art, 25 March–3 June 1975] , New York 1975, pp. 108–09, no. 115
Wilson, G., 'The French Republican Calendar', *Antiquarian Horology*, xvii, 1988, pp. 249–58
Zerubavel, E., *The Seven Day Circle. The History and Meaning of the Week*, Chicago and London 1985

(Scientific Instruments)
Ackermann, S., and Cherry, J., ' Richard II, John Holland and Three Medieval Quadrants', *Annals of Science*, lvi, 1999, pp. 3–23
Arnaldi, M., and Schaldach, K., 'A Roman Cylinder Dial: Witness to a Foreign Tradition', *The Journal of the History of Astronomy*, xxviii, 1997, pp. 107–17
Augarde, J.-D., 'La fabrication des instruments scientifique au XVIIIe siècle et la corporation des fondeurs', *Studies in the History of Scientific Instruments*, eds. C. Blondel *et. al.*, London and Paris 1989, pp. 52–72, esp. p. 68
Bedini, S., 'The 17th century Table Clepsydra', *Physis. Rivista internazionale di storia della scienza*, x,1968, pp. 25–52
Bennett, J.A., *The Divided Circle. A History of Instruments for Astronomy, Navigation and Surveying*, Oxford 1987, pp. 77–79
Bion, N., *The Construction and Principal Uses of Mathematical Instruments*, English transl. by E. Stone, London 1758 (repr. Mendham nj 1995)
Brusa, G., 'Le navicelle orarie di Venezia', *Annali dell'Istituto e Museo di Storia della Scienza di Firenze*, v.i, 1980, p. 55
van Damme, J., 'Michael Piquer, planisferisch astrolabium, in *Sterren in Beelden. Astrologie in de eeuw van Mercator* [exh. catalogue, Sint-Niklaas, Stedelijk Museum, 6 maart–6 juni 1994], Sint Niklaas 1994, pp. 158–62, no. 28
Dekker, E., 'An Unrecorded Medieval Astrolabe Quadrant from c.1300', *Annals of Science*, lii, 1995, pp. 1–49
Dekker, E., 'Epact Tables on Instruments. Their Definition and Use', *Annals of Science*, l, 1993, pp. 303–24
Dekker, E., *Globes at Greenwich*, London 1999, forthcoming
Dekker, E., and Lippincott, K., 'The Scientific Instruments in Holbein's *Ambassadors*', *Journal of the Warburg and Courtauld Institutes*, forthcoming
Diels, H., *Antike Technik*, 2nd edn, Leipzig and Berlin 1920, esp. pp. 154–80
Dohrn–van Rossum, G., *Die Geschichte der Stunde: Uhren und moderne Zeitordnung*, Munich 1992; also, *History of the Hour, Clocks and Modern Temporal Orders*, English transl. by T. Dunlap, Chicago and London 1996
Eckhardt, W., 'Erasmus und Joshua Habermel. Kunstgeschichtliche Anmerkungen zu den Werken der beiden Instrumentenmacher', *Jahrbuch der Hamburger Kunstsammlungen*, xxii, 1977, pp. 13–74
Eckhardt, W., 'Erasmus Habermel: Zur Biographie des Instrumentenmachers Kaiser Rudolfs II', *Jahrbuch der Hamburger Kunstsammlungen*, xxi, 1976, pp. 55–92
Field, J.V., 'European Gearing in the First Millennium: the Archaeological Record', in *Astronomy before the Telescope*, ed. C. Walker, London 1996, pp. 110–22, esp. pp. 113–20
Field, J.V., 'Some Roman and Byzantine Portable Sundials and the London Sundial-Calendar', *History of Technology*, xii, 1990, pp. 103–35
Field, J.V., and Wright, M.T., *Early Gearing: Geared Mechanisms in the Ancient and Medieval World*, London 1985
Field, J.V., and Wright, M.T., 'Gears from the Byzantines: A Portable Sundial with Mechanical Gearing', *Annals of Science*, xlii, no. 2, 1985, pp. 87–138
Focus Behaim-Globus [exh. catalogue, Nuremberg, Germanisches Nationalmuseum, 2 December 1992–20 February 1993], Nuremberg 1992, II, pp. 605–06 (G. Bott)
Folkerts, M., 'Mittelalterliche mathematische Handschriften in westlichen Sprachen in der Berliner Staatsbibliothek. Ein vorläufiges Verzeichnis', *Mathematical Perspectives*, lxvi, 1983, pp. 53–93
Gibbs, S.L., *Greek and Roman Sundials*, New Haven and London 1976
Gingerich, O., 'Zoomorphic Astrolabes and the Introduction of Arabic Star Names into Europe', in *Deferent to the Equant. A Volume of Studies in the History of Science in the Ancient and Medieval Near East in Honor of E.S. Kennedy* [*Annals of the New York Academy of Sciences, vol. 500*], eds. D.A. King and G. Saliba, New York, 1987, pp. 89–104
Gouk, P., *The Ivory Sundials of Nuremberg 1500–1700*, Cambridge 1988, pp. 21–22, 36–45 and 56–58
Gunther, R.T., *Early Science in Oxford*, Oxford 1923, II, pp. 165–69
Gunther, R.T., 'The Great Astrolabe and Other Scientific Instruments of Humphrey Cole', *Archaeologia*, lxxvi, 1927, pp. 273–317, esp. pp. 289–91 and figs. 34–35
Gunther, R.T., *The Astrolabes of the World ...*, Oxford 1932
Higgins, K., 'An Elizabethan Quadrant-Dial in Silver by Humphrey Cole', *Connoisseur*, cxxv, 1950, pp. 118–19
Higton, H., *The Sundials at the National Maritime Museum, London*, forthcoming
Horsky, Z., and Skapova, O., *Astronomy Gnomonics*, Prague 1968, esp. pp. 125–26
Mechanik aus der Wunderkammer, eds. P. Friess and E. Langenstein, Munich 1996, pp. 30–31
Humphrey Cole: Mint, Measurement and Maps in Elizabethan England [British Museum Occasional Papers, no. 126], ed. Silke Ackermann, London 1998
Instrumentos cientificos del siglo XVI. La corte española y la escuela de Lovaina [exh. catalogue, Madrid, Fundación Carlos de Amberes, 26 November 1997–1 February 1998], Madrid 1997, esp. p. 173
Johnston, S., 'The Carpenter's Rule: Instruments, Practitioners and Artisans in 16th-century England', *Proceedings of the Eleventh International Scientific Instrument Symposium*, eds. G. Dragoni, A. McConnell and G.L'E. Turner, Bologna, 1994, pp. 39–45
Jones, A., 'Later Greek and Byzantine Astronomy', in *Astronomy before the Telescope*, ed. C. Walker, London 1997, pp. 98–109
Kirchner, J., *Beschreibende Verzeichnisse der Miniaturen-Handschriften der preussischen Staatsbibliothek zu Berlin*, Leipzig 1926, I, pp. 21–22
Lippincott, K., 'When was Michelangelo born?', *Journal of the Warburg and Courtauld Institutes*, lii, 1989, pp. 228–32
Lippincott, K., 'The Navicula Sundial', *Bulletin of the Scientific Instruments Society*, xxxv, 1992, p. 22
Michel, H., 'The Drum Clepsydra', *Journal Suisse d'Horologerie et de Bijouterie*, 1949, pp. 132–34
Michel, H., 'Astronomical Jades', *Oriental Art*, ii, 1956, pp. 156ff.
Michel, H, 'Les jades astronomiques chinois', *Communications de l'Académie de la Marine*, iv, 1949, pp. 111ff.
Mills, A.A., 'Note: Altitude Sundials for Seasonal and Equal Hours', *Annals of Science*, liii, 1996, pp. 75–84
Noble, J.V., and de Solla Price, D.J., 'The Waterclock in the Tower of the Winds', *American Journal of Archaeology*, lxxii, 1968, pp. 345–55, esp. 351–33
North, J., *The Fontana History of Astronomy and Cosmology*, London 1994, pp. 253–59
La Révolution française et L'Europe, 1789–1799 [exh. catalogue, Paris, Grand Palais], Paris 1989, p. 792
Rohr, R.R.J., *Sundials. History, Theory and Practice*, English transl. by G. Godin, Toronto 1970
Ronan, C., 'Astronomy in China, Korea and Japan', in *Astronomy before the Telescope*, ed. C. Walker, London 1997, pp. 245–68
Rose, V., *Verzeichniss der lateinischen Handschriften der königlichen Bibliothek zu Berlin*, Berlin 1893, I, pp. 308–15
Schätze der Alhambra. Islamische Kunst aus Andalusien [exh. catalogue, Berlin, Kulturforum, 29 October 1995–3 March 1996], Berlin 1996, p. 131, no. 28
Smith, A.H., *A Catalogue of the Sculpture in the Department of Greek and Roman Antiquities. British Museum*, London 1904, III, p. 413
de Solla Price, D.J., 'The Little Ship of Venice – a Middle English Instrument Tract', *Journal for the History of Medicine and Allied Sciences*, xv, 1960, pp. 399-407
de Solla Price, D.J., 'Gears from the Greeks: the Antikythera mechanism – a calendar-computer from c. 80 BC', *Transactions of the American Philosophical Society*, N.S. lxiv, 1975, Part 7, pp. 5–70
Syndram, D., *Wissenschaftliche Instrumente und Sonnenuhren*, Munich 1989
Tesseract. Early Scientific Instruments [sale catalogue, Tesseract, Hastings-on-Hudson ny], LIV, Autumn 1996, no. 17
Turner, A.J., 'The Accomplishment of Many Years: Three Notes towards a History of the Sand-glass', *Annals of Science*, xxxix, 1982, pp. 161–72
Turner, A.J., *The Time Museum, I. Time Measuring Instruments. Part I: Astrolabes; Astrolabe-Related Instruments*, Rockford il 1985
Turner, A.J., *The Time Museum, I. Time Measuring Instruments. Part 3: Water-clocks,, Sand-glasses and Fire-clocks*, Rockford il 1984
Turner, G.L'E., *Antique Scientific Instruments*, Poole 1980
Turner, G.L'E., *Scientific Instruments 1500–1900: An Introduction*, London 1998
Vaughan, D., 'A Very Artificial Workman: The Altitude Sundials of Humphrey Cole', in *Making Instruments Count. Essays on Historical Scientific Instruments presented to Gerard L'Estrange Turner*, eds. R.G.W. Anderson, J.A. Bennett and W.F. Ryan, Aldershot 1993, pp. 191–200
Ward, F.A.B., *A Catalogue of the European Scientific Instruments in the Department of Medieval and Later Antiquities of the British Museum*, London 1981
Zinner, E., *Deutsche und niederländische astronomische Instrumente des 11.–18. Jahrhunderts*, Munich 1956

(Mechanical Time)
Baillie, G.H., *Watches. Their History, Decoration and Mechanism*, London 1929 (reprinted London 1979)
Baillie, G.H., Alan Lloyd, H., and Ward, F.A.B., *The Planetarium of Giovanni de Dondi*, London (Antiquarian Horological Society) 1974
Bedini, S.A., *The Pulse of Time. Galileo Galilei, the Determination of Longitude and the Pendulum Clock* [Biblioteca di Nuncius, Studi e Testi 3], Florence 1991
Benndorf, O., Weiss, E., and Rehm, A., 'Zur Salzburger Bronzescheibe mit Sternbilder' *Jahreshefte der Österreichischen Archäologischen Instituts in Wien*, vi, 1903, pp. 32–49
Breguet, E., *Breguet: Watchmakers since 1775*, Paris 1997
Britten's Old Clocks and Watches and their Makers, 9th edn, ed. C. Clutton, London 1982
Brusa, G. , 'Early Mechanical Horology in Italy', *Antiquarian Horology*, xviii, no. 5, 1990, pp. 485–513
Cardinal, C., *La montre des origines aux XIXe siècle*, Fribourg 1985; also, *The Watch from its Origin to the XIXth Century*, English transl., Secaucus nj 1989
Clutton, C., and Daniels, G., *Watches*, 3rd edn, London, 1979
Collection de montres et automates Maurice et Edouard M. Sandoz, Le Locle n.d.
Coole, P.G., and Neumann, E., *The Orpheus Clocks*, London 1972
Daniels, G., *The Art of Breguet*, London 1975
Dawson, P.G., Drover, C.B., and Parkes, D.W., *Early English Clocks*, Woodbridge 1982
Dickinson, H.W., *Robert Fulton, Engineer and Artist: His Life and Works*, London 1913
Drachmann, A.G., 'The Plane Astrolabe and the Anaphoric Clock', *Centaurus*, iii, 1954, pp. 183–89
Drummond Robertson, J., *The Evolution of Clockwork*, London 1931
Edwards, E., *The Stars of the Pendulum Clock*, Altringham 1977
Haward, J.F., 'The Restoration of the Tudor Clock Salt', *Goldsmiths' Review*, 1972–73
Heger, N., 'Das Kunstwerk des Monats', *Die Monatsblätter des Salzburger Museums*, July 1990
Howse, D., *Greenwich Time and the Longitude*, 2nd edn, London 1997
Hutcheon, W. Jr, *Robert Fulton: Pioneer of Undersea Warfare*, Annapolis md 1981
Jagger, C., *Paul Philip Barraud*, Ticehurst (Antiquarian Horological Society) 1968
Jagger, C., *Royal Clocks. The British Monarchy*

and its Timekeepers, 1300–1900, London 1983
Jaquet, E., and Chapuis, A., *Technique and History of the Swiss Watch*, Boston 1953
Landes, D., *Revolution in Time: Clocks and the Making of the Modern World*, London 1983
Leopold, J.H., *The Almanus Manuscript*, London 1971
Leopold, J.H., *Astronomen, Sterne, Geräte – Landgraf Wilhelm IV. und seine sich selbst bewegenden Globen*, Lucerne 1986
Leopold, J.H., *Die Grosse Astronomische Tischuhr des Johan Reingold, Augsburg, 1581 bis 1592*, Fremersdorf, Lucerne, 1974
Maass, E., 'Salzburger Bronzetafel mit Sternbildern' *Jahreshefte der Österreichischen Archäologischen Instituts in Wien*, V, 1902, pp. 196–97
Maurice, K., *Die Deutsche Räderuhr. Zur Kunst und Technik der mechanischen Zeitmessers im deutschen Sprachraum*, Munich 1976
Maurice, K., and Mayr, O., *Die Welt als Uhr: deutschen Uhren und Automaten, 1550–1650*, Munich 1980
Mody, N.H.N., *Japanese Clocks*, London n.d. [1968]
Mudge, T. Jr, *A Description with Plates ...*, London 1799
Neugebauer, O., 'The Early History of the Astrolabe', *Isis*, XL, 1949, pp. 240–56
Noble, J.V., and de Solla Price, D.J., 'The Waterclock in the Tower of the Winds', *American Journal of Archaeology*, LXXIII, 1968, pp. 345–55, esp. 351–33
Okken, L., 'Wurde die Räderuhr für das abendländische Kloster erdacht?', *Rheinisch-Westfälische Zeitschrift für Volkskunde*, XXXII–III, 1987–88, pp. 117–138
Robertson, J.D., *The Evolution of Clockwork*, London 1931 (reprint Wakefield 1972)
Symonds, R.W., *Thomas Tompion: His Life and Work*, London 1951
Tate, H., *Clocks and Watches*, London 1983
Toesca, I., 'The Royal Clock Salt', *Apollo*, October 1969, pp. 292–97
White, G., *English Lantern Clocks*, Woodbridge 1989
Winter, F.H., *The First Golden Age of Rocketry: Congreve and Hale Rockets of the Nineteenth Century*, Washington, D.C., 1990
Yoder, J.G., *Unrolling Time, Christiaan Huygens and the Mathematization of Nature*, Cambridge 1988

THE DEPICTION OF TIME:
(The Art of Time)

Ades, D., *Dalí*, London 1982
Arndt, K., 'Chronos als Feind der Kunst', *Nederlands Kunsthistorisch Jaarboek*, XXIII, 1972, pp. 329–42
The Autobiography of Thomas Whythorne, ed. J.M. Osborn, Oxford 1961
van Bastelaer, R., *Les Estampes de Peter Bruegel l'Ancien*, Brussels 1908, p. 63, no. 204
Bellioli, L., 'The Evolution of Picasso's Portrait of Gertrude Stein', *The Burlington Magazine*, CXLI, 1999, pp. 12–18
Bellori, G.P., *Le Vite de' Pittori, Scultori et Architettori Moderni*, Rome 1672, pp. 447–8
Beresford, R., *A Dance to the Music of Time by Nicolas Poussin*, London 1995
Bergson, H., *An Introduction to Metaphysics*, English transl. by T.E. Hulme, London 1913
Bilder vom alten Menschen in der niederländischen und deutschen Kunst, 1550–1750 [exh. catalogue, Braunschweig, Herzog Anton Ulrich-Museum, 14. Dezember bis 20. Juni 1994] Brunswick 1994, p. 63
Bivanck, W.G.C., 'Une Impression (Claude Monet)', in *Un Hollandais à Paris en 1891*, Paris 1892
Blunt, A., 'Poussin's "Dance to the Music of Time"', *The Burlington Magazine*, CXVIII, 1976, pp. 844–48
Boccioni, U., 'Manifesto of Futurist Sculpture' (1912) in U. Apollonio, *Futurist Manifestos*, London 1973
Bode, W., 'Lo scultore Bartolomeo Bellano', *Archivio storico dell'arte*, IV, 1891, pp. 397–416, esp. pp. 412 and 415
Bode, W., *Die italienischen Bronzestatuetten der Renaissance*, Berlin 1936, p. 25
Bresciani Alvarez, G., 'Bassorilievo in marmo', in *Alvise Cornaro e il suo tempo* [exh. catalogue, ed. L. Puppi], Venice 1980
Breton, A., *Manifeste du Surréalisme* (1924), in *Manifestoes of Surrealism*, ed. and English transl. by R. Seaver and H.R. Lane, Ann Arbor MI 1969
Breton, A., 'Max Ernst' (1921), in M. Ernst, *Beyond Painting*, New York 1948
Brockhaus, H., 'Ein edles Geduldspiel: "Die Leitung der Welt oder der Himmelsleiter", die sogennanten Taroks Mantegnas vom Jahre 1459–60', in *Miscellanea di storia dell'arte in onore di I.B. Supino*, Florence 1933, pp. 398–400
Brown, C., *Van Dyck*, Oxford 1982
Browne, T., *Urne Buriall* (1658), ed. J. Carter, Cambridge 1967
Burke, J., and Caldwell, C., *Hogarth. The Complete Etchings*, London 1968, no. 267
Campbell, L., *Renaissance Portraits*, New Haven and London 1990
Campbell, L.A., *Mithraic Iconography and Ideology*, London 1968
Clark, A.M., *Pompeo Batoni. A Complete Catalogue of his Works with an Introductory Text*, Oxford 1985, p. 239, cat. 108
Clark, A.M., 'Some Early Subject Pictures by P.G. Batoni', *The Burlington Magazine*, CI, 1959, pp. 232–38
Cohen, C.E., *The Drawings of Giovanni Antonio da Pordenone*, Florence 1980, pp. 64–65 and 84–85, figs. 78–79 and 82
Cook, A.B., *Zeus. A Study in Ancient Religion*, Cambridge 1914–40, II (1925), Part 2, App. A, pp. 859–68
Correspondence de Courbet, ed. P. Ten-Doesschate Chu, Paris 1996
Cumont, F., *Les Mystères de Mithra*, 3rd edn, Brussels 1913
Dalí, S., *The Secret Life of Salvador Dalí*, London 1968
Dieckmann, L., *Hieroglyphics. The History of a Literary Symbol*, St Louis 1970
Disegni veneti di collezioni inglesi [exh. catalogue, Venice, Fondazione Giorgio Cini], ed. J. Stock, Venice 1980, p. 59, cat. 79
Duchamp, M., 'A l'infinitif' (1914), in *The Writings of Marcel Duchamp*, ed. and English transl. by M. Sanouillet and E. Petersen, Oxford 1963
prince d'Essling and Müntz, E., *Pétrarque, ses études d'art, son influence sur les artistes ...*, Paris 1902
prince d'Essling, 'Etudes sur les Triomphes de Pétrarque', *Gazette des Beaux-Arts*, 2e pér., XXXV, 1887, pp. 331, and XXXVI, p. 25
Etudes Mithraiques. IV [Actes de 2e Congrès International. Téhéran, du 1er au 8 septembre 1975], Leiden 1978 (J. Hansman, ' A Suggested Interpretation of the Mithraic Lion-Man Figure', pp. 215–28, and H. Von Gall, 'The Lion-headed and the Human-headed God in Mithraic Mysteries', pp. 511–26)
Fabiny, T., 'Veritas filia Temporis. The Iconography of Time and Truth in Shakespeare', *Shakespeare and the Emblem. Studies in Renaissance Iconography and Iconology*, Szeged 1984, pp. 215–71
Félibien, A., *Entretiens sur les vies et sur les ouvrages des plus excellens peintres, anciens et modernes...*, Paris 1666–88, V (1688), pp. 377–78
Fink, A., *Die Schwarzischen Trachtenbücher*, Berlin 1963
Francken, D., *L'oeuvre gravé des van de Passe*, Amsterdam and Paris 1881
Freedberg, D., *The Prints of Pieter Bruegel the Elder* [travelling exh. catalogue], Tokyo 1989, pp. 186–87
Fremantle, K., *The Baroque Town Hall of Amsterdam*, Utrecht 1959
Fremantle, K., 'Themes from Ripa and Rubens in the Royal Palace of Amsterdam', *The Burlington Magazine*, CIII, 1961, pp. 258–64
Gabriels, J., *Artus Quellien, de Oude: 'Kunstryck Belthouwer'*, Antwerp 1930
La Galleria Estense di Modena. Guida illustrata, ed. J. Bentini, Bologna 1987
Garlick, K., *Sir Thomas Lawrence. A Complete Catalogue of the Oil Paintings*, Oxford 1989
van Gelder, J.G., and Jost, I., *Jan de Bisschop and his Icones & Paradigmata. Classical Antiquities and Italian Drawings for Artistic Instruction in Seventeenth-century Holland*, ed. K. Andrews, Doornspijk 1985
Gemar-Koeltzsch, E., *Holländische Stillebenmaler im 17. Jahrhundert*, Lingen 1995
The Genius of Venice, 1500–1600 [exh. catalogue, London, the Royal Academy, 25 November 1983–11 March 1984], eds. J. Martineau and C. Hope, London 1983, pp. 198 and 335–336 (DL = David Landau)
The Glory of Venice. Art in the Eighteenth Century [exh. catalogue, London, The Royal Academy, 15 September–14 December 1994, and Washington, D.C., The National Galley of Art, 29 January–23 April 1995], edd. J. Martineau and A. Robison, New Haven and London 1994, pp. 210–11 and 502, no. 126
Goldschmidt, F., *Die italienischen Bronzen der Renaissance und das Barock...*, Berlin 1914–22, I (1914), pp. 10–11 , pl. 38
Gurland, P., *Robert Braithwaite Martineau's "The Last Day in the Old Home"*, unpublished M.A. report, Courtauld Institute of Art, 1986
Guthrie, W.K.C., *Orpheus and the Greek Religion. A Study of the Orphic Movement*, London 1935
Haak, B., *The Golden Age: Dutch Painters of the Seventeenth Century*, English transl. by E. Willems-Treeman, London 1984
von Hadeln, D., 'Some Little-Known Works by Titian.', *The Burlinton Magazine*, XLV, 1924, pp. 179–80
Hare, A., *The Story of My Life*, London 1896–1900
Harrington, J., *Aritosto's Orlando Furioso in English Heroical Verse*, London 1591
Hind, A.M., *Early Italian Engraving. A Critical Catalogue with Complete Reproduction of all the Prints Described.....*, London 1938, I, pp. 32–36 (A.I.22) and pp. 131–34 (B.II.5)
Hind, A.M., *Catalogue of the Early Italian Engravings ... in the British Museum*, London 1910
Hornbostel, W., *Sarapis. Studien zur Überlieferungs-geschichte, den Erscheinungsformen und Wandlungen der Gestalt eines Gottes*, Leiden 1973, esp. pp. 90–95
House, J., *Monet: Nature into Art*, New Haven and London 1986
Howard, H., *A Course of Lectures on Painting Delivered at the Royal Academy of Fine Arts*, London 1848
Iversen, E., *The Myth of Egypt and its Hieroglyphs*, Copenhagen 1961
Jasper Johns, ed. J.M. Faerna, Barcelona 1995
de Jongh, E., van Leewen, T., Gasten, A. and Sayles, H., *Dutch Still Life Painting in the Age of Rembrandt*, Auckland 1982
Katalog der Staatsgalerie Stuttgart, Stuttgart 1957, p. 19 and fig. 21
Kiefer, F., 'The Conflation of Fortuna and Occasio in Renaissance Thought and Iconography', *The Journal of Renaissance Studies*, IX, 1979, pp. 1–27
Kirby Talley, M., '"All Good Pictures Crack". Sir Joshua Reynolds's Practice and Studio', in *Reynolds*, ed. N. Penny, London 1986, pp. 55–70
Kirby Talley, M., *Portrait Panting in England: Studies in the technical Literature before 1700*, London 1981
Klotz, G., 'Hogarth und Addison: "Time Smoking a Picture" und "The Spectator"', *Zeitschrift für Kunstgeschichte*, XXII, 1959, pp. 102–07
Knox, G., *Giambattista and Domenico Tiepolo. A Study and Catalogue Raisonné of the Chalk Drawings*, Oxford 1980
Kolakowski, L., *Bergson*, Oxford 1985
Lambert, J.J., *Records of the Skinners of London*, London 1923
Langedijk, K., *The Portraits of the Medici, 15th–18th Centuries*, Florence 1983
Lectures on Painting by the Royal Academicians: Barry, Opie and Fuseli, ed. R.N. Wornum, London 1889
Lessing, G.E., *Laocoon* (1766), English transl., London 1914
Lettres de Paul Gauguin à Daniel de Monfried, Paris 1930
Levenson, J.A., Oberhuber, K. and Sheehan, J.L., *Early Italian Engravings from the National Gallery*, Washington, D.C.,1973
Levey, M., *The National Gallery Catalogues. The Seventeenth and Eighteenth Century Italian Schools*, London 1971, pp. 9–10
Levey, M., *Giambattista Tiepolo. His Life and Art*, New Haven and London 1986
Levy, D., 'Aion', *Hesperia*, XII, 1944, pp. 269ff.
Levy, D., 'Il attraverso la letteratura greca', *Rendiconti della R. Acc. Naz. Dei Lincei, Classe di Scienze Morali*, ser. V, XXXII, 1923, pp. 260ff.
Lippincott, K., 'Mantegna's Tarocchi', *Print Quarterly*, III, 1986, pp. 357–60
Longford, E., *Wellington: Pillar of State*, London 1972
Lübbeke, I., *The Thyssen-Bornemisza Collection. Early German Painting, 1350–1550*, English transl. by M. Thomas Will, London 1991
Mann, T., *Buddenbrooks*, English transl. by H.T. Lowe-Porter, London 1996
Marinetti, F.T., 'Founding and Manifesto of Futurism' (1909), in U. Apollonio, *Futurist Manifestos*, London 1973
Marz, K., *Das Kapital*, English transl. by E. and C. Paul, London 1942
Matisse, H., 'Notes of a Painter' (1908), in J. Flam, *Matisse on Art*, Oxford 1978
Meiss, M., *Painting in Florence and Siena after the Black Death*, Princeton 1951
Merkelbach, R., *Mithras*, Königstein/Ts 1984
Michel, H., 'L'Horloge de Sapence et l'histoire de horlogerie', *Physis. Rivista di Storia delle Scienze*, II, 1960, pp. 291ff.
Miller, H., *First Impression of England and its People*, 11th edn, Edinburgh 1870
Mitchell, W.J.T., *Iconology: Image, Text, Ideology*, Chicago 1986
Morassi, A., *Giambattista Tiepolo: A Complete Catalogue of the Paintings*, London 1962, fig. 324
van Os, H., 'The Black Death and Sienese Painting: A Problem of Interpretation', *Art History*, IV, no. 3, 1981, pp. 237–49
Nicolas Poussin. 1594–1665 [exh. catalogue, Paris, Grand Palais, 24 September 1994–2 January 1995, and London, the Royal Academy, 19 January–9 April 1995], Paris 1994, pp. 195–96 and 278–9
Nilsson, M., 'The Syncretist relief at Modena', *Symbolae Osloensis*, XXIV, 1945, pp. 1ff.
Panofsky, E., 'Father Time', *Studies in Iconology. Humanist Themes in the Art of the Renaissance*, Oxford 1939
Panofsky, E., *Hercules am Scheidewege und andere antike Bildstoffe in der neueren Kunst*, 1930, esp. pp. 12ff.
Panofsky, E., 'Titian's *Allegory of Prudence*: A Postscript', in *Meaning in the Visual Arts: Papers in and on Art History*, New York 1955
Panofsky, E., *Tomb Sculpture. Four Lectures on its Changing Aspects from Ancient Egypt to Bernini*, London 1992
Panofsky, E., and Saxl, F., 'A Late Antique Religious Symbol in Works by Holbein and Titian', *The Burlington Magazine*, XLIX, 1926, pp. 177–81
Parkinson, R., *John Constable, The Man and his Art*, London 1998
Paulson, R., *Hogarth's Graphic Works*, revised edn, London 1970
Paulson, R., *Hogarth (III. Art and Politics, 1750–64)*, Cambridge 1993
Peltzer, R.A., 'Der Hofmaler Hans von Aachen. Seine Schule und seine Zeit', *Jahrbuch der Kunsthistorischen Sammlungen in Wien*, XXX, 1912, pp. 59–82, p. 132
Penny, N., *Catalogue of the European Sculpture in the Ashmolean Museum, 1540 to the Present Day, II. French and other European Sculpture (excluding Italian and British)*, Oxford 1992, p. 80, no. 314
de Piles, R., *Cours de Peinture par Principes*, Paris 1708
Pope-Hennessy, J., *Catalogue of Italian Sculpture in the Victoria and Albert*

Museum, London 1964, I, pp.144–45
Proust, A., 'Edouard Manet (Souvenirs)', *Revue Blanche*, 1 March 1897, pp. 202–03
Puppi, L. 'Tiziano tra Padova e Vicenza', in *Tiziano e Venezia* [Atti del Convegno], Vicenza 1980
Rosenberg, P., and Prat, L-A., *Nicolas Poussin. 1594–1665. Catalogue raisonné des dessins*, Milan 1994, I, pp. 278–79, no. 144
Saxl, F., 'Veritas filia Temporis', *Philosophy and History. Essays presented to Ernst Cassirer*, eds. R. Klibansky and H.J. Paton, Oxford 1936, pp. 197–222
Scheurl, C., *Oratio doctoris Scheurli attingens litteratu presentantiam necnon laudem Ecclesie Collegiate Vittenburgensis*, Leipzig 1509
Schivelbusch, W., *The Railway Journey: Trains and Travel in the 19th Century*, Oxford 1980
Schwartz, A., *The Complete Works of Marcel Duchamp*, London 1997
Seznec, J., *La Survivance des dieux antiques*, London 1940; also, English transl., *The Survival of the Pagan Gods. The Mythological Tradition and Its Place in Renaissance Humanism and Art*, Princeton 1953
Simon, E., 'Sol, Virtus und Veritas im Würzburger Treppenhausfresko des Giovanni Battista Tiepolo', *Pantheon*, XXIX, 1971, pp. 483–96
Simon, R., *The Portrait in Britain and America*, Oxford 1987
Spencer, E.P., 'L'Horloge de Sapience',*Scriptorium*, XVII, 1963, pp. 277ff.
Strong, R.C., *The English Icon*, London and New York 1969
Taylor, L., *Mourning Dress: A Costume and Social History*, London 1983
Titian. Prince of Painters [exh. catalogue, Venice, Palazzo Ducale, 2 June–7 October 1990, and Washington, D.C., National Gallery of Art, 28 October 1990–27 January 1991], pp. 347–48 (LP = Lionello Puppi)
The Triumphs of Petrarch, English transl. by E.H. Wilkins, Chicago 1962
Veldman, I., 'Images of Labour and Diligence in Sixteenth-century Netherlandish Prints: the Work Ethic rooted in Civic Morality or Protestantism?', *Simiolus*, XXI, 1993, pp. 227–64
Venturi, A., *La R. Galleria Estense in Modena*, Modena 1883
Vermassen, M.J., *Corpus Inscriptionum et Monumentorum Religionis Mithriacae*, The Hague 1960
Volkmann, L., *Bilderschriften der Renaissance. Hieroglyphik und Emblematik in ihren Beziehungen und Fortwirkungen*, Leipzig 1923
Walters, H.B., *Catalogue of the Bronzes, Greek, Roman and Etruscan, in the Department of Greek and Roman Antiquities in the British Museum*, London 1899, p. 174, no. 949
Westhoff-Krummacher, H., *Barthel Bruyn der Ältere als Bildnismaler*, Berlin 1965
Wheelock, A.K., Barnes, S.J., and Held, J.S, *Anthony van Dyck* [exh. catalogue, Washington, D.C., National Gallery of Art, 11 November 1990–24 February 1991], Washington, D.C.1990, pp. 231–32, no. 57
Williams, R., *The Country and the City*, London 1973
Wildenstein, D., *Monet: Biographie et catalogue raisonné*, Lausanne and Paris 1979
Wind, E., 'The Trinity of Serapis', in *Pagan Mysteries in the Renaissance. An Exploration of Philosophical and Mystical Soures of Iconography in Renaissance Art*, London 1958
Wittkower, R, 'Chance, Time and Virtue', *Journal of the Warburg and Courtauld Institutes*, I, 1937, pp. 312–31

THE EXPERIENCE OF TIME
(Human Time)
Allan, N., *Ever the Twain Shall Meet: the Interaction of Medical Science, East and West*, London 1993
Arnold, K., Hurwitz, B., McKee, F., and Richardson, R., *Doctor Death: Medicine at the End of Life*, London 1997
Boll, F., 'Die Lebensalter', *Neue Jahrbücher für das klassische Altertum*, XVI (XXXI), 1913, pp. 89ff.
Bylebyl, J.J., 'Galen on the non-Natural causes in the Variation of the Pulse', *Bulletin of the History of Medicine*, XLV, 1971, pp. 482–85
Companion Encyclopaedia of the History of Medicine, eds. W.F. Bynum and R. Porter, London and New York 1993
De Fourcaud, L., 'Les Arts décoratifs au Salon 2ème', *Revue des Arts Dècoratifs*, I, 1893, pp. 377–86, esp. p. 385
Dove, M., *The Perfect Age of Man's Life*, Cambridge 1986, pp. 30–32
Deutsche Weihnacht. Ein Familienalbum, 1900–1945, ed. B. Jochens, Berlin 1996
Dixon, L.S., *Perilous Chastity. Woman and Illness in pre-Enlightenment Medicine*, Ithaca NY and London 1995, esp. pp. 79–86
Floyer, J., *The Physician's pulse-watch; or, An essay to explain the old art of feeling the pulse, and to improve it by the help of a pulse-watch*, London 1707–10
French, R., 'Astrology in medical practice', in *Practical Medicine from Salerno to the Balck Death*, eds. L. García-Ballester, R. French, J. Arrizabalaga and A. Cunningham, Cambridge 1994, pp. 30–59
Horine, E.F., 'An Epitome of Ancient Pulse Lore', *Bulletin of the History of Medicine*, X, 1941, pp. 209–49
Imhof, A.E., 'The Implications of Increased Life Expectancy for Family and Social Life', in *Medicine in Society: Historical Essays*, ed. A. Wear, Cambridge 1992
Kern, S., *The Culture of Time and Space: 1880–1918*, London 1983
Pingeot, A., Le Normand-Romain, A., and de Margerie, L., *Musée d'Orsay. Catalogue sommaire illustré des sculptures*, Paris 1986, p. 116
Porter, R., *The Greatest Benefit to Mankind: A Medical History of Humanity from Antiquity to the Present*, London 1997
Rabinbach, A., *The Human Motor: Energy, Fatigue, and the Origins of Modernity*, Berkeley and Los Angeles 1992
Reiser, S.J., 'Technology and the Use of the Senses in Twentieth-century Medicine', in *Medicine and the Five Senses*, eds. W.F. Bynum and R. Porter, Cambridge 1993
Rusnock, A., 'Necrologues: Classifications of Death in the Enlightenment', talk presented at 'Ordering Nature in the Enlightenment' conference at William Andrews Clark Memorial Library, 15–16 May 1998
Saxl, F., 'Microcosm and Macrocosm in Medieval Pictures', in F. Saxl, *Lectures*, ed. G. Bing, London 1957, pp. 58–72
Thane, P., 'Geriatrics', in *Companion Encyclopaedia of the History of Medicine*, ed. W.F. Bynum and R. Porter, London and New York 1993, pp. 1092–111

(The Creation of History)
Ackerman Smoller, L., *History, Prophecy and the Stars. The Christian Astrology of Pierre d'Ailly, 1350–1420*, Princeton 1994, esp. pp. 88ff.
Blankert, A., *Amsterdams Historisch Museum. Schilderijen daterend van voor 1800 voorlopige catalogus*, Amsterdam 1975–79
Carey, S.W., *Theories of the Earth and Universe. A History of the Dogma of the Earth Sciences*, Stanford CA 1988, pp. 70–71
Dawson Adams, F., *The Birth and Development of the Geological Sciences*, London 1938, pp. 77–136, esp. pp. 114–16
Edson, E., *Mapping Time and Space. How medieval Mapmakers viewed their World*, London 1997
Gould, S.J., *Time's Arrow and Time's Cycle: Myth and Metaphor in the Discovery of Geological Time*, Harmondsworth 1988
Graeser, S., 'Alpine Minerals', *Rocks and Minerals*, LXXIII, no. 1, 1998, pp. 14–32
How Fragrant the Rose. Samplers and Historic Embroideries, 1650–1850, Whitney 1998, p. 32
Landes, R., 'Lest the Millennium be Fulfilled', *The Use and Abuse of Eschatology in the Middle Ages*, eds. W. Verbeke, D. Verhelst and A. Welkenhuysen, Louvain 1988, pp. 137–211
McCoy, R., 'Swift Dog: Hunkapa Warrior, Artist and Historian', *American Indian Art*, XIX, no. 3, 1994, pp. 68–75
Philips, J., *A Treatise of Geology*, London 1837
Pointon, M., 'The Representation of Time in Painting: A Study of William Dyce's *Pegwell Bay, Kent: A Recollection of October 5th, 1858*', *Art History*, I, no. 1, 1978, pp. 100–03
Pointon, M., *William Dyce 1806–1864: A Critical Biography*, London 1979
te Rijdt, R.J.A., 'Een ontwerp voor een zaalstuk doot Jacob de Wit', *Leids. Kunsthistorisch Jaarboek*, X, 1995, pp. 241–57
Ruskin, J., *The Complete Works of John Ruskin*, London 1906
Samplers: 'All creatures great and small', Whitney 1996, pp. 33–34
Schnapp, A., *The Discovery of the Past. The Origins of Archaeology*, English edn, London 1996
Staring, A., *Jacob de Wit, 1695–1754*, Amsterdam 1958, p. 157, fig. 99
Watson, A., *The Early Iconography of the Tree of Jesse*, Oxford 1934

(Wrinkles in Time)
Art Treasures of England. The Regional Collections [exh. catalogue, London, The Royal Academy, 22 January–13 April 1998], London 1998, no. 272
Boskovits, M., *Pittura fiorentina alla vigilia del Rinascimento, 1370–1400*, Florence 1975, p. 365
Captain James Cook [exh. catalogue, London, The National Maritime Museum, 1988], ed. D. Cordingly, London 1988, esp. pp. 42–3 and 105
Clark, C., *Shakespeare and Science*, Birmingham 1929
Coleman, J.A., *Relativity for the Layman*, New York 1954
Cryer, F.H., *Divination in Ancient Israel and its Near-Eastern Environment. A Socio-historical Investigation*, Sheffield 1994, esp 174–77
Eade, J.C., 'Marcantonio Michiel's Mercury Statue: astronomical or astrological?', *Journal of the Warburg and Courtauld Institutes*, XLIV, 1981, pp. 207–09
English Romantic Writers, ed. D. Perkins, New York 1967, pp. 275–78
Fischer, R., 'The Biological Fabric of Time', typescript text (London, The Warburg Institute [HAF 60])
Hartner, W., 'The Mercury Horoscope of Marcantonio Michiel of Venice. A Study in the History of Renaissance Astrology and Astronomy', in *Vistas in Astronomy*, ed. A. Beer, London and New York, I, 1952, pp. 84–138
Hawking, S., *A Brief History of Time*, London 1988
Hoskinson, J., *Roman Sculpture from Cyrenaica in the British Museum* [Corpus Signorum Imperii Romani, II, 1], London 1975, p. 53, no. 93
Imagining Rome: British Artists and Rome in the Nineteenth Century [exh. catalogue, Bristol Museum and Art Gallery, 3 May–23 June 1996], London 1996
Lippincott, K., 'A Masterpiece of Renaissance Drawing: A *Sacrificial Scene* by Gian Francesco de' Maineri', *Museum Studies (Art Institute of Chicago)*, XVII, no. 1, 1991, pp. 7–21 and 88–89
Miola, R.S., *Shakespeare's Rome*, Cambridge 1983
Mr Holman Hunt's Picture 'The Shadow of Death', London (Thomas Agnew's & Sons) 1874
Pattie, T., *Astrology as Illustrated in the Collections of the British Library and the British Museum*, London 1980, pp. 13–16
Pope-Hennessy, J., 'A Statue by Antonio Minelli', *The Burlington Magazine*, XCIV, 1952, pp. 24–28
Pope-Hennessy, J., *Catalogue of the Italian Sculpture in the Victoria and Albert Museum*, London 1964, pp. 510–12, no. 539
Quinton-Smith, M., 'The Winter Exhibition at the Royal Academy. I: Paintings of St Fina and of Eve Recumbent', *The Burlington Magazine*, CIV, 1962, pp. 62–66
Starr, I., *The Rituals of the Diviner*, Malibu CA 1983

THE END OF TIME
Beresford, R., *Dulwich Picture Gallery. Complete Illustrated Catalogue*, London 1998
Hirschfeld, G., *The Collection of Ancient Greek Inscriptions in the British Museum, IV. Knidos, Halicarnassus and Branchidae*, London 1893, p. 224, no. 1114
Lane, A., *Later Islamic Pottery: Persia, Syria, Egypt, Turkey*, London 1957, pls. 75–79
Lucian, *Opera*, ed. and transl. by M.D. MacLeod, London and Cambridge (Loeb editions), 1961, VII, pp. 170–75
The Print in Germany, 1880–1933. The Age of Expressionism [exh. catalogue, London, The British Museum], eds. F. Carey and A.Griffiths, London 1984, p. 86
Smith, A.H., *A Catalogue of Sculpture in the Department of Greek and Roman Antiquities in the British Museum*, London 1904, III, pp. 363–64, no. 2391
Toynbee, J.M.C., *Death and Burial in the Roman World*, London 1971
Welcher, F.G., *Syllogie Epigrammatum Graecorum Marmoris et Libris*, Bonn 1828, no. 67

WORLD CULTURES AND TIME
(Aboriginal Australians)
Dreamings Tjukurrpa. Aboriginal Art of the Western Desert. The Donald Kahn Collection, ed. J.-A. Birnie Danzker, Munich 1994
Kupka, K., *Peintres aboriginales d'Australie*, Paris 1972, p. xxxv
Morphy, H., *Aboriginal Art*, London 1998
Morphy, H., 'Landscape and the Reproduction of the Ancestral Past', in *Anthropology of Landscape: Perspectives on Place and Space*, eds. E. Hirsch and M. O'Hanlon, Oxford 1995, pp. 194–209
Munn, N., 'The Cultural Anthropology of Time: A Critical Essay', *Annual Review of Anthropology*, XXI, 1992, pp. 93–123
Spencer, W.B., and Gillen, F.J., *The Arunta*, London 1927
Stanner, W.E.H., 'Relgion, Totemism and Symbolism', in *White Man Got No Dreaming*, Canberra 1979

(Aztecs)
Aveni, A., *Empires of Time. Calendars, Clocks and Cultures*, New York 1989
Circa 1492: Art in the Age of Exploration [exh. catalogue, Washington, D.C., 12 October 1991–12 January 1992], ed. by J. Levenson, New Haven and London 1991, esp. p. 552 (M.D.C. = Michael D. Coe)
León-Portilla, M., *Aztec Thought and Culture*, Washington, D.C.,1979
Townsend, R.F., *The Aztecs*, London 1992

(Babylonians)
Aro, J., 'Remarks on the Practice of Extispicy in the Time of Esahaddon and Assurbanipal', *La Divination en Mésopotamie ancienne et dans les régions voisines* [XIVe Rencontre Assyriologique Internationale], Paris 1966, pp. 109–117
Budge, E.A.W., *Facsimiles of Egyptian Hieratic Papyri in the British Museum*, London 1923, pp. 34–38, no. 10184 (Sallier IV)
Budge, E.A.W., *British Museum. A Guide to the Babylonian and Assyrian Antiquities*, London 1900, esp. pp.128–29
Contenau, G., *La Divination chez les Assyriens et les Babyloniens*, Paris 1940, esp. pp. 235–83
Frankfort, H, *Kingship and the Gods: A Study of Ancient Near-Eastern Religion as the Integration of Society and Nature*, Chicago 1948
Grayson, A.H., 'Königslisten und Chroniken', *Reallexxikon der Assyriologie und vorderasiatische Archaeology* , Berlin 1980, p. 90
Heidel, A., *The Babylonian Genesis. The Story of the Creation*, 2nd edn., Chicago 1951

Jacobsen, T., 'Mesopotamia' in H. Frankfort *et al., The Intellectual Adventure of Ancient Man. An Essay on Speculative Thought in the Ancient Near East*, Chicago and London 1946, pp. 125–219

Jacobsen, T., *The Treasures of Darkness: A History of Mesopotamian Religion*, New Haven 1976, esp. pp. 167–91

Jeyes, U., 'The "Palace Gate" of the Liver. A Study of Terminology and Methods of Babylonian Extispicy', *Journal of Cuneiform Studies*, XXX, 1978, pp. 209–233

Koch, J., *Neue Untersuchungen zur Topographie des babylonischen Fixsternhimmels*, Weisbaden 1989, esp. pp. 111–13

Langdon, S., *Enuma elish. The Babylonian Epic of Creation. Restored from the recently recovered Tablets of Assur. Transcription and Commentary*, Oxford 1923

Neugebauer, O., *Astronomical Cuneiform Texts. Babylonian Ephemerides of the Seleucid Period for the Motion of the Sun, the Moon and the Planets*, Princeton 1955, I, pp. 144–46 (no. 122)

Nougayrol, J., 'Le Foie "d'orientation", BM 50494', *Revue d'Assyriologie*, LXII, 1968, pp. 31–50

Britton, J., and Walker, C., 'Astronomy and Astrology in Ancient Mesopotamia', in *Astronomy before the Telescope*, ed. C. Walker, London 1996, esp. pp. 42–67

Sollberger, E., 'New Lists of Kings of Ur and Isin', *Journal of Cuneiform Studies*, VIII, 1954, pp. 134–35

Thureau-Dangin, F., *Rituels accadiens*, Paris 1921, esp. pp. 127–54

(China)

Bedini, S., *The Trail of Time. Time Measurement with Incense in East Asia (Shih-chien ti tsu-chi)*, Cambridge 1994

Birrell, A., *Chinese Mythology: An Introduction*, Baltimore 1993

Bulling, A., *The Meaning of China's Most Ancient Art. An Interpretation of the Pottery Patterns from Kansu (Ma Ch'ang and Pan-Shan) and their development in the Shang, Chou and Han Period*, Leiden 1952

Cullen, C., and Farrer, A.S.L., 'On the term *hsüan-chi* and the flanged trilobate discs', *Bulletin of the School of Oriental and African Studies*, xlvi, no. 1, 1988, pp. 52–76

Doré, H., *Recherches sur les superstitions en Chine*, Shanghai 1914–29

Feuchtwang, S.D.R., *An Anthropological Analysis of Chinese Geomancy*, Laos 1974

Forke, A., *The World-Conception of the Chinese. Their Astronomical, Cosmological and Physico-Philosophical Speculations*, London 1925

Goodrich, A.S., *Peking Paper Gods. A Look at Home Worship*, Nettetal 1991

Du Halde, J. B., *A Description of the Emperor of China and Chinese-Tartary, together with the Kingdoms of Korea and Tibet, containing the Geography and History (natural as well as civil) of these Countries*, London 1741, esp. pp. 184–207

Karlgren, B., 'The Book of Documents (Shu Ching)', *Bulletin of the Museum of Far Eastern Antiquities (Stockholm)*, XXII, 1950 pp. 1ff.

Loewe, M., 'China', in *Divination and Oracles*, eds. M. Loewe and C. Blacker, London 1981, pp. 38–62, esp. pp. 56–59 and pl. 4

Loewe, M., *Ways to Paradise. The Chinese Quest for Immortality*, London 1979, esp. pp. 60–85 and Appendix I

Needham, J., and Wang Ling, *Science and Civilisation in China, III. Mathematics and the Sciences of the Heavens and the Earth*, Cambridge 1959

Rawson, J., *Chinese Jade from Neolithic to the Qing*, London 1995, esp. pp. 160–63

Schaefer, E.H., *Pacing the Void. T'ang approaches to the Stars*, Berkeley CA 1977

Twichett, D., *Printing and Publishing in Medieval China*, London 1983

Unearthing China's Past, eds. J. Fontein and Tung Wu, Boston 1973, esp. pp. 100–02

Werner, E.T.C., *Myths and Legends of China*, London 1922

Wood, F., *Chinese Illustration*, London 1985

(Egypt)

Borchardt, L., *Die altägyptische Zeitmessung*, Berlin and Leipzig 1920

Borchardt, L., *Die Mittel zur zeitlichen Festlegung von Punkten der ägyptischen Geschichte und ihre Anwendung*, Cairo 1935

The British Museum Book of Ancient Egypt, eds. S. Quirke and J. Spencer, London 1992

Budge, E.A.T.W., *The Gods of the Egyptians: or, Studies in Egyptian Mythology*, London 1904

Clagett, M. *Ancient Egyptian Science. A Source Book*, Philadelphia 1989–95

Cotterell, B., Dickson, F.P., and Kamminga, J., 'Ancient Egyptian Water-clocks: A Reappraisal', *The Journal of Archaeological Sciences*, XIII, 1986, pp. 31–50

Cumont, F., *Les religions orientales dans le paganisme romain*, 1929

Isler, M., 'The Merkhet', *Varia Aegyptiaca*, VII, no. 1, 1991, pp. 53–67

Quirke, S., *Ancient Egyptian Religion*, London 1992

Shaw, I., and Nicholson, P., *British Museum Dictionary of Ancient Egypt*, London 1995

Sloley, R.W., 'Primitive Methods of Measuring Time, with special reference to Egypt', *Journal of Egyptian Archaeology*, XVII, 1930, pp. 166–78, esp. p. 169

Walker, S., and Bierbrier, M. (with Roberts, P., and Taylor, J.), *Ancient Faces. Mummy Portraits from Roman Egypt* [exh. catalogue, London, The British Museum, 14 March–20 July 1997], London 1997, p. 184, no. 246

Wells, R.A., 'Astronomy in Egypt', in *Astronomy before the Telescope*, ed. C. Walker, London 1996, pp. 28–41, esp. p. 37

Witt, R.E., *Isis in the Graeco-Roman World*, London 1971

(Hindu)

Balslev, A.N., *A Study of Time in Indian Philosophy*, Wiesbaden 1983

Barrett, D., 'The Dancing Siva in Early South Indian Art', *Studies in Indian Sculpture and Painting*, London 1990, pp. 327–67

Burton, T. Richard, *Hindu Art*, London 1992

Daniélou, A., *The Myths and Gods of India*, 3rd edn, Rochester VT 1991

Gopinatha Rao, T.A., *Elements in Hindu Iconography*, Madras 1914–16

In the Image of Man. The Indian Perception of the Universe through 2000 years of Painting and Sculpture [exh. catalogue, London, Hayward Gallery, 25 March–13 June 1982], London 1982

Mandal, K.K., *A Comparative Study of the Concepts of Space and Time in Indian Thought*, Varanasi 1968

Tantra [exh. catalogue, London, Hayward Gallery, 30 September–7 November 1971], London 1971

Zimmer, H., *The Art of Indian Asia. Its Mythology and Transformations*, ed. and completed by J. Campbell, New York 1955

(Inuit)

Boas, F., 'The Central Eskimo', *6th Annual Report of the Bureau of American Ethnology for the Years 1884–1885*, Washington, D.C., 1888, pp. 399–669

Driscoll, B., *Inuit Myths, Legends and Songs*, Winnipeg 1982, p. 45

Gad, F., *The History of Greenland, II: 1700–1782*, Montreal 1973

Giddings, J.L., *Kobuk River People*, Fairbanks AL 1961

Kane, E.K., *Arctic Explorations in the Years 1853, '54, '55*, 2 vols., Philadelphia 1856

MacDonald, J., *The Arctic Sky. Inuit Astronomy, Lore and Legend*, Toronto 1998

Parry, W. E., *Journal of a Second Voyage for the Discovery of a North-West Passage from the Atlantic to the Pacific: Performed in the Years 1821–22–23, in His Majesty's Ships Fury and Hecla*, London 1824

Rasmussen, K., *Intellectual Culture of the Iglulik Eskimos. Report of the 5th Thule expedition, 1921–24*, vol. 7, no. 1, Copenhagen 1929

Rasmussen, K., *Iglulik and Caribou Eskimo Texts. Report of the 5th Thule expedition*, vol. 7, no. 3, Copenhagen 1930

Rasmussen, K., *The Netsilik Eskimos: Social Life and Spiritual Culture. Report of the 5th Thule Expedition, 1921–24*, vol. 8, nos. 1 and 2, Copenhagen 1931

Stefánsson, V., *Hunters of the Great North*, New York 1922

(Islam)

Ackermann, S., and King, D., 'An Exceptional Quadrant: Timekeeping and Calendrical Calculations in Late 19th-century Ottoman Turkey', forthcoming

L'Apparence des cieux. Astronomie et astrologie en terre d'Islam [exh. catalogue, Paris, Musée du Louvre, 18 juin–21 Settembre 1998], Paris 1998

Chebel, M., *Symbols of Islam*, Paris 1997

The Encyclopaedia of Islam, eds. Houfsma, M. Th., *et al.*, Leiden 1913–36; new edn, Leiden 1960 to present

von Grunebaum, G.E., *Muhammadan Festivals*, London 1951

Heinen, A., *Islamic Cosmology – A Study of as-Suyt's al-Hay'a as-sanya f l-hay'a as-sunnya, with critical edition, translation and commentary*, Beirut 1982

Ilyas, M., *Astronomy of Islamic Calendar*, Kuala Lumpur 1998

Ilyas, M., *Astronomy of Islamic Times for the Twenty-First Century*, London 1988

King, D.A., 'Kibla. The sacred direction', in *The Encyclopedia of Islam*, Leiden 1979, V, pp. 83–88

King, D.A. , *Islamic Mathematical Astronomy*, London 1986; 2nd edn, Aldershot 1993

King, D.A., *Astronomy in the Service of Islam*, Aldershot 1993

King, D.A., 'The Orientation of Medieval Islamic Religious Architecture and Cities', *Journal for the History of Astronomy*, XXVI, 1995, pp. 253–74

King, D.A., 'Islamic Astronomy', in *Astronomy before the Telescope*, ed. C. Walker, London 1996, pp. 143–174

King, D.A., *World-Maps for Finding the Direction and Distance to Mecca – Innovation and Tradition in Islamic Science*, Leiden 1999

Lazarus-Yafeh, H., *Some Religious Aspects of Islam*, Leiden 1981

Mayer, L.A., *Islamic Astrolabists and their Works*, Geneva 1956, p. 75

Mernissa, F., '*The Muslim and Time*', in *The Veil and the Male Elite*, Reading 1991, pp. 15–24

Peters, F.E., *The Hajj: The Muslim Pilgrimage to Mecca and the Holy Places*, Princeton 1994

Peters, F.E. , *Mecca – A Literary History of the Muslim Holy Land*, Princeton 1994

Turner, H.R., *Science in Medieval Islam. An Illustrated Introduction*, Austin TX 1995

(Japan)

Bedini, S., *The Trail of Time. Time Measurement with Incense in East Asia (Shih-chien ti tsu-chi)*, Cambridge 1994, esp. pp. 54 and 58–64

Caillet, L., 'Time in the Japanese Ritual Year', in *Interpreting Japanese Society*, ed. J. Hendry, London 1998, pp. 15–30

Crump, T., 'The Pythagorean View of Time and Space in Japan',in *Interpreting Japanese Society*, ed. J. Hendry, London 1998, pp. 42–56

Hendry, J. 'The pre-Gregorian Calendar and some Implications for Marriage' in *Marriage in Changing Japan*, London 1981, appendix

Hendry, J., '*Tomodachi ko*: Age-Mage Groups in Northern Kyushu', *Proceedings of the British Association for Japanese Studies*, VI, no. 2, 1981, pp. 43–56

Jingû Shikinen Sengû. Explanation of the 61st Regular Removal of the Grand Shrine of Ise, Tokyo 1993

Lewis, D., 'Years of Calamity: *Yakudoshi* Observances in Modern Japan', in *Interpreting Japanese Society*, ed. J. Hendry, London 1998, pp. 196–212

Suzuki, D.T., *Zen and Japanese Culture. Zen Buddhism and its Influence on Japanese Culture*, New York 1958

Utamaro. Estampes. Livres Illustrées, Paris 1977

Ukiyo-e. Die Kunst der heiteren vergänglichen Welt Japan 17.–19. Jahrhundert [exh. catalogue, Essen, Villa Hügel, 17 March–30 June 1972], Essen 1972, p. 106

(Judaism)

Burnaby, S.B., *Elements of the Jewish and Muhammadan Calendars, with Rules and Tables and Explanatory Notes on the Julian and Gregorian Calendars*, London 1901

Catalogue of the Permanent and Loan Collection of the Jewish Museum, London, ed. R.D. Barnett, London 1974

Encyclopaedia Judaica. Das Judentum in Geschichte und Gegenwart, Berlin 1928–34

Levi, L., *Jewish Chrononomy. The Calendar and Times of Day in Jewish Law*, New York 1967

The Jewish Encylopedia, London and New York 1967

Spier, A., *The Comprehensive Jewish Calendar*, New York 1952

Stioui, R., *Le Calendrier Hébraïque*, Paris 1988

La vida Judía en Sefarad [exh. catalogue, Toledo, Sinagoga del Tránsito, November 1991–January 1992], Toledo 1992, p. 297, no. 146

(Maori)

Binney, J., 'A lost drawing of the Nukutawhiti', *New Zealand Journal of History*, XIV, no. 1, 1980, pp. 3–24, esp. p. 21

Hakiwai, A.T., 'Maori Society Today', in *Maori Art and Culture* [exh. catalogue, London, The British Museum], ed. D. C. Starzecka, London 1996, pp. 50–68, esp. pp. 52–54

King, J.C.H., *New Zealand 1990*, Auckland 1990, esp. p. 10

Neich, R., 'Wood-carving', in *Maori Art and Culture* [exh. catalogue, London, The British Museum], ed. D.C. Starzecka, London 1996, pp. 69–113, esp. pp. 106–07

(Native American)

Baatsoslanii Joe, E., and Bahti, M., *Navajo Sandpainting Art*, Tucson AZ 1978

Beautyway: A Navajo Ceremonial, ed. L.C. Wyman, New York 1957

Burlan, C., *North American Indian Mythology*, Feltham MX 1965

Dorsey, G.A., *The Cheyenne. II, The Sun Dance*, Chicago 1905

Feest, C.F., *The Native Arts of North America*, London 1980

Gill, S.D., *Songs of Life* [Iconography of Religions, X, 3], Leiden 1979

Halpin, D., *The Raven Rattle* (University of British Columbia Museum of Anthropology, VI)

Hultkrantz, A., *Prairie and Plains Indians* [Iconography of Religion, X, 2], Leiden 1973

Holsbeke, M., 'Husk Faces. Bringers of Fertility among the Onondaga', in *The Object as Mediator. On the Transcendental Meaning of Art in Traditional Cultures* [exh. catalogue, Antwerp, Etnografisch Museum, 30 November 1996–15 March 1997], ed. M. Holsbeke, Antwerp 1996, pp. 103–07

Klah, H., *Navajo Creation Myth. The Story of Emergence*, Santa Fe NM 1942

McKenney, T.L., and Hall, J., *History of the Indian Tribes of America with Biographical Sketches and Anecdotes of the principal Chiefs ...*, Philadelphia 1837–44, esp. II (1842), pp. 225 and 237

Merrill, R.H., *The Calendar Stick of Tshi-zun-hau-kau*, Bloomfield Hills MI n.d.

Riechard, G.A., *Navajo Religion. A Study of Symbolism*, New York 1950

Visions of the People: A Pictorial History of Plains Indian Life [travelling exh. catalogue], Minneapolis 1992, p. 275

Wardwell, A., *Objects of Bright Pride. Northwest Coast Indian Art from the American Museum of Natural History*, New York 1978

Index

Credits

This book has been published to accompany the exhibition held at The Queen's House, National Maritime Museum, Greenwich, London, 1 December 1999 – 24 September 2000

First published in 1999 by
Merrell Holberton Publishers Ltd
42 Southwark Street
London SE1 1UN

Distributed in the USA and Canada by
Rizzoli International Publications, Inc.
through St Martin's Press
175 Fifth Avenue
New York, NY 10010

British Library Cataloguing in Publication Data
The story of time
1. National Maritime Museum 2. Time – Exhibitions
3. Time 4. Time – Philosophy 5. Time – Social aspects
I. Lippincott, Kristen II. National Maritime Museum
529

ISBN: 1 85894 072 9 (hardback)
ISBN: 1 85894 073 7 (paperback)

Produced by Merrell Holberton Publishers Ltd
Designed by Karen Wilks
Printed and bound in Italy

Jacket/cover illustration: *White clouds in sky*, photograph by Doug Armand, © Tony Stone Worldwide

Photographic Credits

The Royal Collection, © Her Majesty The Queen 179, 201, 214, 218. © English Heritage Photographic Library/ Stonehenge p. 34 below, p. 37; © English Heritage Photographic Library/ Down House 277.© The National Trust, Anglesey Abbey/ photo James Austin 195. © The National Trust,Powys Castle/ photo: Paul Highnam 171. Amherst MA: Mead Art Museum, Amherst College: Gift of Samuel H. Kress Foundation 22. Amsterdam: Amsterdams Historisch Museum 262; Amsterdams Historisch Museum/ The City of Amsterdam/ Rijksmuseum 204; Koninklijk Istituut voor de Tropen (Irene de Groot, KIT) 48; Rijksmuseum 229. Arizona: © Dr Emily Umberger p. 55. Baltimore MD: AURA/ ST ScI(with support from NASA, contract NA S5-26555) J. Hester (Arizona State University) and NASA p. 284; AURA/ ST ScI (with support from NASA, contract NA S5-26555) W.N. Colley, E. Turner (Princeton University), J.A. Tyson (AT&T Bell Labs) and NASA p. 287. Basel: Basel-Naturhistorisches Museum 276. Berlin: Staatliche Mussen zu Berlin, Preußischer Kulturbesitz Ägyptische Museum und Papyrussammlung/ photo Margarete Büsing 87, 101, 264, p. 270; Antikensammlung, Staatliche Museen zu Berlin - Preußischer Kulturbesitz 40; Archiv des Heimatsmuseums, Charlottenburg/ © Nicolaische Verlagsbuchhandlung - Beuermann GmbH, Berlin 247; Staatliche Museen zu Berlin Preußischer Kulturbesitz Skulpturensammlung/ PKB 93, 99, 225; Staatliche Museen zu Berlin Preußischer Kulturbesitz Gemäldegalerie/ photo Jörg P. Anders 79; Staatliche Museen zu Berlin Preußisher Kulturbesitz Museum für Islamische Kunst/ BPK 26, 44; Staatliche Museen zu Berlin Preußisher Kulturbesitz Museum für Volkerkunde/ BPK 9. Bern: Bernisches Historisches Museum, Ethnography Dept/ photo: S. Rebsamen 21. Bielefeld: Stiftung Huelsmann/ photo: von Uslar Bielefeld 114, 123. Bismarck ND: State Historical Society of North Dakota 273. Bloomfield Hills MI: Tim Thayer, Cranbrook Institute of Science 53. Boston: Museum of Fine Arts: Gift of Julia Bird, Madeline Kidder and Sybil Wolcott 146. Bradford: © National Museum of Photography, Film and Television/ Science and Sound Picture Library p. 220. Braunschweig: Stadtbibliothek 59. Brescia: Musei Civici d'arte e storia 94. Brussels: Musées Royaux d'Art et d'Histoire 85. Bury St Edmunds: Private Collection/ photo National Maritime Museum 60, 184. Cambridge: The Syndics of the Fitzwilliam Museum 205; Cambridge University, The Museum of Archaeology and Anthropology: 261, 263, 294; © Needham, Ling and de Solla Price/ Cambridge University Press 180; Whipple Museum of the History of Science/ © photograph National Maritime Museum 103, 283, p. 96 top; Whipple Museum of the History of Science/ St John's College, Cambridge/ © photograph National Maritime Museum 69. Cambridge MA: © President and fellows of Harvard College, Harvard University/ photo David Mathews p. 186. Chesham: © Peter Sanders p. 56 top. Chicago: Collection of Mr and Mrs E.A. Bergman, Chicago/ © The Joseph and Robert Cornell Memorial Foundation p. 205. Dorchester: © The Dorset Natural History and Archaeological Museum 115. Dresden: Gemäldegalerie Alte Meister 161; Staatliche Kunstsammlungen Dresden, Mathematisch-Physikalischer Salon 176,199. Dublin: The Trustees of the Chester Beatty Library 68; © National Museum of Ireland 16. Edinburgh: © National Gallery of Scotland/ photo: Antonia Reeve 74, 208; © Trustees of the National Museums of Scotland 1999, 80, 254. Faversham: © Harris Belmont Charity/ © photograph National Maritime Museum 153, 154, 156, 157, 160, 172. Fleurier: Parmigiani Fleurier 198. Florence: Biblioteca Nazionale Centrale 174. St Petersburg, FLA: © Salvador Dalí Museum Inc./ Salvador Dalí-Foundation Gala-Salvador Dalí/ DACS 1999 235. Furtwagen: Deutsches Uhrenmuseum 159. Geneva: © Musée d'Horlogerie/ photo: Maurice Aeschimann 200; © Musée d'Histoire des Sciences/ Ville de Genève 28. Graz: Landesmuseum Joannem, Kunstgewerbliche Sammlungen 104. The Hague: Koninklijke Bibliotheek 66. Hamburg: Hamburger Kunsthalle/ photo: © Elke Walford 210; Museum für Kunst und Gewerbe 89, 130. Hamilton, NY: Colgate University 280, © Linda Schele p. 52 below. Heidelberg: © Universitätsbibliothek 25. Hildesheim: Archives of the Pelizaeus-Museum Hildesheim p. 271. Ipswich: © Haward Horological Ltd/ © phtograph National Maritime Museum 149. Jerusalem: © The Israel Museum, Avshalom Avital/ Albert Einstein™ licenced by The Hebrew University/ Represented by the Roger Richman Agency, Inc., Beverly Hills CA. 298. Kendal: C.H. Bagot, Esq. p. 198. Lancaster: The Ruskin Foundation (Ruskin Library, University of Lancaster) p. 198. Leeds: City Art Gallery, Leeds Museums and Galleries/ Bridgeman Art Library, London/ New York 290; Brotherton Library Special Collections, Leeds University Library 267. Leiden: Rijksmuseum voor Volkenkunde 251; Stedelijk Museum de Lakenhal 230. LeLocle: Château des Monts, Musée d'Horologie 196, 197. Liège: Musée de la Vie Wallonne 111. Linschoten: Private Collection 92. Liverpool: © The Board of Trustees, The National Museums & Galleries on Merseyside 305, p. 52 top. London: © The British Museum 8, 10, 13, 18, 19, 24, 29, 32, 36, 38, 39, 41, ,43, 45, 46, 49, 50, 58, 61, 62, 67, 70, 78, 82, 84, 86, 95, 96, 100,102, 131, 133, 144, 155, 162, 189, 207, 209, 211, 212, 215, 217, 219, 220, 222, 224, 227, 257, 258, 272, 282, 287, 292, 299, 301, 303, 306, 312, p. 34 top, p. 48 below, p. 59, p. 245; On loan from the V&A © The British Museum 150; The British Library 1, 2, 5, 14, 34, 42, 55, 71, 81, 97,134, 141, 175, 223, 242, 255, 266, 270, 285, 286, p. 13, p. 42, p. 48 top, p. 79, p. 246 below, p. 250 top; © Christie's Image Ltd, 1999 p. 56 below; The Helly Namad Gallery/ © ADAGP, Paris and DACS, London, 1999 p. 197; The Horniman Museum & Gardens/ photograph National Maritime Museum , 248, 311; The Horniman Museum & Gardens/ photograph National Maritime Museum/ © Fred Stevens 7. © The Jewish Museum/ photograph National Maritime Museum 143, 250; © The Jewish Museum 51, 145, 152. The Museum of London 163, 187; © National Gallery 213, 221, p. 214; National Portrait Gallery/ Private Collection p. 190. The Natural History Museum 77; Jonathan Betts/ photograph National Maritime Museum 182, 183;.© National Maritime Museum 30, 31, 90, 98, 101, 112, 113, 116, 117, 119, 120, 124, 125, 128,135, 136, 137, 140, 164, 166, 169, 170, 177, 178, 181, 185, 194, 295, p. 10, p. 12, p. 134 top,p. 102, p. 103,p. 110,p. 114, p. 137; Private Collection/ photograph National Maritime Museum 158,269 Courtesy of Rebecca Hossack Gallery 296; Royal Academy of Arts 233; The Royal Artillery Historical Trust, Woolwich/ © photograph National Maritime Museum 180; © Royal Geographical Society/ photo: John Miles p. 264 top; Science Museum/ Science and Society Picture Library 63, 88, 91, 110, 122, 126, 127, 132, 142; © Sotheby's, London 121; © Tate Gallery, London 1998/ ARS, NY and DACS, London, 1999 237. © Tate Gallery, London 1988 232; © Tate Gallery, London 1998/ photo John Webb 231,274, p. 194; By permission of The Trustees of the Dulwich Picture Gallery 300; Dr Denys Vaughan/ illus. J.M. Farrant, British Museum p. 121; V&A Picture Library 12, 17, 23, 76, 191, 206, 226, 234, 243, 259, 260, 271, 288, 308, p. 246 top; © The Trustees of the Wellcome Trust 245, 246,; © The Trustees of the Wellcome Institute Library p. 223; The Worshipful Company of Goldsmiths. 186 Madrid: Museo Naçional del Prado 256, p. 189; © Museo Thyssen-Bornemisza p. 190. Manchester: By permission of The John Rylands University Library of Manchester 1; Manchester City Art Galleries 278, 284; Helen Rees/ © University of Manchester 309; The Manchester Museum, The University of Manchester 279, 310; The Whitworth Art Gallery, The University of Manchester 6. Melbourne: National Gallery of Victoria, Felton Bequest, 1988/ © Anthony Wallis, Aboriginal Artists Agency, PO Box 282, Cammeray, NSW 2062 p. 264 bottom. Modena: Ministerio per i Beni e le Attività Culturali, Biblioteca Estense Universitaria 244; Ministerio per i Beni e le Attività Culturali, Galleria Estense 203. Munich: Bayerische Staatsgemäldesammlungen/ photo: Joachim Blauel, ARTOTHEK p. 194; © Bayerisches Nationalmuseum 109, 188, 190, 192, 193; © Deutsches Museum 106, 107. New York: © Department of Library Services, American Museum of Natural History/ photo Denis Finnin 54; © 1999 The Museum of Modern Art/ Katherine S. Dreier Bequest p. 202 top; © 1999 The Museum of Modern Art/ Gift of James Thrall Soby p. 202 below; New York Public Library, William Frederick Allen Papers, Manuscripts and Archives Division 165, p. 134 below; ULAE/ © Robert Rauschenberg/ VAGA, NY/ DACS, London, 1999 238; ULAE/ © estate of Jasper Johns/ VAGA, NY/ DACS, London 1999 236; Private Collection/ ULAE/ © estate of Jasper Johns/ VAGA NY/ DACS, London 1999 75. Nuremberg: © Germanisches Nationalmuseum 118. Oslo: © Munch Museet/ Svein Andersen & Sidsel de Jong, 1998/ Munch-Ellingsen Group/ DACS, London 1999 252. Oxford: © Ashmolean Museum 20; Bodleian Library, University of Oxford p. 139; © Bill Coaldrake p. 80; © Joy Hendry p. 80. Paris: Cliché Bibliothèque Nationale de France 3, 4, 35, 241; © Photothèque des Musées de la Ville de Paris 37, 73, 167,168; Musée de l'Institute du Monde Arabe/ photo Philippe Maillard/ IMA 302; © Photo RMN, Paris p. 15; © RMN Paris/ photo H. Lewandowski 15; © RMN Paris/ photo J.G. Berizzi 249; RMN Paris/ photo J.G. Berizzi 297;Musée Nationale d'Histoire Naturelle 281; Musée du Moyen-Age, Cluny/ © Photo RMN/ J.G. Berizzi 125. Piacenza: © Galleria d'Arte Moderna Ricci-Oddi/ photo F. Ili Manzetti 291. Prague: National Technical Museum/ © Mgr. Rudolf Jung. Clen Asoclace 108. Rhode Island: Museum of Art, Rhode Island School of Design/ Helen M. Danforth Fund/ photo Erik Gould p. 1; © The Provost, Brown University/ Neugebauer and Parker, Brown University Press, 1969 p. 268. Rimini: Biblioteca Civica Gambalunga/ Photo by Paritani 253. Rotterdam: © Institut Collectie Nederland/ photo Museum Boijmans Van Beuningen 239. Salzburg: Carolino Augusteum, Salzburger Museum für Kunst und Kulturgeschichte 147. Siena: © Archivio di Stato 72; Private Collection p. 240. Spikkestad: © The Schøyen Collection/ © photograph National Maritime Museum 47, 56, 64, 65,240, 265. St Florian: Stiftsbibliothek St Florian/ Augustiner Chorherrenstift p. 243. St Germain -en-Laye: Cliché L. Hamon, Musée des Antiquités Nationales 148. Stalybridge: © Astley Cheetham Art Gallery/ Tameside MBC Leisure Services 289. Stuttgart: Staatsgalerie Stuttgart 216. Toronto: The Royal Ontario Museum 27, p. 76 top; The Royal Ontario Museum/ © T. Paningajek 293. Verona: Biblioteca Civica p. 250 below. Vienna: Gesellschaft der Musikfreunde in Wien p. 214; Kunsthistorisches Museum 228, p. 193; Österreichische Nationalbibliothek: Globe Museum/ Rudolph Schmidt Collection 129. Washington DC: Dumbarton Oaks Research Library and Collection 57, 83; National Museum of American History/ Smithsonian Institution (Negative CS 82-13347) 105; Smithsonian Institution Libraries 1999 © Smithsonian Institution 173; Silvio A. Bedini (Smithsonian Institution)/ photo Hugh Talman 138, 139; Courtesy of The National Museum of the American Indian, Smithsonian Institution 11. Winnepeg: © Ernest Mayer, Winnepeg Art Gallery 307. Winterthur: The Konrad Kellenberger Clock and Watch Collection, Gewerbemuseum Winterthur 152. Witney (Oxon): Witney Antiques 268. Wupperta: Wupperthaler Uhrenmuseum 151. York: The Yorkshire Museum 275. Zurich: © 1999 by Kunsthaus Zurich. All rights reserved 304; Uhrenmuseum Beyer 202.